Inorganic Membranes for Energy and Environmental Applications

Arun C. Bose
Editor

Inorganic Membranes for Energy and Environmental Applications

 Springer

Editor
Arun C. Bose
U.S. Department of Energy
National Energy Technology Laboratory
Pittsburgh, PA, USA
arun.bose@netl.doe.gov

ISBN: 978-0-387-34524-6   e-ISBN: 978-0-387-34526-0
DOI 10.1007/978-0-387-34526-0

Library of Congress Control Number: 2008935100

Printed on acid-free paper

springer.com

Preface

The study of inorganic membrane materials and processes is a rapidly expanding research interest area. This is due to the recognition that research results from recent years enabling membrane-based technology modules have the potential to improve the cost, efficiency, and environmental performances of energy production systems beyond current benchmarks. The chapters document progress in inorganic membranes, especially in advanced materials and novel separation concepts, and uniformly convey an inspired expectation that the information will help identify R&D opportunities for providing engineering and technological solutions towards an affordable, efficient, and environmentally progressive energy sector. The book provides a single source reference for researchers contemplating continued enhancement of inorganic membranes with novel capabilities for applications to future energy and fuel production systems, and also offers ideas for technology-based responses to address greenhouse gas emission concerns.

The chapters on Ion-Transport Membrane (ITM) oxygen production technology update readers on the current status and the commercial potential of the non-cryogenic ITM air separation process. The chapters on metal and alloy membranes for separating hydrogen from coal or natural gas derived shifted synthesis gas will allow the readers to stay abreast of recent developments and challenges in Hydrogen Transport Membrane (HTM) technology, and provide ideas and opportunities to further build on today's HTM technology knowledge base and technical know-how. The chapters on porous membranes have been written to provide the readers with information on the existing art about the subject.

This book is meant for scientists, engineers, and industry R&D personnel engaged in the development, engineering scale-up, and applications of inorganic membrane materials and processes; non-technical management staff delegated with programmatic and resource allocation decision authorities; and academics involved in generating inorganic membrane R&D products. By avoiding deep theoretical foundations, the book's content is meant to be readily accessible to readers who want a look at the current membrane technology state-of-the-art. The editor intends this material to advance the culture of science involving inorganic membranes and hopes that students, researchers, and program officials will equally find the information useful to advance their own work.

Arun C. Bose
Editor

Contents

Part I   Ion Transport Membranes and Ceramic-Metal Membranes

1   **Progress in Ion Transport Membranes for Gas Separation
    Applications**   3
    Arun C. Bose, Gary J. Stiegel, Phillip A. Armstrong, Barry J. Halper,
    and E. P. (Ted) Foster

2   **Viability of ITM Technology for Oxygen Production and Oxidation
    Processes: Material, System, and Process Aspects**   27
    Marcel J. den Exter, Wim G. Haije, and Jaap F. Vente

3   **Oxygen-Ion Transport Membrane and Its Applications in Selective
    Oxidation of Light Alkanes**   53
    Weishen Yang and Rui Cai

4   **Ceramic Membrane Devices for Ultra-High Purity Hydrogen
    Production: Mixed Conducting Membrane Development**   67
    S. Elangovan, B. Nair, T. Small, B. Heck, I. Bay, M. Timper,
    J. Hartvigsen, and M. Wilson

5   **Decomposition of Yttrium-Doped Barium Cerate
    in Carbon Dioxide**   83
    Theodore M. Besmann, Robert D. Carneim,
    and Timothy R. Armstrong

6   **Mixed-Conducting Perovskite Reactor for High-Temperature
    Applications: Control of Microstructure and Architecture**   95
    Grégory Etchegoyen, Thierry Chartier, Alain Wattiaux,
    and Pascal Del Gallo

7 **Mixed Protonic-Electronic Conducting Membrane for Hydrogen
   Production from Solid Fuels** 107
   Shain J. Doong, Francis Lau, and Estela Ong

Part II   Metal and Alloy Membranes

8 **Hydrogen Separation Using Dense Composite Membranes:
   Part 1 Fundamentals** 125
   Michael V. Mundschau

9 **Hydrogen Separation Using Dense Composite Membranes
   Part 2: Process Integration and Scale-Up for H$_2$ Production
   and CO$_2$ Sequestration** 155
   David H. Anderson, Carl R. Evenson IV, Todd H. Harkins,
   Douglas S. Jack, Richard Mackay, and Michael V. Mundschau

10 **Gasification and Associated Degradation Mechanisms Applicable
   to Dense Metal Hydrogen Membranes** 173
   Bryan Morreale, Jared Ciferno, Bret Howard, Michael Ciocco John
   Marano, Osemwengie Iyoha, and Robert Enick

11 **Un-supported Palladium Alloy Membranes for the Production of
   Hydrogen** 203
   Bruce R. Lanning, Omar Ishteiwy, J. Douglas Way, David Edlund,
   and Kent Coulter

12 **Palladium-Copper and Palladium-Gold Alloy Composite
   Membranes for Hydrogen Separations** 221
   Fernando Roa, Paul M. Thoen, Sabina K. Gade, J. Douglas Way,
   Sarah DeVoss, and Gokhan Alptekin

13 **Composite Pd and Pd/Alloy Membranes** 241
   Yi Hua Ma

14 **Model-Based Design of Energy Efficient Palladium Membrane
   Water Gas Shift Fuel Processors for PEM Fuel Cell Power Plants** . . . 255
   Mallika Gummalla, Thomas Henry Vanderspurt, Sean Emerson,
   Ying She, Zissis Dardas, and Benoît Olsommer

Part III   Porous, Silica and Zeolite Membranes

15 **Zeolite Membranes** 275
   Weishen Yang and Yanshuo Li

**16 Advanced Materials and Membranes for Applications in Hydrogen
     and Energy Production** 287
     Balagopal N. Nair, Y. Ando, and H. Taguchi

**17 Catalytic Dehydrogenation of Ethane in Hydrogen Membrane
     Reactor** 299
     Jan Galuszka, Terry Giddings, and Ian Clelland

Index 313

Contributors

Gokhan Alptekin TDA Research, Inc., Wheat Ridge, CO 80033, USA

David H. Anderson Eltron Research & Development Inc., 4600 Nautilus Court South, Boulder, CO 80301-3241, USA

Y. Ando R&D Center, Noritake Company LTD., Miyoshi, Aichi 470-0293, Japan

Phillip A. Armstrong Air Products and Chemicals, Inc., Allentown, PA 18195, USA

Timothy R. Armstrong Metals and Ceramics Division, Oak Ridge National Laboratory, Oak Ridge, TN 37831, USA

I. Bay Ceramatec, Inc., 2425 South 900 West, Salt Lake City, UT 84119-1517, USA

Theodore M. Besmann Metals and Ceramics Division, Oak Ridge National Laboratory, Oak Ridge, TN 37831, USA

Arun C. Bose U.S. Department of Energy, NETL, Pittsburgh, PA 15236, USA

Rui Cai Satate Key Lab of Catalysis, Dalian Institute of Chemical Physics, Chinese Academy of Sciences, Dalian 116023, China

Robert D. Carneim Metals and Ceramics Division, Oak Ridge National Laboratory, Oak Ridge, TN 37831, USA

Thierry Chartier Laboratoire de Science des Procédés Céramiques et Traitements de Surface (SPCTS), CNRS/UMR 6638, ENSCI, Limoges, France

Jared Ciferno US DOE National Energy Technology Laboratory, 626 Cochrans Mill Road, PO Box 10940, Pittsburgh, PA 15236, USA

Michael Ciocco NETL Support Contractor, Parsons, Inc., Pittsburgh, PA, USA

Ian Clelland Natural Resources Canada, CANMET Energy Technology Centre-Ottawa, 1 Haanel Drive, Ottawa, Ontario, Canada K1A 1M1

Kent Coulter Southwest Research Institute, TX 78228, USA

Zissis Dardas United Technologies Research Center, 411 Silver Lane, East Hartford, CT 06108, USA

Marcel J. den Exter Energy research Centre of the Netherlands, PO Box 1, 1755 ZG Petten, The Netherlands

Sarah DeVoss TDA Research, Inc., Wheat Ridge, CO 80033, USA

Shain J. Doong Gas Technology Institute, 1700 South Mount Prospect Road, Des Plaines, IL 60018-1804, USA; UOP, 25 E. Algonquin Rd. Des Plaines, IL 60016, USA

David Edlund Idatech, LLC, Bend, OR 97701, USA

S. Elangovan Ceramatec, Inc., 2425 South 900 West, Salt Lake City, UT 84119-1517, USA

Sean Emerson United Technologies Research Center, 411 Silver Lane, East Hartford, CT 06108, USA

Robert Enick NETL Research Associate, University of Pittsburgh, Department of Chemical & Petroleum Engineering, Pittsburgh, PA, USA

Grégory Etchegoyen Centre de Transfert de Technologies Céramiques, Parc d'Ester, Rue Soyouz, BP36823, 87068 Limoges Cedex

Carl R. Evenson IV Eltron Research & Development Inc., 4600 Nautilus Court South, Boulder, CO 80301-3241, USA

E.P. (Ted) Foster Air Products and Chemicals, Inc., Allentown, PA 18195, USA

Sabina K. Gade Colorado School of Mines, Golden, CO 80401, USA

Pascal Del Gallo Air Liquide, Centre de Recherche Claude-Delorme, Jouy-en-Josas, France

Jan Galuszka Natural Resources Canada, CANMET Energy Technology Centre-Ottawa, 1 Haanel Drive, Ottawa, Ontario, Canada K1A 1M1

Terry Giddings Natural Resources Canada, CANMET Energy Technology Centre-Ottawa, 1 Haanel Drive, Ottawa, Ontario, Canada K1A 1M1

Mallika Gummalla United Technologies Research Center, 411 Silver Lane, East Hartford, CT 06108, USA

Wim G. Haije Energy research Centre of the Netherlands, PO Box 1, 1755 ZG Petten, The Netherlands

Barry J. Halper Air Products and Chemicals, Inc., Allentown, PA 18195, USA

Todd H. Harkins Eltron Research & Development Inc., 4600 Nautilus Court South, Boulder, CO 80301-3241, USA

J. Hartvigsen Ceramatec, Inc., 2425 South 900 West, Salt Lake City, UT 84119-1517, USA

B. Heck Ceramatec, Inc., 2425 South 900 West, Salt Lake City, UT 84119-1517, USA

Bret Howard US DOE National Energy Technology Laboratory, 626 Cochrans Mill Road, PO Box 10940, Pittsburgh, PA 15236, USA

Omar Ishteiwy Colorado School of Mines, Golden, CO 80401, USA

Osemwengie Iyoha NETL Research Associate, University of Pittsburgh, Department of Chemical & Petroleum Engineering, Pittsburgh, PA, USA

Douglas S. Jack Eltron Research & Development Inc., 4600 Nautilus Court South, Boulder, CO 80301-3241, USA

Bruce R. Lanning ITN Energy Systems, Inc., 8130 Shaffer Parkway, Littleton, CO 80127-4107, USA

Francis Lau Gas Technology Institute, 1700 South Mount Prospect Road, Des Plaines, IL 60018-1804, USA; GreatPoint Energy, 1700 South Mt. Prospect Rd., Des Plaines, IL 60018, USA

Yanshuo Li State Key Laboratory of Catalysis, Dalian Institute of Chemical Physics, Chinese Academy of Sciences, Dalian 116023, China

Yi Hua Ma Center for Inorganic Membrane Studies, Department of Chemical Engineering, Worcester Polytechnic Institute, Worcester, Massachusetts 01609, USA

Richard Mackay Eltron Research & Development Inc., 4600 Nautilus Court South, Boulder, CO 80301-3241, USA

John Marano NETL Support Contractor, Technology and Management Service, Inc., Pittsburgh, PA, USA

Bryan Morreale US DOE National Energy Technology Laboratory, 626 Cochrans Mill Road, PO Box 10940, Pittsburgh, PA 15236, USA

Michael V. Mundschau Eltron Research & Development Inc., 4600 Nautilus Court South, Boulder, CO 80301-3241, USA

B. Nair Ceramatec, Inc., 2425 South 900 West, Salt Lake City, UT 84119-1517, USA

Balagopal N. Nair R&D Center, Noritake Company LTD., Miyoshi, Aichi 470-0293, Japan

Benoît Olsommer United Technologies Research Center, 411 Silver Lane, East Hartford, CT 06108, USA

Estela Ong Gas Technology Institute, 1700 South Mount Prospect Road, Des Plaines, IL 60018-1804, USA

Fernando Roa Colorado School of Mines, Golden, CO 80401, USA; Current Address: Intel Corporation, Chandler, AZ 85223, USA

Ying She United Technologies Research Center, 411 Silver Lane, East Hartford, CT 06108, USA

T. Small Ceramatec, Inc., 2425 South 900 West, Salt Lake City, UT 84119-1517, USA

Gary J. Stiegel U.S. Department of Energy, NETL, Pittsburgh, PA 15236, USA

H. Taguchi R&D Center, Noritake Company LTD., Miyoshi, Aichi 470-0293, Japan

Paul M. Thoen Colorado School of Mines, Golden, CO 80401, USA

M. Timper Ceramatec, Inc., 2425 South 900 West, Salt Lake City, UT 84119-1517, USA

Thomas Henry Vanderspurt United Technologies Research Center, 411 Silver Lane, East Hartford, CT 06108, USA

Jaap F. Vente Energy research Centre of the Netherlands, PO Box 1, 1755 ZG Petten, The Netherlands

Alain Wattiaux Institut de Chimie de la Matière Condensée de Bordeaux (ICMCB), CNRS/UPR 9048, Pessac, France

J. Douglas Way Colorado School of Mines, Golden, CO 80401, USA

M. Wilson Ceramatec, Inc., 2425 South 900 West, Salt Lake City, UT 84119-1517, USA

Weishen Yang Satate Key Lab of Catalysis, Dalian Institute of Chemical Physics, Chinese Academy of Sciences, Dalian 116023, China

Part I
Ion Transport Membranes
and Ceramic-Metal Membranes

Chapter 1
Progress in Ion Transport Membranes for Gas Separation Applications

Arun C. Bose, Gary J. Stiegel, Phillip A. Armstrong, Barry J. Halper, and E. P. (Ted) Foster

Prologue

This chapter describes the evolution and advances of ion transport membranes for gas separation applications, especially separation of oxygen from air. In partnership with the US Department of Energy (DOE), Air Products and Chemicals, Inc. (Air Products) successfully developed a novel class of mixed ion–electron conducting materials and membrane architecture. These novel materials are referred to as ion transport membranes (ITM). Generically, ITMs consist of modified perovskite and brownmillerite oxide solid electrolytes and provide high oxygen anion and electron conduction typically at high temperatures driven by an oxygen potential gradient without the need for external power. The partial pressure ratio across the ITM layer creates the driving force for oxygen separation.

These oxide ion conductors can be fabricated in desired module configurations for commercial reaction or separation applications. ITMs can separate oxygen from air at high temperatures and pressures at high flux and purity and have the potential to dramatically reduce the cost of oxygen production compared to conventional cryogenic processes. ITM reactors allow low-cost, single-step approaches to conduct partial oxidation chemistries, such as the partial oxidation of methane to synthesis gas, by matching oxygen transport rate with the natural gas reforming kinetics.

Technology research, development, and demonstration (RD&D) is currently in progress to develop, scale up, and deploy large-scale tests of the ITM technology to achieve significant advancements of the oxygen production process. The ITM oxygen R&D efforts will result in the development of a novel air separation technology for large-scale production of oxygen from air with the co-production of power and for the integration of ITM oxygen systems with combined cycle coal gasification plants and other advanced power generation systems. An overriding goal of the RD&D is to cut the cost of oxygen production by approximately one-third compared

A.C. Bose (✉)
US Department of Energy, NETL, Pittsburgh, PA 15236, USA
e-mail: arun.bose@netl.doe.gov

A.C. Bose (ed.), *Inorganic Membranes for Energy and Environmental Applications*,
DOI 10.1007/978-0-387-34526-0_1, © Springer Science+Business Media, LLC 2009

to conventional cryogenic air separation technologies and demonstrate all necessary technical and economic requirements for ultimate commercial deployment.

Motivation

Oxygen is a key requirement for many advanced industrial processes with an annual US market approximating several billion dollars. It is the third largest volume chemical produced worldwide. The bulk of the commercial oxygen supply is extracted from the atmosphere and purified by cryogenic distillation. Since the invention of over a century ago, cryogenic distillation for oxygen production has gone through numerous advances in the twentieth century, including improvements in column design, column packing, and the thermodynamic cycle. It is generally recognized that cryogenic oxygen is not affordable for low-cost, next-generation energy production and other oxygen-intensive industries. Cost-effective, environmentally progressive power production requires a novel oxygen production process beyond current benchmarks of technology performance and cost. Energy scenario analyses indicate that future technology concepts and environmental mandates for clean energy production could be techno-economically enabled by a low-cost air separation technology. Lower cost oxygen would broaden the applicability of clean energy processes, such as the oxygen-blown gasification of coal to the integrated gasification combined cycle power plants, oxygen-enhanced combustion (oxycombustion) of coal to produce power, and the conversion of coal to transportation fuels and hydrogen. Oxygen-intensive industries such as steel, glass, non-ferrous metallurgy, refineries, and pulp and paper would realize cost and productivity benefits as a result of the ITM oxygen. Process economic analyses illustrate that the benefits of the ITM oxygen technology relative to cryogenic air separation technology is significant for nearly all oxygen-intensive industries.

Non-cryogenic air separation processes use physical property differences rather than boiling points to separate oxygen from nitrogen and other constituents of air. Two alternative air separation technologies, adsorption and polymeric membrane separations, are also commercially important technologies that have been practiced for many years. A common technology is pressure swing adsorption (PSA). PSA units produce oxygen by passing air through a vessel containing adsorbent materials and, for oxygen service, include a variety of materials that selectively adsorb nitrogen, moisture, and carbon dioxide. Efficiency limitations inherent in adsorption technology restrict its application to relatively small plants of less than approximately 150 tons-per-day (TPD) oxygen production capability, while polymer-based membranes simply do not provide the separation factor and flux required for economical large-scale production of oxygen. Highly active, oxygen-selective, and stable solid-state materials, such as reported by Teraoka et al. [1], could play a significant role in membrane-based commercial separations of oxygen from air. ITMs can be considered as a medium or a barrier to effect selective permeation of a desired gas species at practical flux, and the driving force is a function of the inlet and outlet partial pressures of the permeate species [2].

Evolution of Ion-Transport Membranes

The world's first patent on cryogenic air separation plant for the production of oxygen was issued on February 27, 1902. Cryogenic air separation technology, now proven and reliable, has gone through many improvements since its commercial entry about a century ago. As an example, the early processes required air compression to 200 atmospheres, whereas the present-day process requires lower air compression, as low as 10–15 atmospheres. The industrial gas industry continues to refine and improve cryogenic air separation units (ASU) through technology research and development and various execution strategies. However, the overall thermodynamic efficiency of the modern-day cryogenic ASU is likely reaching its theoretical limits; significant technical and cost breakthroughs are unlikely to result in a step-change reduction in the cost of producing oxygen. The ITM oxygen technology is a radically different approach to producing low-cost, high-purity tonnage oxygen at temperatures synergistic with power production and many other oxygen-intensive applications.

DOE-sponsored R&D resulted in the development of ITMs that can separate oxygen from air at high temperatures and pressures with high flux and purity in a single-stage operation. In power cycles generically, the work output for a given heat input depends on the high to low pressure ratio and the capital efficiency depends on the temperature ratio. The operating conditions of ITMs are complementary to the efficient and low-cost power generation economics: high pressure and high temperature. DOE-supported studies established the technical feasibility of the ITM air separation process and confirmed a step-change cost reduction. ITMs are oxygen conductors and, therefore, 100% oxygen selective. The separation factors achieved by ITMs are several orders of magnitude higher than can be achieved by other separation methods (Fig. 1.1).

Initial test results demonstrating the oxygen separation capacity of mixed ion–electron conducting membranes [3] led DOE to recognize that membrane-based technology offers a near-term realistic opportunity to manufacture oxygen at lower cost. Although many engineering challenges have to be met first, significant commercial application potential was evident. This led to the ITM Oxygen Cooperative Research and Development Agreement between the DOE and Air Products [4, 5]. This government–industry partnership has resulted in several inventions of ITM materials and processes to advance the oxygen production technology beyond the current state of the art [6] and, ultimately, industrial commercialization, attesting to the DOE leadership in a crucial long-term technology.

ITM Materials

ITM oxygen technology is based on a novel class of engineered ceramics that have both electronic and oxygen ionic conductivity properties when operated at high temperatures, typically 800–900°C. The mixed conductors are complex

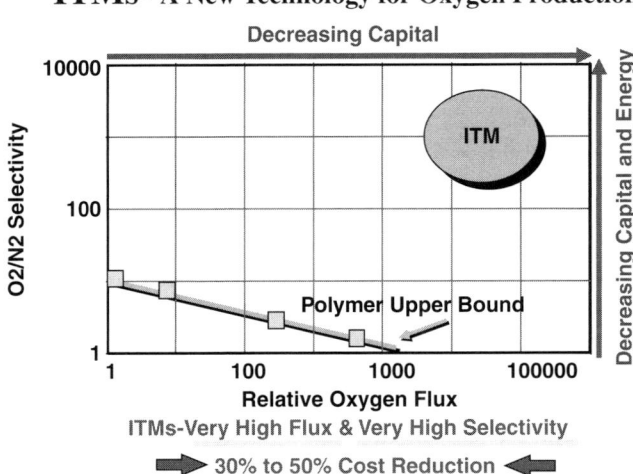

Fig. 1.1 ITM Provides Commercially Relevant Flux and Selectivity for Oxygen Separation at Lower Cost (Figure provided by Air Products and Chemicals, Inc., under a US Department of Energy Cooperative Agreement)

formulations of inorganic mixed-metal oxides, for example, perovskites such as $(La,Sr)(Fe,Co,Cu)O_{3-\delta}$ whose crystal structures are deficient in oxygen to achieve charge neutrality causing a distribution of oxygen vacancies in their lattices. The brownmillerites can be viewed as perovskites with one-sixth of the oxygen atoms removed and include $SrFeO_{2.5}$, $CaFe_{2.5}$, $BaInO_{2.5}$, $SrGd_2O_5$, $Ca_3Fe_2TiO_8$, and $Ba_3In_2ZrO_8$.

It has been reported [7] that La $Ni_{1-x}Fe_xO_3$ behaves as a mixed conductor at x >0.2. In 1985, Teraoka [1] reported oxygen permeability of $\bar{L}a_{1-x}Sr_xCo_{1-y}Fe_yO_{3-\delta}$ in connection with its defect structure and oxygen-extractive properties. Other works [8, 9] by these researchers indicated that Ba and Cu substituted $LnCoO_3$-based perovskite-type oxides (where Ln is an element from the lanthanides series) are oxygen-permeable membranes and Ln also affected the oxygen permeability, which increased with a decrease of Ln ionic radius. Ionic and electronic conductivities of $La_{1-x}Sr_xCo_{1-y}Fe_yO_{3-\delta}$ were measured and a vacancy mechanism for oxygen ion transport was suggested. Teraoka et al. concluded that $La_{1-x}Sr_xCo_{1-y}Fe_yO_{3-\delta}$ is a promising oxygen permeating material above 500 °C and could be used as an oxygen separation membrane.

Preferred mixed conducting oxides for ITM oxygen compositions are generically represented by the formula $La_xA'_xCo_yFe_{y'}Cu_{y''}O_{3-z}$ where A' is selected from Sr, Ba, Ca, or Mg, and z is a number that makes the composition charge neutral. Selection of A is based on achieving low expansion of the dense phase with temperature. Copper provides improved control over dimensional changes when used in the mix with cobalt and iron, and a coefficient of thermal expansion matches with the substrate. This composition also provides superior resistance when used in the presence of moisture. A US patent describes A site-rich and B site-rich multi-component metallic oxides that are suitable for use in oxygen separation devices in the presence of high-pressure carbon dioxide and moisture [10]. A choice of a temperature regime to overcome the problems associated with the degradation of Ca, Sr, or Mg oxides caused by carbon dioxide has been described [11]. A generic variation of mixed conducting metal oxides can be represented by the formula $(Ln_{1-x}A_x)_z(B_{1-y}B'_y)O_{3-\delta}$. Ln represents elements selected from the lanthanides block of the periodic table; A and B are cationic substituents. 'A' represents one or more elements selected from Mg, Ca, Sr, and Ba; B and B' each represent one or more elements selected from Sc, Ti, V, Mn, Fe, Co, Ni, Cu, Cr, Al, Zr, Mg, and Ga; δ is a number that renders the compound charge neutral.

Impurities such as sulfur dioxide can react with ion-conducting membranes and reduce performance, and the reactivity depends on the partial pressure of SO_2 in oxygen-containing stream and certain reactant species present in the ITM. It has been observed [12] that each membrane composition has a critical threshold SO_2 partial pressure, above which SO_2 will react with the ITM materials to impact oxygen flux. This result helps to determine the required level of sulfur removal from oxygen-containing feeds.

Membrane Stability and Mechanical Issues

The lattice volume in ITMs may change with oxygen partial pressure. For example, the membrane may experience expansion on the oxygen-lean side and contraction on the oxygen-rich side, leading to the potential for stresses within the membrane. At constant operating temperature, oxygen will be incorporated into the solid ITM lattice structure when the oxygen partial pressure increases and the membrane material will contract. Also at constant temperature, oxygen will be evolved from the permeate side of the solid lattice structure when the oxygen partial pressure decreases and the membrane material will expand. These problems have been partly circumvented by the development of modified perovskite ITMs, such as brownmillerites and transition metal-doped dual phase perovskites, and are still under further development, partly by operational control as described later.

Although oxygen flux increases inversely with membrane thickness, if membrane thickness becomes smaller than a few hundred microns, it must be supported on a porous substrate. This requires novel methods to fabricate thin membranes mounted on porous supports, membrane sealing materials and techniques, and optimal membrane geometry.

A key technology challenge of the ITM wafer fabrication is, therefore, to deposit a thin dense coating on the surface of a porous substrate. The dense layer is comprised of the ITM metallic oxides, and functions as solid state electro-chemical cells to support flow-through of oxygen ions relying on a partial pressure induced driving force. The layer on the air side of the membrane generally experiences little or no expansion, but on the inside where the oxygen partial pressure is lower, the substrate may expand. When this occurs, the membrane can delaminate from its porous support and eventually fracture. In addition, membrane reactor design strives to minimize the thermal expansion mismatch between the membrane and the substrate and controlling the overall expansion by, for example, choosing support and membrane materials with the same coefficient of expansion, preferably using the same materials. The porous support may be formed of the same or similar composition as the multi-component metal oxide dense layer, a compatible multi-component metal oxide layer with surface exchange reaction catalytic activity, or non-ion–electron conducting inert materials [13]. Several preparation techniques for the dense and support layers are documented in the referenced patent.

A long operating life of ITM oxygen membranes requires good material stability in the industrial process environments to which the membranes will be routinely subjected in practice. Thermodynamic modeling work with a model compound, lanthanum cobalt oxide ($LaCoO_{3-\delta}$), assessed the stability of ITM-like materials. Experimental stability data verified the first principle thermodynamic stability model results and the technique was applied to predict the stability of the ITM oxygen materials in practical operating conditions [14] and for process control system designs.

Quantitative, tensile creep measurements for an ITM oxygen material provide physical property data required for detailed thermal and mechanical modeling of ITM oxygen structures as well as the design of commercial modules. A study determined the tensile creep behavior of $(La_{1-x}Sr_x)(Co_{1-y-z}Fe_yCu_z)_aO_{3-\delta}$ and found power dependence of creep rate on applied stress and on the oxygen partial pressure. The researchers believe that the mechanism causing and controlling creep is grain boundary sliding and derived a steady-state creep rate equation as a function of stress, temperature, and P_{O2}. This study provides some insight into how to control creep failure modes and improve ITM material stability [15].

Temperature gradients in a mixed conducting metal oxide ceramic part create differential strains due to differential thermal expansion and contraction. Similarly, oxygen stoichiometry gradients in a ceramic part can create differential strains due to differential chemical expansion and contraction. If the thermal transients in an operational system occur too quickly, thicker parts of the ITM stack may not equilibrate rapidly enough with oxygen feed and the permeate sides of the membrane, and the membrane material will tend to expand or contract near the surfaces at a different rate than the material in the membrane interior. A gradient in oxygen stoichiometry may be sufficiently large to create a correspondingly large differential chemical expansion, and therefore large mechanical stresses, and can cause a failure of the part. When an ITM wafer is housed within a module structure its ability to change dimensions is restricted.

During ITM operation, the feed and permeate sides are at different oxygen partial pressures and activities. An equilibrated oxygen activity can prevent build-up of differential strains on the feed and permeate sides of the membrane and minimize or prevent stress-induced failure, especially during heating and cooling cycles, and planned and emergency shutdowns.

Isocompositional heating and cooling is a process control technique [16] to improve ITM stack operational reliability and comprises a method of controlling the differential strain below a maximum allowable level between the feed side and the permeate side of an operating oxygen separation ceramic membrane. Isocompositional heating and cooling requires that the feed and permeate sides of the membrane be at the same oxygen activity.

Sealing and Joining Techniques for ITM Modules

For high-temperature gas separation applications, leak-free sealing of the ITM module components and parts is essential and requires chemically resistant ceramic–metal and ceramic–ceramic seals with similar mechanical, chemical, and expansion characteristics as the membrane material. Little prior art exists for sealing and joining designs for tonnage-quantity ITM modules. ITM ceramics are susceptible to breakage and will have to be joined preferably without any joint interface property difference, possibly as a routine plant maintenance procedure.

An innovative seal described in the US patent [17] between a metal header and a ceramic, full-sized oxygen product pipe performed successfully during engineering tests at 200 psig, opening a novel approach for a ceramic–metal joining and sealing technology.

A transient liquid phase (TLP) method has been developed for joining ITMs [18] using TLP liquids made from model compounds similar to that used in making ITM ceramic materials. The joining method produces no interfacial phase and no property disparities with the joined parts. A TLP material is applied at the interface of the parts to be joined. The assembly is heated to the joining temperature at which point a liquid phase is formed at the interface. The liquid phase fills gaps at the interface and diffuses into parent materials being joined. As the liquid phase diffuses into the surrounding materials, the interface solidifies isothermally into a material of similar or compatible composition to that of the joined materials, leaving behind either no interfacial phase or a compatible, refractory interfacial phase. TLP joints can be used at temperatures that approach or can exceed the joining temperature.

A lanthanum calcium ferrite (LCF) perovskite system ($La_xCa_{1-x}FeO_3$) was joined to itself to demonstrate the feasibility of the TLP method. LCFs can be used as a component of air separation membranes, so the assemblies need to operate at high temperatures. Therefore the joint interface needs to resist creep, avoid problems with thermal cycling, and be chemically stable in the high-temperature environment. Ideally, the interface in such systems is the same as the parts in the assembly.

A model material selected for the joining experiment was $La_{0.9}Ca_{0.1}FeO_3$, so the preferred interface would also be $La_{0.9}Ca_{0.1}FeO_3$. The flux and refractory phase amounts are chosen so that when combined they yield a stoichiometry of $La_{0.9}Ca_{0.1}FeO_3$. For example, a combination that yields the correct stoichiometry is when the flux consists of 78 wt% Fe_2O_3 and 22 wt% CaO and the refractory phase consists of 29 wt% Fe_2O_3 and 71 wt% La_2O_3. Monolithic specimens of

$La_{0.9}Ca_{0.1}FeO_3$ were joined using the processed powders. Many successful joints were produced. The joints were ideal and scanning electron microscopic and X-ray analyses have shown that the TLP joints are essentially pore-free ideal joints. Additional sealing and joining methods are described in US patents 7,011,898 and 7,094,301 [19, 20].

ITM Operating Principles and Performance Enhancements of ITMs

The ITMs are non-porous, mixed ion and electron-conducting materials operating typically at 800–900°C. Ion and electron flow paths occur through the membrane layer and are driven by the relative oxygen partial pressure ratio across the membrane, typically 100–300 psig on the feed side and low to sub-atmospheric pressure on the permeate side. Sub-atmospheric permeate side pressure operations for a given feed side pressure improves separation flux. A compressor pulls the oxygen away from the permeate side of the membrane and compresses it to its final delivery pressure. Electrons flow across the membrane layer counter-current to the oxygen anions to complete an internal electronic circuit (Fig. 1.2). This approach can be used for producing oxygen in large quantities. A process variable is the trans-membrane partial pressure ratio that can be varied to an extent to achieve a desired oxygen throughput rate.

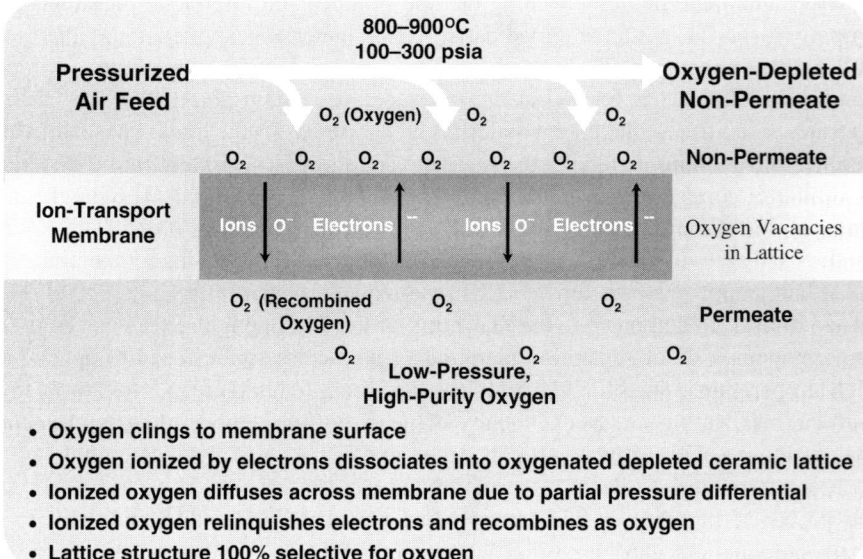

Fig. 1.2 Oxygen Separation Using Ion Transport Membrane (Courtesy of the US Department of Energy and Air Products and Chemicals, Inc.)

The flux of oxygen through an ITM oxygen membrane is dependent on three mechanisms: the first is the dissociation of oxygen molecules into oxygen anions at the membrane surface, known as surface oxygen exchange kinetics. The second is the migration of oxygen anions through the membrane, while the third is the recombination of oxygen anions back to oxygen molecules on the other side of the membrane. Each mechanism has the potential to become the rate-controlling step.

The addition of surface catalysts depending on the specific ITM composition and modifications of the membrane surface may improve flux, provided gas phase mass transfer resistances are negligible, that is, the separation operation is not proceeding in mass transfer controlling modes. Oxygen separation from air proceeding via interfacial catalysis on the membrane surface could promote the initial dissociative adsorption of oxygen at the reducing surface leading to the formation of oxygen anions (O^{2-}) and, as oxygen molecules are reformed, electron transfer takes place counter-currently from the permeate side. The mechanism may be represented by [21]

$$O_2 + 4e^- + 2V_{\ddot{o}} \rightarrow 2O_o^{2-} \tag{1.1}$$

where $V_{\ddot{o}}$ represents an oxide ion vacancy in a normal lattice position where such vacancy sites may be expected to be compatible towards stabilization of O^{2-} species. Subsequently, O^{2-} migrates from the catalytic surface into the mixed (O^{2-}/e^-) conducting membrane bulk phase for later reformation into molecules or for partial oxidation chemistry.

Ideally, catalysis for promoting the oxygen reduction reaction should possess both mixed ionic and electronic conduction properties. This is most conveniently achieved by a catalyst having a lattice closely related to the membrane bulk. This requirement can be met by species, such as Fe^{2+}, Co^{3+}, or Ni^{4+}, within perovskite B lattice sites. The oxygen species, being energetically close to the perovskite-related catalyst conduction band edge, can facilitate subsequent net electron transfer from the metal to the dissociatively adsorbed oxygen ion, especially under conditions where this band edge is negative to that of the dissociatively adsorbed oxygen phase [21].

It has been reported [22] that micro/nano surface reactive layers introduced on the surface of oxygen ion transport membranes enhanced the oxygen permeability of dense oxygen-permeable perovskite-type ceramic membranes, $La_{0.7}Sr_{0.3}Ga_{0.6}Fe_{0.4}O_{3-\delta}$. The results also showed that the oxygen permeation flux is influenced by the surface area of the surface-reactive layer.

Staged separations at controlled pressure ratios have been shown to produce more oxygen from a specific amount of air and at lower power consumption when compared with the single-stage operation [23]. However, staging in ITM devices is not necessary to improve product purity.

Flux Expression for Oxygen Separation

The flux of oxygen through an ITM device is given by the Nernst–Einstein formula,

$$j_{O_2} = \frac{\sigma_i RT}{4Ln^2F^2} \ln\left(\frac{p'_{O_2}}{P''_{O_2}}\right)$$

where j_{O2} is the oxygen flux, F is Faraday's constant, L is the membrane thickness, n is the charge of the charge carrier ($= 2$), R is the ideal gas constant, T is the absolute temperature, P'_{O_2} is the oxygen partial pressure at the feed surface of the membrane, P''_{O2} is the oxygen partial pressure at the permeate surface of the membrane, and σ_i represents the material conductivity.

This expression clearly identifies the natural logarithm of the oxygen partial pressure ratio as the driving force for the oxygen flux. The oxygen flux through an electrochemical device can be made quite high, simply by manipulating the partial pressure ratio of oxygen on the two sides of a membrane. It is even more pronounced by the thickness of the dense layer. In a conventional membrane (e.g., polymeric), the driving force involves the difference between the two partial pressures, but yields a relatively low range of potential fluxes due to poor separation selectivities. The kinetic diameters of oxygen and nitrogen are too close to separate by size exclusion filtration at commercially relevant production rates with desired product purities.

Integration of ITM Oxygen with Energy Cycles

The operation of the ITM oxygen technology is synergistic with power cycle integration applications. The co-production of oxygen and power with lower capital cost and parasitic power loads compared to cryogenic processes is an incentive to develop the ITM oxygen technology for energy and oxygen-intensive industries in order to improve the cost, efficiency, and environmental performances. A US patent [24] describes recovery of the useful work by expansion of the pressurized oxygen-depleted, non-permeate stream at a temperature below the ITM device operating temperature. A simple schematic of one implementation approach of an ITM oxygen membrane is shown in Fig. 1.3a and illustrates the effectiveness of the membrane separation in this application [25].

The approach in Fig. 1.3a is based on an ITM-specific developmental gas turbine. With some gas turbines, the entire air stream required for oxygen production can be extracted from the compressor section of a gas turbine and fed to the ITM unit. Because this stream experiences relatively little pressure drop as it passes over the membrane, the energy contained in the hot, pressurized non-permeate stream can be recovered by reinjecting the hot pressurized oxygen-depleted, non-permeate gas into the combustion turbine for conversion of its thermal and hydraulic energy to shaft work. By contrast, if the air is fed to a cryogenic ASU instead of the ITM oxygen unit, the heat in the air stream would be rejected and much of the pressure energy expended during oxygen extraction.

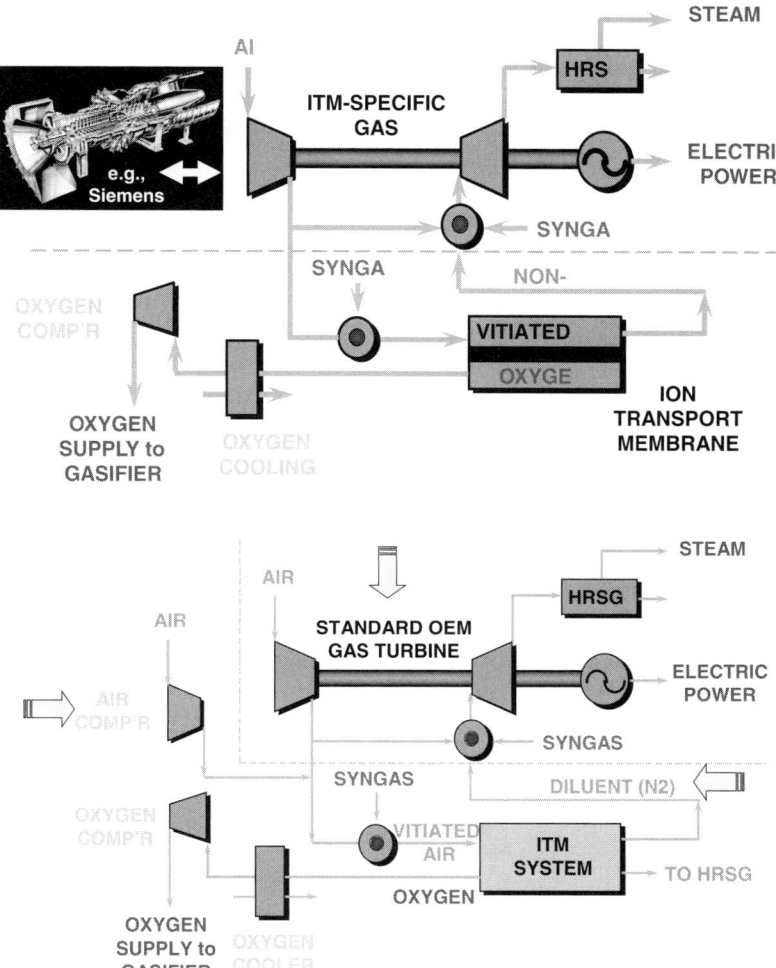

Fig. 1.3 a Schematic of ITM Oxygen Integration With Conventional IGCC Using ITM-Specific Turbines (Figure provided by Air Products and Chemicals, Inc., under a U.S. Department of Energy Cooperative Agreement). **b** Schematic of ITM Oxygen Integration with Conventional IGCC Using Commercially Available Turbines (Figure provided by Air Products and Chemicals, Inc., under a US Department of Energy Cooperative Agreement)

ITM oxygen is also amenable to other commercially available gas turbines where only a portion of the air required for the ITM air separation unit can be extracted from the gas turbine. In this case, as shown in Fig. 1.3b, a supplemental air compressor is used to provide the remaining air. This approach has been shown to preserve much of the cost benefits of the technology.

ITM Oxygen Technology RD&D Program Overview

The goal of the ITM oxygen technology RD&D efforts is to develop and scale-up the technology to tonnage-quantity scales for stand-alone oxygen plants and integration with coal gasification and other advanced power generation systems. A collateral goal is to demonstrate nearly one-third lower capital cost and parasitic energy requirement compared to conventional cryogenic air separation processes. The work scope is extensive, long term, and high risk and requires a high degree of technical sophistication needed to address the specific scientific and technical objectives of the program mandates. The intended result is to overcome the technology, market, economic, and environmental barriers to implementing ITM oxygen technology in the nation's energy portfolio.

Phase I involved materials and process R&D, material selection for scale-up, engineering fundamental studies, module design approaches, ITM process and ceramic infrastructure development, scale-up techniques, and technical risk reduction. Based on the successful outcome of Phase I, Phase II scaled up the technology to a 5 TPD engineering prototype to obtain performance data on full-scale membrane modules. In Phase III, the technology is being scaled to a 150 TPD Intermediate Scale Test Module to obtain materials, cost, engineering, fabrication, operating, and performance data necessary for the 1,500–2,500 TPD unit as early-entrance commercial plants. The three-phase technology RD&D effort will demonstrate all necessary technical and economic requirements for scale-up en route to eventual industrial commercialization.

RD&D Results and Accomplishments Summary

Phase I objectives, focused on materials and process R&D and the design, construction, and operation of a 0.1 TPD Technology Development Unit (TDU), have been successfully achieved. The TDU test data allowed establishing the cost and performance targets for stand-alone, tonnage-quantity commercial ITM oxygen plants and integration schemes of ITM oxygen with advanced power generation systems. As shown in Fig. 1.4, the commercial cost and performance targets for the ITM were achieved. Studies on turbo-machinery integration schemes and pathways have been completed.

Thin cost-optimized, planar ceramic wafers have been designed and fabricated as the building blocks of membrane separation modules, and currently an inventory is being built for constructing scaled-up prototypes. Statistical wafer samples undergo quality control tests and a technique has been developed to re-use the rejected wafers. This breakthrough enables the development of high-performance ITM modules at low cost and meets the lower cost goal of the ITM oxygen process. The research team successfully built the first 0.5–1.0 TPD commercial-scale individual ITM oxygen separation modules.

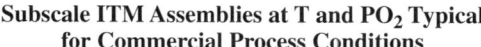

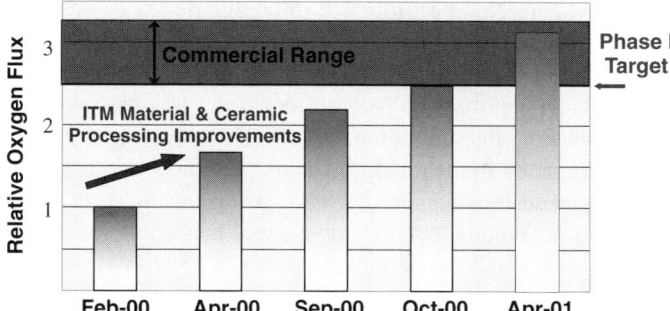

Fig. 1.4 Phase I Cost and Performance Targets for Scale-Up and Commercialization Decisions (Figure provided by Air Products and Chemicals, Inc., under a US Department of Energy Cooperative Agreement)

The Phase II activities focused on the design, construction, commissioning and testing of a 5 TPD sub-scale engineering prototype (SEP) facility. A schematic is shown in Fig. 1.5. The prototype facility is testing the performance of the full-size ITM oxygen modules housed in a pressure vessel at a variety of operating conditions and generating process information for further scale-up to a 100–150 TPD engineering test facility in Phase III.

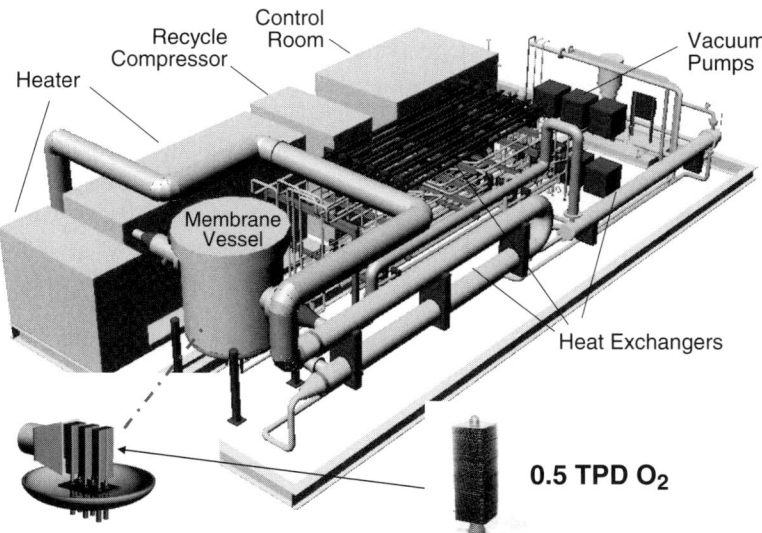

Fig. 1.5 Schematic of the Five TPD Subscale Engineering Prototype for Testing Commercial-Scale ITM Oxygen Modules (Figure provided by Air Products and Chemicals, Inc., under a US Department of Energy Cooperative Agreement)

The 5 TPD prototype achieved target performance for system pressure and temperature at an airflow rate approaching commercial operating conditions. Designed experiments assessed module performance through a series of scheduled process variable changes producing oxygen of more than 99% purity. Two commercial-scale modules in the SEP have been successfully cycled between operating and idle conditions without any material and performance degradations. This success attests that the materials and architecture of the new technology can withstand the stresses accompanying anticipated operational transients and allow successful steady-state operations to produce designed quantities of oxygen product. Thus significant milestones enabling the technology towards commercial maturity have been achieved.

In Phase III, it is planned to operate the 100–150 TPD ITM facility integrated with a gas turbine. The facility will provide the engineering, scale-up, cost, operating, and performance data for building an oxygen supply module for early-entrance commercial plants.

ITM Oxygen Module Design

DOE has evaluated two different architectural approaches for ITM-based oxygen separation, planar and tubular; however, engineering scale-up of the planar approach is currently being pursued. The planar design allows desired gas phase mass transfer and is amenable to standard ceramic processing technologies.

Structures composed of an ITM dense layer and the contiguous support are called wafers. Several wafers, when joined together and placed inside a pressure vessel forming a separation device, are the ITM air separation module. ITM module is the separation devices and the ITM wafers form the building blocks of the separation device that allows the selective permeation of oxygen at commercially relevant flux and purity.

Inverse relationship of oxygen flux with membrane thickness requires that membrane devices are composed of thin film structures. Mechanistically, the flux is inversely proportional to the membrane thickness when surface oxygen exchange kinetics is not controlling, and reducing membrane thickness can have a beneficial effect. Ideally, oxygen flux should be dominated by transport through the dense layer. The thin film membrane devices should be able to support the pressure and mechanical loads, and a multi-layer structure has been developed (Fig. 1.6). The porous layer is a continuum of the dense layer and preferably is made of the same ITM oxygen material. Composite structures, where both the dense and porous support layers are made of ion–electron conducting phases, have been reported in a patent literature [26]. The pore diameters of the porous layer are much lower than the usual substrate pore size and oriented to minimize kinetic limitations thus substantially improving oxygen separation flux. The oxygen flux can be improved by conduction through the walls of the pores.

Several solid state membrane wafer units are joined to form a membrane module and channels are incorporated between the wafers for mediating the air flow.

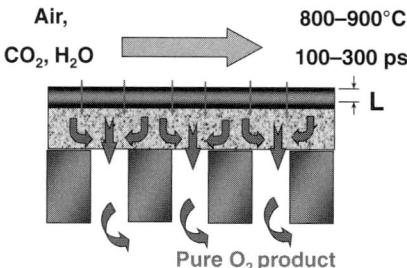

Fig. 1.6 Supported Thin-Film ITM Structure (Figure provided by Air Products and Chemicals, Inc., under a US Department of Energy Cooperative Agreement)

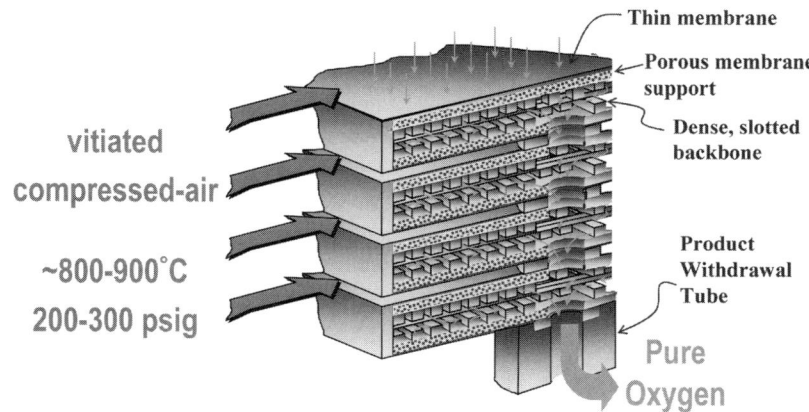

Fig. 1.7 ITM Oxygen Module Architecture - Planar Wafer Stack Design (Figure provided by Air Products and Chemicals, Inc., under a US Department of Energy Cooperative Agreement)

The wafer can consist of internal layers with different porosity or slots leading to a central hole that, when joined with other wafers in a module stack, allow recovering oxygen product from the module oxygen pipe (Fig. 1.7). Deposition of dense oxygen separation layer on channel-free graded support but with connected porosity reduces kinetic limitations to oxygen transport and, thus, significantly improves oxygen flux [27].

Planar solid state membrane module for separating oxygen from air that provides improved pneumatic and structural integrity and with ease of manifolding has been fabricated. The modules are formed from a plurality of planar membrane units where each membrane unit is composed of a porous support layer in contact with a dense mixed conducting oxide layer in flow communication with air on the feed side from which oxygen will be separated. A class of multi-component metallic oxides particularly suited for fabricating parts and components used in ITM oxygen separation devices have been described in a patent [28]. These materials provide a favorable balance of oxygen flux, resistance to degradation at high oxygen partial pressures, favorable sintering properties and thermal expansion match with other separation parts. This material is represented by the formula $Ln_x A'_x Co_y Fe_{y'} Cu_{y''} O_{3-z}$, where Ln is selected from the lanthanide block.

Commercial Production of ITM Oxygen Modules

The planar wafers described in a previous section have been scaled to their full commercial dimensions and produced in volume on a pilot production line using standard ceramic manufacturing techniques. The production activities established the feasibility of achieving low-cost production required to meet overall economic targets. Multi-wafer modules were constructed and statistically tested; the modules produced oxygen exceeding commercial flux and purity targets in high-temperature, high-pressure, pilot-scale experiments. A schematic of a planar wafer module design is shown in Fig. 1.7, in which four wafers are shown joined to a common oxygen withdrawal tube.

Each wafer consists of a thin, outer ITM dense layer through which the oxygen ions diffuse. The thin layer is supported by a porous layer, which is itself supported by a slotted layer. Hot, high-pressure air flows between the wafers. Oxygen passes from the air outside each wafer, through the thin dense outer layer, through the pores of the porous layer, and into the slots of the innermost layer. Oxygen is collected at the center of each wafer in a tubular region formed by the joined wafers and passes out of the module through a ceramic tube sealed to a metal pipe. The high-pressure air on both sides of each wafer serves to compress the wafer equally. The planar design also makes for a very compact separation device while facilitating good gas phase mass transfer. All the layers are made of the same ceramic material and, therefore, provide thermal and chemical expansion match.

Stacking and joining a number of commercial-size wafers is used to construct multi-wafer modules. Figure 1.8 illustrates the ITM oxygen submodule construction

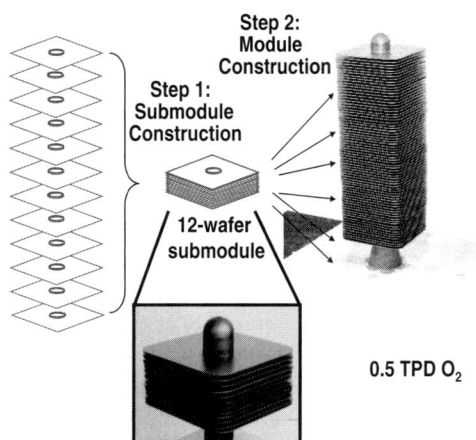

Fig. 1.8 Commercial Module Construction Process Steps (Figure provided by Air Products and Chemicals, Inc., under a US Department of Energy Cooperative Agreement)

protocol. Individual wafers, each fitted with a ceramic spacer ring to maintain a consistent gap between adjacent wafers, are joined together to form a 12-wafer submodule. Fitted with a ceramic tube and cap for testing, such submodules have succeeded in producing on the order of 0.1 TPD oxygen each in pilot tests. Multiple submodules are joined to make commercial-size modules. Also shown in Fig. 1.8 is a photograph of a 0.5 TPD commercial-scale module formed from joining multiple submodules. The 0.5 TPD module is fitted with a conical tube and end cap. The conical tube facilitates connection of the module to an internally insulated metal pipe and provides structural stability to the module. Scale-up to 1.0 TPD oxygen production capacity modules by joining 0.5 TPD submodules has also been completed.

The engineering criteria used in the design of the SEP would apply to the design of scaled-up modules and, thus, can predict performance of commercial plants. A commercial concept of an ITM oxygen production vessel is shown in Fig. 1.9, which shows an array of multi-wafer modules arranged in a common flow duct and connected through a series of manifolds to an oxygen header below. Each commercial-scale module produces approximately 1 TPD oxygen, and many modules are arrayed in parallel banks and the banks are arrayed in series to the gas flow. The scale-up scheme for the modules is summarized in Fig. 1.10. The commissioned 5 TPD prototype described in a previous section is shown in Fig. 1.11.

ITM Oxygen Technology Validation and Technology Benefits

The ITM oxygen technology has been validated at both laboratory and pilot scales showing a record oxygen flux exceeding the commercial target by over 30%. Although the commercial flux is considered business-sensitive information, some researchers [29] have reported oxygen flux through brownmillerite materials of approximately 10 ml/min-cm^2.

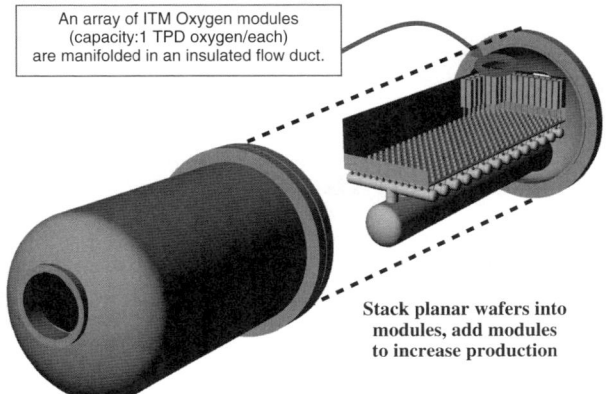

Fig. 1.9 Pilot-Scale Vessel Is a Prototype for Commercial-Scale Concept (Figure provided by Air Products and Chemicals, Inc., under a US Department of Energy Cooperative Agreement)

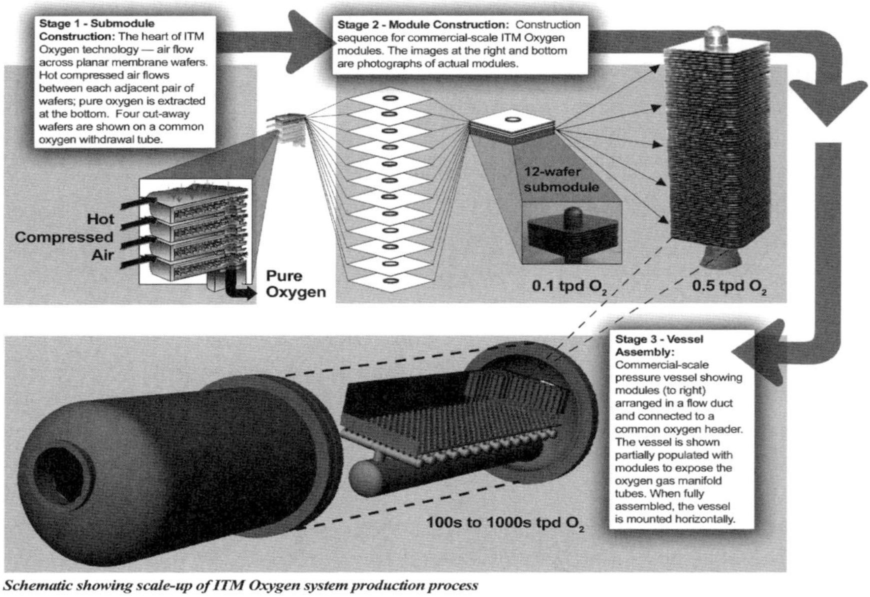
Stage 1 - Submodule Construction: The heart of ITM Oxygen technology — air flow across planar membrane wafers. Hot compressed air flows between each adjacent pair of wafers; pure oxygen is extracted at the bottom. Four cut-away wafers are shown on a common oxygen withdrawal tube.
Stage 2 - Module Construction: Construction sequence for commercial-scale ITM Oxygen modules. The images at the right and bottom are photographs of actual modules.
Hot Compressed Air
Pure Oxygen
12-wafer submodule
0.1 tpd O2
0.5 tpd O2
Stage 3 - Vessel Assembly: Commercial-scale pressure vessel showing modules (to right) arranged in a flow duct and connected to a common oxygen header. The vessel is shown partially populated with modules to expose the oxygen gas manifold tubes. When fully assembled, the vessel is mounted horizontally.
100s to 1000s tpd O2
Schematic showing scale-up of ITM Oxygen system production process

Fig. 1.10 Summary of Scale-Up Scheme (Courtesy of the US Department of Energy and Air Products and Chemicals, Inc.)

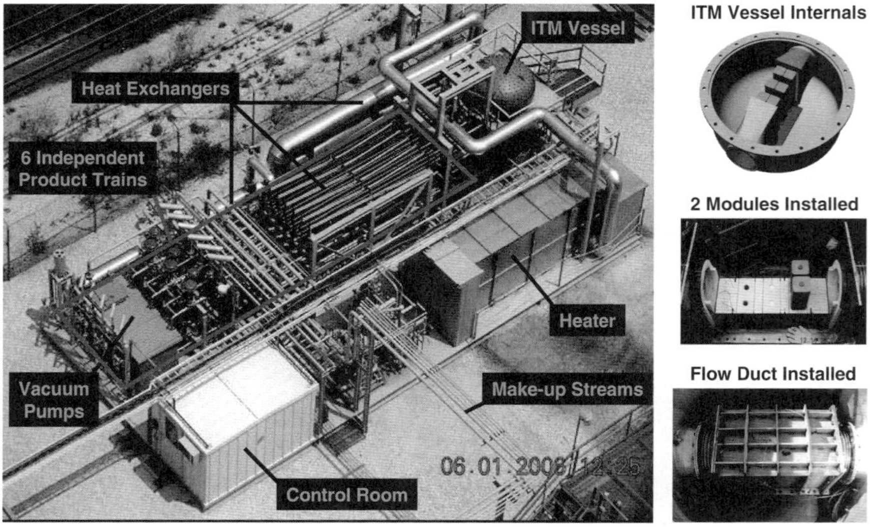
ITM Vessel
Heat Exchangers
6 Independent Product Trains
Heater
Vacuum Pumps
Make-up Streams
Control Room
06.01.2006 12:25
ITM Vessel Internals
2 Modules Installed
Flow Duct Installed

Fig. 1.11 Sub-Scale Engineering Prototype Commissioned: Oxygen Production from Commercial-Scale Modules (Figure provided by Air Products and Chemicals, Inc., under a US Department of Energy Cooperative Agreement)

Full-size wafers have been successfully operated at 200–400 psig pressure loads at full operating temperatures for several thousand hours without material, seals, or performance degradation. Oxygen was produced at > 99% purity at conditions typically expected in commercial operations. The planar design minimizes the effects of material creep and the ITM material used is also resistant to unwanted chemical reactions at anticipated commercial process conditions. The design provides for uncomplicated fabrication of compact modules containing multi-layered ITM oxygen wafers for tonnage-quantity oxygen production.

Process engineering and economic evaluations of IGCC power plants, complemented with experimental data have compared ITM oxygen with a state-of-the-art cryogenic air separation unit [30]. The study projects decreases in the power requirement of the air separation unit by nearly one-third, while increasing the power plant output improves the installed capital of the IGCC facility by 7% and the efficiency by approximately 2% by full integration with a power producing turbine. Studies confirm the suitability of Siemens existing SGT6-6000G platform for use in supporting significant air extraction, simulating full integration from the compression side of the gas turbine for feed to the ITM vessel. A high volume of air is a key parameter in reducing the amount of membrane required and lowering overall cost. Full Integration means that the ITM ASU gets all of its air from the compressor section of the turbine and supplies the turbine the high-pressure, high-temperature non-permeate stream.

Partial integration system studies with other commercially available turbines preserves the ITM-IGCC integration benefits including a reduction of the IGCC plant specific cost ($/kW) by 9% with a net power (MWe) output increase of 15% and plant efficiency increase of 1.2% with over 25% cost ($/short TPD) savings in oxygen production. Partial Integration is when the ASU gets no or partial air from the compression side of the turbine, but still supplies the non-permeate stream to the turbine. This integration scheme needs a compressor.

It is now well recognized that use of oxygen in combustion processes would also provide potentially cost-effective emission reduction and carbon management opportunities, but the cost of oxygen has been a barrier to the widespread application of coal Oxycombustion. ITM oxygen offers the potential for retrofitting the existing coal-boiler fleet to Oxycombustion modes to improve power plant efficiencies and mitigate pollutant emissions. Preliminary studies indicate that, for Oxycombustion of coal, the power needed to operate the ITM ASU and the required capital expenditures are lower compared to cryogenic ASU.

Oxygen-intensive industries such as steel, glass, non-ferrous metallurgy, refineries, and pulp and paper would realize cost and productivity benefits as a result of ITM oxygen. Process economic analyses illustrate that the benefits of the ITM oxygen technology relative to cryogenic air separation technology are significant for nearly all oxygen-intensive industries. Process intensification and lower plant foot prints are other benefits of integrated ITM oxygen-energy production processes.

ITM Oxygen Technology Commercialization Timeline

The Phase III of the ITM oxygen technology RD&D efforts include performance testing of an intermediate scale test facility to produce 100–150 TPD of oxygen from a prototype ITM oxygen pressure vessel integrated with a gas turbine. This work will lead to an understanding of the engineering development and operational issues of an integrated energy system and generate cost, engineering design, scale-up, operational and performance data supporting the development and marketing of integrated energy production systems.

Small commercial gas turbines may be used for integration with ITM oxygen for co-production of power and moderate amounts of oxygen. This platform will perform as an integration-ready machine for early low tonnage (100s) commercial opportunities. Then the much larger, the first-of-a-kind, facility will lead ultimately to an IGCC-ready ITM oxygen technology product (1,000's TPD oxygen). The technology development plan is presented in Fig. 1.12.

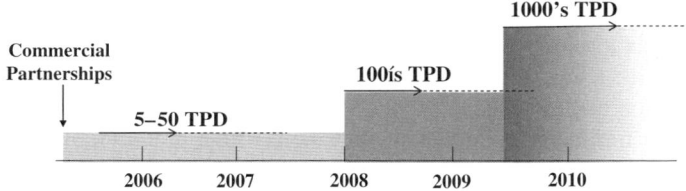

Fig. 1.12 Commercialization Pathway for ITM Oxygen (Figure provided by Air Products and Chemicals, Inc., under a US Department of Energy Cooperative Agreement)

The ITM oxygen technology's potential to serve industrial markets is expected toward the end of the next decade. The anticipated commercialization time line is summarized as follows:

2001–2007	Build and test 1–5 TPD prototype plant;
2010+	Build and test up to 150 TPD plant;
2014+	Build first early-entrance commercial plant (1,000–2,000 TPD).

Epilogue

DOE is supporting the development of the ITM air separation technology for large, tonnage-quantity oxygen as stand-alone plants and for integration of ITM oxygen with power generation systems. A multi-disciplinary effort was launched that required the simultaneous development of ceramic materials and manufacturing methods, engineering fundamental studies, energy cycle analyses, ceramic oxygen conductor module design, construction, and operation. The project has since reached and is moving beyond the prototype phase. ITM oxygen will enable advancement of energy production systems by achieving step-change improvements in efficiency, environmental performance, and cost. It has broad commercial potential to energy production applications, such as coal gasification and combustion systems, coal-to-hydrogen and electricity co-production plants, and has the potential to provide a technology-based response to global climate change concerns by improving the CO_2 emission performance and capture economics of coal-based power plants [31].

The compositions of ITM materials are complex and application dependent, and are based on required operating temperature and environment. Generically, ITMs are engineered ceramics that operate at $700–1,000°C$ and at high pressures, and can be fabricated in tubular or planar configurations. Depending on the applications, ITMs can be pressure driven or chemical potential driven. In general, these materials allow the rapid transfer of oxygen ions, achieving very high flux which is orders of magnitude higher than polymeric membranes with theoretically infinite selectivity. This property enables compact and efficient gas separator equipment designs. ITMs produce oxygen from air in a single separation stage under the gradient of oxygen partial pressure.

The application of ITMs to important industrial needs requires the design and fabrication of membrane modules that address both mechanical and process conditions. The ITM effort has led to compact, ceramic separator module and equipment designs and fabrication methods. Ceramic module architecture is typically either tubular or planar; however, for oxygen applications the full potential of ITMs is obtained through planar designs that rely on advanced lamination fabrication techniques. Laminated structures composed of layers of featured ceramic materials are joined to form an engineered structure. Layers are used to fabricate large, multi-passage membrane architectures satisfying all flow distribution, heat exchange, and mechanical strength requirements.

In partnership with DOE, Air Products has developed high-flux, compact oxygen ion conductors in planar configurations to separate oxygen from air and achieved production flux exceeding commercial performance targets with excellent product purity. The research team built the first commercial-scale ITM oxygen air separation module. A 5 TPD subscale engineering prototype facility has been successfully commissioned. The prototype achieved target performance for flux, oxygen purity, and system pressure, temperature, and airflow rates typical of the commercial operating conditions.

ITM oxygen integrates well with turbine-based power cycles. Successful embodiment of ITM oxygen with advanced power cycles will increase efficiencies and environmental performance while reducing the capital and operating costs of coal-based energy plants. System and experimental studies confirm the economic benefits of ITM oxygen technology. Process economic analyses to date have been mainly focused on the coal-based IGCC power plant application. Other studies have detailed the use of ITM oxygen in other applications, such as oxycombustion of coal. These process economic analyses illustrate that the benefits of the ITM oxygen technology relative to cryogenic air separation technology are significant. For example, an ITM oxygen plant integrated with an IGCC facility could reduce the air separation plant capital and power requirement by nearly one-third each compared to IGCC-cryogenic facilities. ITM oxygen can reduce the parasitic power load of coal boilers by over 30%.

Advanced energy production systems and future environmental mandates would require use of more oxygen rather than air. Today's cryogenic technologies are challenged to produce affordable oxygen for energy and other oxygen-intensive industries. The planned accomplishments of the current efforts include the commissioning a nominal 150 TPD test facility comprising an integrated ITM oxygen separator with a gas turbine. The nominal 150 TPD unit will provide scale-up data to allow the design and construction of a nominal 2,000 TPD plant planned for testing as the oxygen supply module of early-entrance commercial plants for stand-alone oxygen production, integration with IGGC other advance power generation systems, and co-producing hydrogen and electricity from coal.

References

1. Teraoka Y, Zhang HM, Furukawa S, Yamazoe N. Oxygen permeation through perovskite type oxides. Chem Lett. 1985;14(11):1743–46.
2. Burggraaf AJ, Cot L. Fundamentals of inorganic membrane science and technology. Elsevier, Netherlands; 1996.
3. Bose AC, Stiegel GJ, Sammells AF. Proceedings of the Fifth International Conference on Inorganic Membranes, Nagoya, Japan, 22–28 June 1998.
4. DOE Begins Research Effort to Revolutionize Oxygen Production for Future Energy, Industrial Processes, DOE Fossil Energy Techline, 7 October 1998.
5. Bose AC, Richards RE, Sammells AF, Schwartz, M. Beyond state-of-the-art gas separation processes using ion transport membranes. Desalination. 2002;144:91–2.
6. Dyer PN, Richards RE, Russek SL, Taylor DM. Ion transport membrane technology for oxygen separation and syngas production. Solid State Ionics. 2000;134:21–33.
7. Rao CNR, Prakash O, Ganguly P. Electronic and magnetic properties of $LaNi_{1-x}Co_xO_3$, $LaCo_{1-x}Fe_xO_3$, and $LaNi_{1-x}Fe_xO_3$. J Solid State Chem. 1975;15:186–92.
8. Teraoka Y, Nobunaga T, Yamazoe N. Effect of cation substitution on the oxygen semipermeability of perovskite type oxides. Chem Lett. 1988;503–6.
9. Teraoka Y, Zhang HM, Okamoto K, Yamazoe N. Mixed ionic-electronic conductivity of $La_{1-x}Sr_xCo_{1-y}Fe_yO_{3-\delta}$ perovskite type oxides. Mat Res Bull. 1988;23:51–8.
10. Carolan MF, Dyer PN, Motika SA, Alba PB. Compositions capable of operating under high carbon dioxide partial pressures for use in solid state oxygen producing devices. U.S. Patent 5,712,220, 27 June 1998.
11. Carolan MF, Dyer PN, Thorogood RM. Process for recovering oxygen from gaseous mixtures containing water or carbon dioxide which process employs ion transport membranes. U.S. Patent 5,261,932, 16 Nov 1993.
12. Carolan MF, Dyer PN, Fine SM, Makitka A, Richards RE, Schaffer LE. Inorganic membranes. U.S. Patent 5,683,797, 4 Nov 1997.
13. Stein VE, Richards RE, Brengel DD, Carolan MF. Sulfur control in ion conducting membrane systems. U.S. Patent 6,602,324,B2, 5 Aug 2003.
14. Hutchings KN. Tensile creep behavior of a potential ceramic oxygen separation material. M.S. Thesis, The Pennsylvania University, Oct 2002.
15. Hutchings KN, Shelleman DL, Tressler RE. Tensile creep behavior of $(La_{1-x}Sr_x)(Co_{1-y-z}Fe_yCu_z)_aO_{3-\delta}$: stress, temperature, and oxygen partial pressure dependence. Proceedings of the Material Science and Technology Conference, Pittsburgh, Pennsylvania, 25–28 September 2005.
16. Control of differential strain during heating and cooling of mixed conducting metal oxide membranes. Document Type and Number: United States Patent 7311755. Publication Date: 12/25/2007.Link to this page:http://www.freepatentsonline.com/7311755.html

17. Rynders SW, Minford E, Tressler RE, Taylor DM. Compliant high temperature seals for dissimilar materials. U.S. Patent 6,302,402, B1, 16 Oct 2001.
18. Butt Darryl P. Joining ion-transport membranes using a novel transient liquid phase process. Final Technical Report, DOE UCR Contract DE-FG26-05NT42536.
19. Butt DP, Cutler RA, Rynders SW, Carolan MF. Method of joining ion transport membrane materials using partially or fully transient liquid phase. U.S. Patent 7,011,898, 2006.
20. Butt DP, Cutler RA, Rynders SW, Carolan MF. Method of forming a joint. U.S. Patent 7,094,301, 2006.
21. Sammells AF, Mundschau MV, editors. Dense inorganic membranes. New York: Wiley-VCH; 2006.
22. Lee KS, Shin TH, Lee S, Woo SK. Enhancement of oxygen permeability by the introduction of a micro/nano surface reactive layer on the oxygen ion transport membrane. J Ceram Process Res. 2004;5(2):143–7.
23. Yee TF, Srinivasan R, Thorogood RM. Oxygen production by staged mixed conductor membranes. U.S. Patent 5,447,555, 5 Sep 1995.
24. Russek SL, Knopf JA, Taylor DM. Oxygen production by ion-transport membranes with non-permeate work recovery. U.S. Patent 5,753,007, 19 May 1998.
25. Armstrong PA, Foster EP, Sorensen JC, Stein VE, Suzuki H. ITM oxygen for energy intensive applications. Proceedings of the Twenty-First Annual Pittsburgh Coal Conference, Osaka, Japan, 13–17 Sep 2005.
26. Thorogood RM, Srinivasan R, Yee TF, Drake MP. Composite mixed-conductor membranes for producing oxygen. U.S. Patent 5,240,480, 31 Aug 1993.
27. Taylor DM, Bright JD, Carolan MF, Cutler RA, Dyer PN, Minford E, et al. Planar solid state membrane module. U.S. Patent 5,681,373, 28 Oct 1997.
28. Carolan MF, Dyer PN, Motika SA. Compositions capable of operating under high oxygen partial pressures for use in solid state oxygen producing devices. U.S. Patent 5,817,597, 6 Oct 1998.
29. Sammells AF, Barton TF, Peterson DR, Harford ST, Mackay R. Methane conversion to syngas in mixed-conducting membrane reactors. Proceedings of the 4th International Conference on Catalysis in Membrane Reactors, Zaragoza, Spain, 3–5 July 2000.
30. Stein VE, Juwono E, Demetri EP. Improving IGCC economics through ITM oxygen integration. 18th International Pittsburgh Coal Conference, New Castle, Australia, 4–7 Sep 2001.
31. Bose AC, Armstrong PA. AIChE Annual Meeting, Nov 2007.

Chapter 2
Viability of ITM Technology for Oxygen Production and Oxidation Processes: Material, System, and Process Aspects

Marcel J. den Exter, Wim G. Haije, and Jaap F. Vente

Abstract This chapter is devoted to the state of the art of the most important aspects of high temperature ceramic air separation membranes for oxygen production and oxidation processes. Alternative technologies, operational principle, fields of application, energy efficiency and cost aspects, materials science, module design, and sealing will be discussed.

Introduction

The threat of global warming due to increasing carbon dioxide (CO_2) concentrations has been recognized as one of the main environmental challenges of this century. To limit atmospheric CO_2 concentrations to acceptable levels of about 550 ppm [1] major changes in energy consumption are required in the coming decades. Still, fossil fuels are widely expected to remain the world's major source of energy for well into the twenty-first century. Although supply of oil and gas is under threat due to political instability and uncertainties on reserves, the use of coal is increasing, with concomitant higher CO_2 emissions [2]. To meet this target set for atmospheric CO_2 concentrations, the development of breakthrough technologies is essential. Otherwise, it will prove to be impossible to reach the dramatic decrease of the CO_2 emission into the atmosphere during the conversion of fossil fuels to other forms of energy, for example, electricity or hydrogen. The three main routes for mitigation of CO_2 emissions in electricity plants can be defined as:

1. Post-combustion processes: CO_2 is captured from the flue gases.
2. Pre-combustion processes: The fuel (natural gas or coal) is converted into hydrogen and CO_2. The CO_2 is separated and hydrogen is combusted in a gas turbine.
3. Oxyfuel processes: Combustion is carried out using pure oxygen, resulting in a flue gas that mainly contains water (H_2O) and CO_2.

M.J. den Exter (✉)
Energy Research Centre of the Netherlands, PO Box 1, 1755 ZG Petten, The Netherlands
e-mail: denexter@ecn.nl

A.C. Bose (ed.), *Inorganic Membranes for Energy and Environmental Applications*,
DOI 10.1007/978-0-387-34526-0_2, © Springer Science+Business Media, LLC 2009

These routes are connected with carbon capture with subsequent sequestration. Another approach is to avoid the production of CO_2 emissions altogether through increased industrial energy efficiency and thus a lower energy consumption. The topic of this chapter, oxygen production, is related to the last two points. In the first three routes mentioned above, three different methods can be used for the separation of CO_2 and the other gases: absorption in solvents, separation by membranes and adsorption or absorption on or in a sorbent.

This chapter is devoted to the state of the art of the most important aspects of high temperature ceramic air separation membranes for oxygen production and oxidation processes. Alternative technologies, operational principle, fields of application, energy efficiency and cost aspects, materials science, module design, and sealing will be discussed.

Alternative Technologies

Polymeric membranes are used for oxygen enrichment or depletion of air. Not only are large-scale applications of the oxygen enrichment found in the industry but also small-scale applications are possible which are much closer to the individual consumer, such as for wellness and health branch. Oxygen depletion is often related to safety as fire prevention in buildings and future blanketing over airplane fuel tanks. The polymers are commonly based on among others polyphenyloxide (Parker Gas Separation) and polyimides (Air Products). The separation is based on preferred dissolution and diffusion of oxygen over nitrogen in these materials. Operating temperature is limited by the thermal stability of the polymers used. The membrane unit consists of bundles of hollow fibers, a porous polymer support with a dense polymer separating layer, or spiral wounds. The technology is mature and due to its modular nature applicable both for small and large production volumes. Pressure Swing Adsorption (PSA) and Vacuum Swing Adsorption (VSA) technologies are based on the preferred absorption of nitrogen in, for example, zeolite beds. Purities of oxygen of about 90–94 percent are obtainable. Plant size is up to about 1,500 Nm^3 per hr. Cryogenics is currently the most developed mature large-scale oxygen production technology, with a typical plant size being 30,000–50,000 Nm^3 per hr or 9,000–15,000 tons per day. The produced oxygen is of very high purity $> 99\%$. In addition, this technology can also be used to produce nitrogen and argon. However, oxygen is believed to be the main incentive for the separation. Relatively new developments that are currently being investigated are the chemical looping combustion [3, 4] cycle and the so-called Ceramic Autothermal Recovery (CAR) [5] process originally devised by British Oxygen Company (BOC), now a part of Linde Gas. These processes are very similar. In CLC a metal or metal oxide is oxidized by air in one reactor and transferred to a reduction reactor that is fed with, for example, syngas. The reduced material is transferred back to the oxidation reactor. The CAR concept uses a set of two or more batch reactors alternating between oxidation and reduction cycles, containing a perovskite material $ABO_{3-\delta}$ that is cycled between two oxygen contents δ_1 and δ_2.

General Aspects of ITM Technology

The process that is the subject of this article consists of the use of high temperature ceramic membranes that selectively transport oxygen. They are referred to with several acronyms of which ITM (Ion Transport Membranes), OTM (Oxygen Transport Membranes) and MIEC (Mixed Ionic Electronic Conducting) membranes prevail. We will use ITM throughout this article.

ITM technology can be used as a replacement of energy-demanding cryogenic distillations or PSA/VPSA plants. ITM technology in ASU's can result in high purity oxygen streams provided that defect-free membranes can be prepared on industrial scales and high quality sealing technology can be developed. The main drawback of the ITM technology is that it does not produce nitrogen. Advantages can be found in local decentralized production of oxygen to small-scale on-demand production in places where oxygen is actually needed in chemical processes. This would result in abandoning transport of oxygen over public roads, while storage facilities and associated safety measures are no longer required.

PSA/VPSA technology provides oxygen ranging from small-scale (e.g., small portable devices) to large-scale plants. It is anticipated that ITM technology can be introduced in the medium to large-scale production range following normal economical considerations while introduction in the "portable" sector requires comparison of module sizes as a function of oxygen output, additionally.

A schematic representation of the air separation process using ITM as a whole is presented in Fig. 2.1. The driving force for the transport of oxygen is a difference in partial oxygen pressure (pO_2) between the feed and permeate side of the membrane. As a result, oxygen permeates from the side with the high pO_2 to the side with the

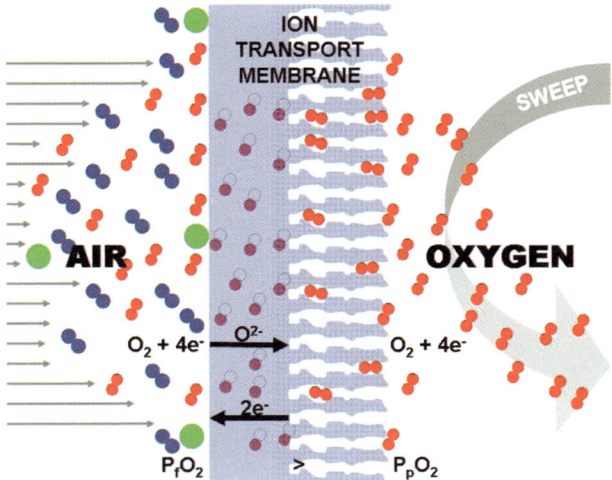

Fig. 2.1 Schematic representation of the transport of oxygen through an ion transport membrane consisting of a dense (*left*) and porous (*right*) part. The sweep can be either "inert," for example, CO_2 or H_2O, or "fuel," for example, CH_4

low pO_2. On the feed side the oxygen molecules reduce to two oxide ions (O^{2-}). This pair of oxide ions permeate through a thermally activated hopping mechanism to the permeate side and recombine. Simultaneously, two pairs of electrons move from the permeate side to the feed side.

The majority of the ITM compounds showing mixed ionic electronic conductivity belong to the structural family of Ruddlesden–Popper compounds [6] $A_{n+1}B_nO_{3n+1}$. The end members of this family are the cubic perovskite $SrTiO_3$ for $n = \infty$ and the layered K_2NiF_4 structure for $n = 1$. These end members form two very versatile families of compounds and are generally referred to as ABO_3 and A_2BO_4. The A cations are typically relatively large cations from trivalent lanthanide series or divalent earth metals (large spheres in Fig. 2.2), the smaller B cations are in general transition metals and are found in octahedral coordination. Ionic conduction in cubic perovskites is achieved when oxygen vacancies are formed as a result of a too low total formal valence on the cations. In the formed $ABO_{3-\delta}$ compounds oxygen vacancies can move from one site to another through a thermally assisted hopping mechanism. Typically this movement is quenched at temperatures below $\sim 450°C$, and it becomes significant in terms of technical applicability at temperatures well above $700°C$. In A_2BO_4 compounds, the oxide conduction can be quite different from that of the perovskites. One option is similar to perovskites and leads to sub-stoichiometric oxygen content. In the second option extra oxide ions are intercalated in between the subsequent perovskite-like layers resulting in an over-oxidized state ($A_2BO_{4+\delta}$), which may eventually lead to higher reduction stability. $ABO_{3-\delta}$ is clearly an isotropic oxide conductor whereas $A_2BO_{4+\delta}$ is a 2D anisotropic oxide conductor. Depending on the composition, both types can show high electronic conductivity, and even super conductivity [7–9]. Well-known examples of perovskites with a high oxygen conductivity are $La_{1-x}SrCo_{1-y}Fe_yO_{3-\delta}$ [10], and for the layered systems $La_2Ni_{0.9}Co_{0.1}O_{4-\delta}$ [11] can be mentioned.

Fig. 2.2 Archetypical structures of the cubic perovskite, ABO_3, (*left*) and K_2NiF_4, A_2BO_4, (*right*). Green balls: large A cations, blue octahedra of small B cation in the center surrounded by six oxide ions

Techno-economic evaluations [12] have shown that oxygen selective membranes are required that can deliver oxygen at a rate of over $10\,\text{ml/cm}^2\text{min}$. This value will be used as a base case for all our considerations.

Applications

The two main applications of high temperature membrane technology are the production of high purity oxygen as valuable chemical and as a reactant for various (partial) oxidation processes such as methane conversion [13]. For the first application, this technology is in direct competition with the well established and energy intensive cryogenic distillation. For the second application this, in comparison, is much less clear. Now the (partial) oxidation and the separation of air take place within the same unit operation, leading to a membrane reactor. These applications may look very similar at first sight, but difference start to appear on closer inspection. The most important ones are the necessity of catalysts and the very different demands that are being put on the membrane materials. For the production of pure oxygen a partial oxygen pressure of about 0.3 bar is anticipated on the permeate side. In the case of partial oxidations this value can be as low as 10^{-18} bar. In the chemical industry, oxygen is used in a large number of processes. In Table 2.1 the main chemical processes are listed where air and/or pure oxygen is used, divided into worldwide production figures in megaton per year.

The main use of oxygen in industry is for production of ethylene oxide, ethylene dichloride, propylene oxide, and acetic acid. Notice that the sequence of product production quantities is slightly different since different amounts of oxygen are required for each ton of product. This especially applies for polyethylene, requiring only a small oxygen quantity per ton of product. Most of these production processes are carried out in plants making use of oxygen but sometimes air is used as oxygen source as well. Additionally, many chemical processes are carried out using oxygen in the form of air only. Examples are the production of nitric acid, formaldehyde, terephtalic acid, and carbon black as the main air-consuming production processes. Production of these 4 compounds consumes an additional 106.5 Mton/a of oxygen, 74% of the total oxygen consumption (129.5 Mton/a) of the full range of air consuming production processes.

Huge oxygen consumers outside the chemical industry are the steel producers. The oxygen consumption for primary steel production being six times the sum of the major oxygen consuming chemicals mentioned in Table 2.1. The relative contributions to oxygen consumption in the steel industry can be found in blast furnace enrichment (32%), basic oxygen furnaces (43%), electric arc furnaces (19%), and cutting and burning activities (6%).

Next to bulk production of chemicals and steel, other applications exist where oxygen is required. For instance, air enriched with oxygen is used for bleaching purposes in the pulp and paper industry, not putting strains on the purity of the oxygen used. Finally, a market that is likely to show a fast growth over the coming years is that of oxygen of gas-to-liquid (GTL) plants. In this case natural gas is first transformed into syngas, which in turn is used in a Fischer-Tropsch process for

Table 2.1 World use of gaseous oxygen (GOX) in bulk production of chemicals and steel in 2004

Product	Production [Mt/a]	O_2 use [Mt/a]	Process Description
Ethylene oxide	15.1	7.2	Oxidation of ethylene. Air is sometimes used but O_2 preferred in new and larger plants.
Ethylene dichloride	49.1	4.0	Oxichlorination of ethylene. Air is sometimes used
Propylene oxide	5.8	2.0	Epoxidation of propylene: PO/TBA-route. Air is sometimes used
Acetic acid	8.1	1.6	Oxidation of naphta/n-butane (35 %) or acetaldehyde (5 %). Air is sometimes used
Titanium oxide	4.3	0.9	From the ore: chlorination (in presence of O_2) to $TiCl_4$, subsequent oxidation to TiO_2
Vinyl acetate	5.0	0.8	From ethylene, acetic acid and O_2.
Acetaldehyde	2.4	0.7	Oxidation of ethylene (Wacker-Hoechst process). The 1-step process uses O_2; the 2-step process uses air
Perchloroethylene	0.7	0.1	Oxychlorination of ethylene dichloride (PPG process)
Acetic anhydride	1.9	0.1	Oxidation of acetaldehyde. Air is sometimes used
Polyethylene (LDPE)	18.7	0.01	Polymerisation of ethylene in a high pressure tubular reactor. Oxygen is used as initiator (radical formation)
Cyclododecanol Cyclododecanone Crotinic acid	0.01	0.02	Oxidation of cyclododecane. Air is sometimes used
Subtotal		17.4	
Steel [14]	1241.0	104.0	
Total		121.4	

the preparation of highly linear hydrocarbons. The air separation unit (cryogenic distillation) on its own is responsible for about 30–40% of the total capital costs. A similar costing figure can be put on the running cost.

Oxygen is put into the market as gaseous oxygen (GOX) by pipelines and lique-fied oxygen (LOX) in bottles or tanks. Despite supplying both forms, most of the oxygen in chemical processes is used in the gaseous form. Membrane technology directly supplies gaseous oxygen that, depending on the process, only needs to be pressurized for use whereas the temperature level allows for easy integration in high temperature processes.

Pure Oxygen Production

A typical process flow diagram for oxygen production is presented in Fig. 2.3. Our in-house assessments have shown that, if no high temperature thermal integration is possible, the co-generation of oxygen and electric power is required to render this system to be energetically more efficient than conventional cryogenic distillation.

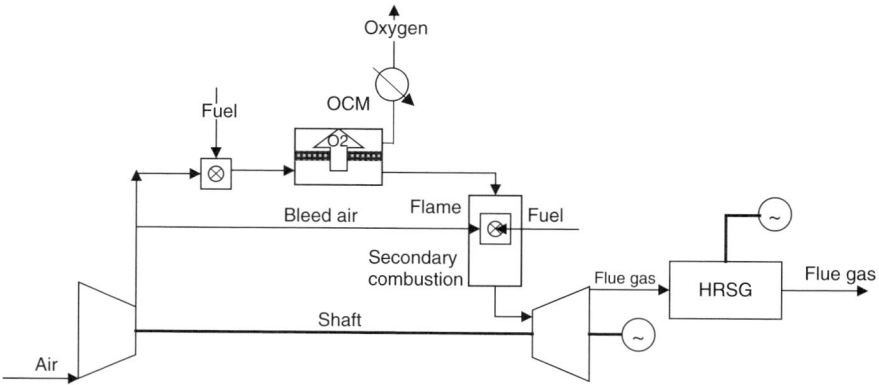

Fig. 2.3 Large-scale oxygen production (30,000 Nm^3/hr) using ITM technology. For small-scale systems (1,500 Nm^3/hr) the Heat Recovery Steam Generator (HRSG) cannot be applied

One specific process scheme, as depicted in Fig. 2.3, consists of first compressing the incoming air followed by internal combustion of natural gas. The hot air is depleted of oxygen in the following separation module, and the retentate is fed to a secondary combustion chamber. The hot off-gas is used to power the turbine, and the flue gas is fed to a HRSG unit. If the energy efficiency of such a process is compared with that of a cryogenic distillation unit in co-junction with the average efficiency of the Dutch electricity park, an energy saving of over 20% can be achieved (Fig. 2.4). However, if the base case includes a state-of-the-art STAG (steam and gas turbine), then the efficiency improvement is limited to $\sim 8\%$.

This comparison can also be made with PSA- and VPSA-based system, as presented in Fig. 2.5. For this case, depending on the purity of the oxygen required, the energy savings can range from -4% (so a small penalty!) to over 50%. It is clear that the production of oxygen using ITM is most promising for small-scale applications where very high purity oxygen is required.

Costs

We have made a rough estimate of the costs for both a large and a small-scale system (Table 2.2). The difference between *Total process equipment* and *Total fixed capital* is caused by including percentages, for example, piping, installation, and engineering, following standard cost engineering procedures [15].

In order to determine the economic viability the running coasts and revenues have been estimated (Table 2.3) as well. The main contribution to the fixed costs can be found in the membrane overhaul that is supposed to take place every three years. The oxygen price for large-scale production is taken to be equal to the price for gaseous oxygen from a cryogenic plant. The small-scale price is taken to be equal to that of a PSA plant.

For economic viability the simple pay out time (SPOT) should be smaller than five years. This gap can be reduced by several factors: cheaper membranes, higher

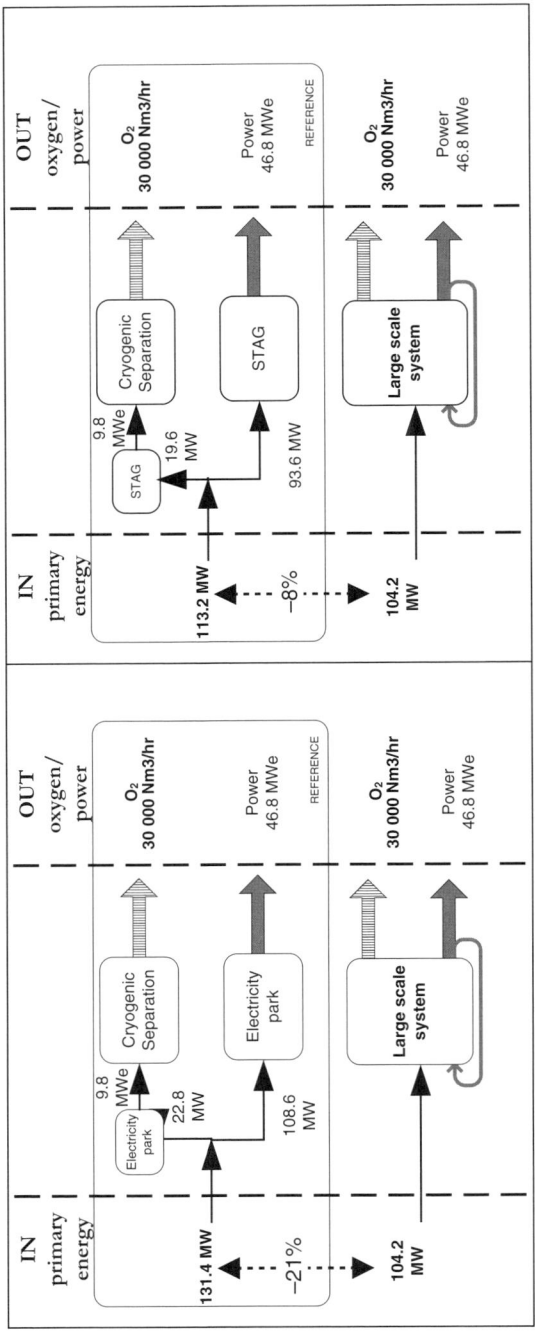

Fig. 2.4 Comparison of cryogenic and ITC technology as energy use for average power plant (*left*) and modern STAG based power plant (*right*)

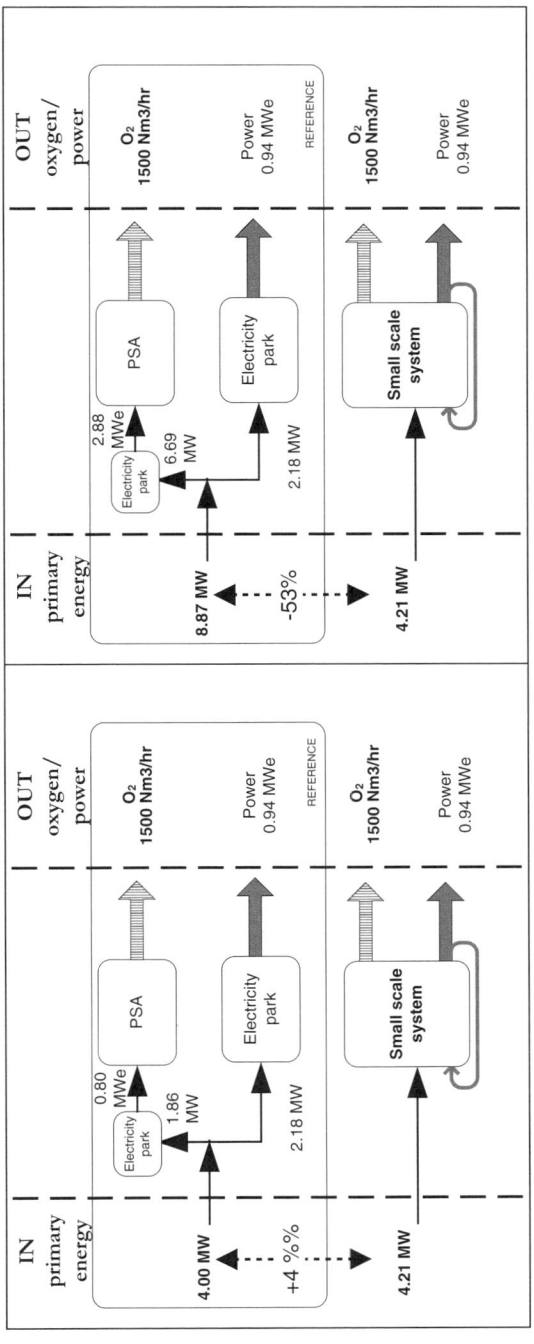

Fig. 2.5 Comparison PSA and ITC technologies for small-scale systems for 90–94% Oxygen purity (*left*) and 100% purity (*right*)

Table 2.2 Breakdown of total fixed capital for large and small-scale ITM systems (the HRSG is not used in the latter system)

Investments	#	k€	#	k€
	Large-scale		Small-scale	
Process equipment				
Heat exchangers	3	362	2	112
Oxygen compressor	1	1530	1	18
Membrane module shells	38	7692	1	385
Burners (or modification)	2	200	2	60
Gas turbine + HRSG	1	33724	1	1100
Membrane tubes*		7500		375
Total process equipment		50809		2050
Total fixed capital		101949		4695

*Oxygen flux taken: $10 \, ml/cm^2 min$

Table 2.3 Economic evaluation of the large and small-scale systems

	M€/year	
Costs	Large-scale	Small-scale
Variable costs	12.3	0.490
Fixed costs	9.2	0.460
Admn. and sales	0.2	0.040
Total	21.5	0.970
Revenues		
Oxygen	13.5	1.09
Electricity	13.6	0.280
Total	27.1	1.360
Cash Flow	5.5	0.400
Simple Pay Out Time	16.8 years	11.0 years

flux through the membranes, longer lifetime of the membranes, higher energy prices and CO_2 trading. Fluxes will have to be at least two to three times higher if this were the only option. We do not believe that this is feasible.

Oxidation Reactions

The processes to be discussed under this topic are all designed to maintain a strict separation between N_2 and carbon containing streams in the generation of electricity. Four different system configurations in which ITM provides the pure oxygen can be distinguished:

- Systems in which all fuel is combusted in the membrane reactor (MR) at the permeate side.
- Systems with partial oxidation in the membrane reactor at the permeate side.
- Systems without combustion anywhere near the membrane, but in a separate unit.
- Systems for afterburning of flue gasses from, for example, fuel cells.

Table 2.4 Viability of ITM based membrane process options in power production

Oxidation Process	Feed Coal, biomass, waste, slurry, etc.	Natural gas, gasoline, syngas, etc.
Full oxidation in membrane reactor	Not viable, mechanical deterioration of membrane	Viable but temperature control difficult and expensive
Partial oxidation in membrane reactor	Not viable, mechanical deterioration of membrane	Viable as a nitrogen free syngas generator
Oxidation and separation not integrated	Viable in principle, but techno-economic uncertainties; not really a membrane reactor For coal gasification it is viable and proven (replaces ASU)	Viable, systems being under investigation by several groups [16–19] (replaces ASU)
Post-oxidation in membrane reactor	Viable if flue gas is N_2 lean, ASU is competitor	Viable, especially in SOFC applications (flue gas N_2 lean)

In summary: the lighter the color of the cells in the table the more viable the option is.

When a distinction is made between fuels that can be present in the form of gas or vapor and those that will be present as solids or liquids, the applicability of an ITM based membrane reactor is found to be limited to a few cases (Table 2.4).

Quite a number of system layouts have been proposed for Zero Emission Power Plants (ZEPP) [16–23]. From these specific systems a general scheme as depicted in Fig. 2.6 can be constructed. The main issue is to keep the nitrogen and carbon side of the system strictly separated. On the carbon side all conversions of the fuel take place in the end leading to predominantly water and CO_2 as combustion products and the production of electricity by single or combined cycle processes. On the nitrogen side air is separated at high temperature and pressure. The oxygen lean hot retentate is expanded in a gas turbine combined cycle. In a few cases fuel is burned on the nitrogen side to adjust the turbine entrance temperature to the optimum value, in which case we have low emission power plants.

Costs

For the ZEPP system presented in Fig. 2.7 the allowable installed costs for the capture plant have been estimated to be in between 80 and 120 M€ for a 1.4 GW$_{\text{thermal input}}$ natural gas combined cycle (NGCC) plant. To meet the cost targets, an ITM based ZEPP power plant should have an electric efficiency of at least 52%. Capture ratios (carbon captured/carbon fed to the process) of 100% result in plant efficiencies lower than 50%. Capture ratios of the order of 85% are accompanied with plant efficiencies of about 52%. The allowable installed costs window is satisfied when the oxygen flux is at least 20 ml/cm^{-2}min^{-1} and the costs of the ITM tubes should be less than 1,500 €/m^2. This clearly sets the targets for materials and turbine development in order for the system to be economically viable within the capture cost boundary conditions.

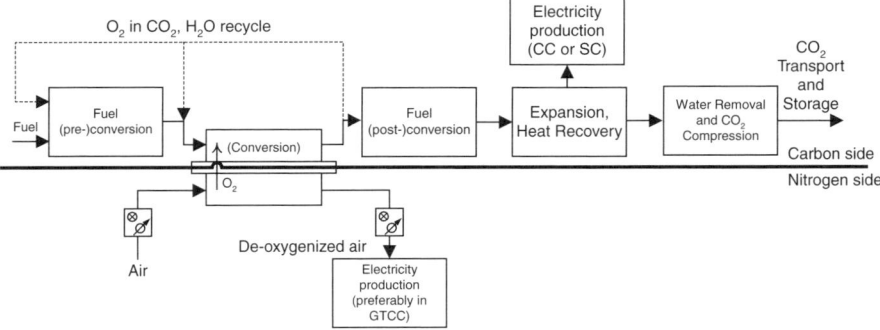

Fig. 2.6 General building blocks for power/ITM /carbon capture systems in which carbon and nitrogen sides are strictly separated (ZEPP)

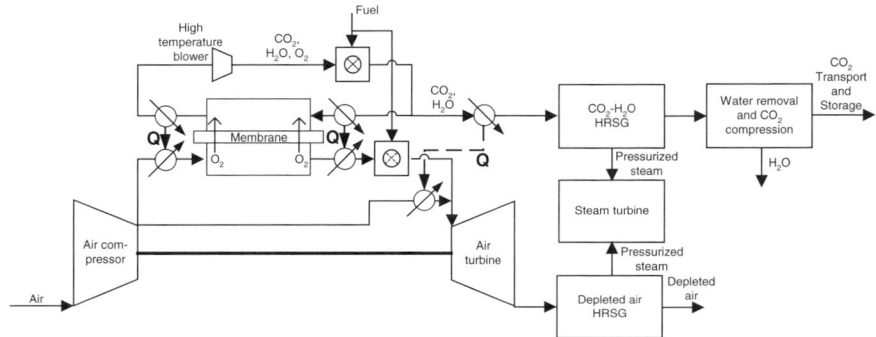

Fig. 2.7 ZEPP using ITM technology [16, 17]

Membrane Materials and Sealing Aspects

Membrane Materials

As mentioned earlier oxygen fluxes of at least $10\,\mathrm{ml/cm^2 min}$ are needed for economic viability of the membrane process. This value is equivalent to an electron current density of $\sim 3\,\mathrm{A/cm^2}$. From this flux, we can also calculate the time it takes for all the oxide ions in the perovskite layer to be replaced by fresh oxide ions. Depending on the thickness of the selective layer, this value is of the order of a few minutes. It is clear that these values put very large demands on the stability of the membrane material. It also means that in one minute all the oxygen has to be removed from an air layer with an equivalent thickness of $\sim 50\,\mathrm{cm}$. We believe that that can only be achieved if the air on the feed side is compressed. The selectivity demands of over 1,000 are likely to be met with defect-free membranes. Further demands include stable performance under real industrial conditions (presence of CO_2, H_2O). As shown in Fig. 2.8, the required flux ($> 10\,\mathrm{ml/cm^2 min}$) can be reached by at least 2 different perovskites $SrCo_{0.8}Fe_{0.2}O_{3-\delta}$ (SCF) and $Ba_{0.5}Sr_{0.5}Co_{0.8}Fe_{0.2}O_{3-\delta}$ (BSCF) in

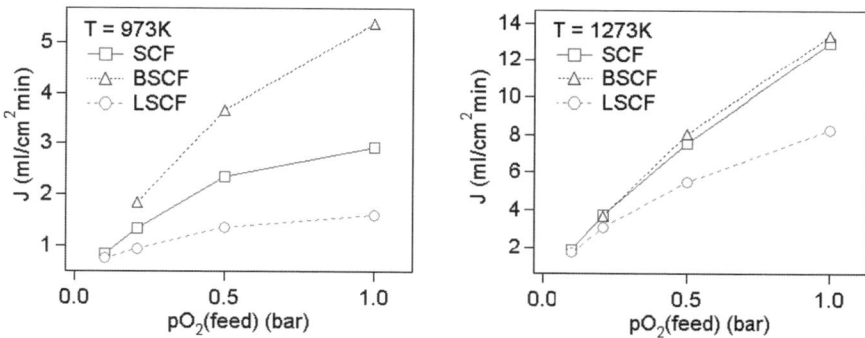

Fig. 2.8 Oxygen fluxes for SCF, BSCF, and LSCF at 973 K and 1273 K for a 5×5 cm^2 and 200 μm thick dense membrane as a function of oxygen feed partial pressure

contrast to, for example, the lanthanum containing $La_{0.2}Sr_{0.8}Co_{0.8}Fe_{0.2}O_{3-\delta}$ (LSCF) for which lower flux values are measured [24].

The two compositions, SCF and BSCF, have very similar oxygen fluxes at high temperature, but at lower temperature, BSCF has a much higher oxygen flux than SCF. This has been ascribed to the occurrence of a low temperature phase with the brownmillerite (see Fig. 2.9) structure in which the oxygen vacancies are ordered in rows [25]. Such ordered phases tend to have a lower permeability for oxygen [26].

Concomitant with the phase transition from brownmillerite to perovskite, SCF shows a negative thermal expansion coefficient over a limited temperature range [24]. This phase transition is present up to a remarkably high partial oxygen pressure of ~0.1 bar [25]. Such a phase transitions may cause stress and enhance degradation of the membrane upon start-up and shut-down cycling. In contrast to SCF, BSCF [27] was shown not to have transitions to brownmillerite and may therefore be preferred over SCF. However, BSCF is believed to be susceptible to kinetic demixing [28] under the influence of a gradient in the oxygen chemical potential, and phase separation may also play a critical role as life time determining factor [24].

SCF or BSCF are suitable candidates for the production of pure oxygen. For more reducing conditions like using syngas as a combustible at the permeate size the high content of Co is detrimental. Based on observations by [29–32] $Sr_{0.97}Ce_{0.03}Fe_{0.8}Co_{0.2}O_{3-\delta}$ (SCFC) and $La_2Ni_{0.9}Co_{0.1}O_{4-\delta}$ (LNC) were selected as promising candidates for operation under reducing conditions and are at present under scrutiny in our laboratories.

In Fig. 2.10, the oxygen flux through LNC and SCFC between 1,073 and 1,173 K is presented. Both LNC and SCFC are measured as a 200 μm thick self-supporting membrane of 5×5 cm^2 surface area, denoted D. The feed is one bar of air and the permeate side is swept with one bar of helium. Both materials have similar fluxes up to about 1,100 K. At higher temperature, SCFC shows a strong increase of oxygen permeation. Asymmetric SCFC membranes have also been made, consisting of a porous support of about 200 μm thickness and a dense top layer of about 10 μm thickness, denoted P-D, the porous part being on the air feed side.

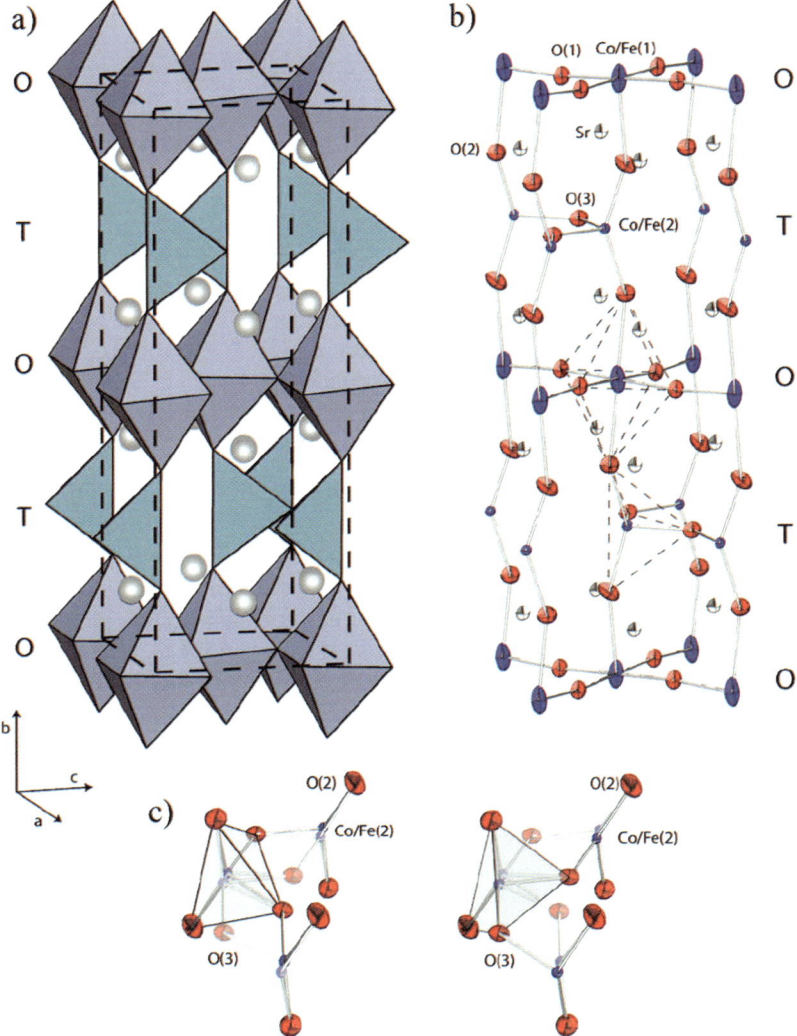

Fig. 2.9 The structure of $SrCo_{0.8}Fe_{0.2}O_{2.5}$ in the brownmillerite structure

In one measurement He is used as a sweep gas and in the other a gas with a typical composition of SOFC off gas (SOFC o.g.), for example, $CO : CO_2 : H_2 = 1 : 2 : 2$ and a relative humidity of about 40% that is clearly a reducing composition. The oxygen flux is usually given by the following expression:

$$J_{O_2} = \frac{RT\sigma_e\sigma_i}{16F^2(\sigma_e + \sigma_i)L} \ln\left(\frac{p_h}{p_l}\right)$$

(2.1)

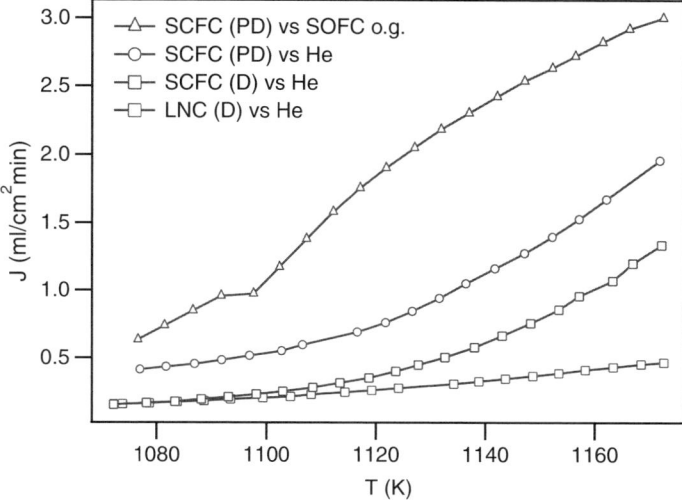

Fig. 2.10 Oxygen flux between 800 and 900°C of a dense LNC membrane and dense and supported SCFC membranes with 1 bar air on the feed side and different sweep gasses on the permeate side (see text)

Here J_{O2} is the oxygen permeation rate, R gas constant, T the absolute temperature, σ_e and σ_i the electronic and oxide ionic conductivities, F the Faraday constant, L the thickness, and ph and pl the partial oxygen pressures on feed and permeate side respectively. Here it is assumed that the bulk diffusion is rate limiting. From this equation, it is clear that for high fluxes, high feed or very low permeate pressures are benevolent as well as small membrane thicknesses. Purely based on thermodynamics, values for the partial oxygen pressure at the permeate side can be as low as 10^{-18} when partial or complete combustion of fuel takes place. Referring to the measurements shown on SCFC it is clear that a thinner membrane gives a higher flux as well as a lower oxygen partial pressure in the case of the reducing SOFC off gas composition. Differences in temperature dependence between He and SOFC off gas stem from the fact that the shift equilibriums also change drastically as a function of temperature whereas in the case of He the exponential behavior of an activated diffusion process is reflected. Clearly the transport in LNC is more difficult than in SCFC at these temperatures.

Membrane Architecture

In order to obtain sufficiently high oxygen flux, a thin (10–40 μm) and dense perovskite layer is required that must be strong and robust enough to withstand typical pressure differences of ~15 bar at elevated temperatures (800–1,000°C) Therefore, a support is needed on which the perovskite layer is deposited. The main advantage

Fig. 2.11 Oxygen-conducting membranes consisting of a porous support and a dense top layer; both parts consist of the same perovskite

of a membrane configuration where the support material is identical to the dense top layer is that the build up of undesirable tensions resulting from differences in the thermal expansion coefficient can be prevented. Furthermore, solid state reactions between the support and top layer are prevented, and no interlayer with a negative effect on the oxygen flux will be formed. Tubular membranes can be prepared by first extrusion of a porous perovskite support. Subsequently, a thin dense selective perovskite layer is deposited. Challenges are still found in the preparation of a defect free top layer that would otherwise reduce the selectivity. At the same time the underlying support must remain porous. Fine tuning of the recipes for the extrusion paste and the dense top layer, in terms of composition, viscosity, and so on. together with appropriate drying and sinter protocols can result in supported tubular membranes as depicted in Fig. 2.11. Our choice for a tubular membrane system has been made on the basis of a conceptual module design as detailed below.

Sealing

A proper sealing system to link the ceramic membranes tubes to the steel module is required.

High demands exist with respect to the seal materials since they have to operate under rather extreme conditions in terms of an oxidizing environment in combination with a high temperature. It is not anticipated that low temperature sealing is a viable option since this would create major temperature differences over the membrane causing stress while parts of the membranes that are present in the low temperature zones are not functional with respect to oxygen conductance, decreasing the economics of the module. Many other aspects of concern affect the choice of material:

– chemical inert towards the perovskite and the steel module;
– physical compatibility such as matching expansion coefficients, low creep rate;
– high temperature resistance;

– resistance against oxidation;
– possibility of adaptation to un-roundness of the membrane tube (e.g. ductility);
– industrial producibility; ease of handling/manifolding.
– economical viability such as cost prize/automation

Apart from materials issues, the engineering aspects have to be addressed. Different type of seal principles can be distinguished such as compression seal, ceramic glue like, and, for example, glasses. Solid state reactions between one of the more promising type of seal, a glass that crystallizes at the operating temperature are rather common [33]. A proper choice of thermal expansion coefficients (TEC) of the sealing materials is a prerequisite for success. The thermal expansions of SCF and BSCF are known from neutron diffraction data [25, 27] and dilatometry [24]. The values are rather high and are similar to that of commonly used stainless steel (e.g. 316). For a compression seal [34] the TEC of the sealing material should, preferably, be just a little higher than that of both the steel parts and the membrane material as it will be able to close the cap between both materials. For seal materials with a lower TEC, end-cap materials are required with a smaller TEC. Different steel types with lower TEC's exist such as 410S, 430, and 444 or materials like Monel or Inconel. The required oxygen resistance is, however, limiting the choice of steel types.

In Fig. 2.12, the principle of compression sealing of membranes with a tubular configuration is depicted. The membrane tube is mounted in a metal end-cap and the space in between end-cap and membrane filled by the sealing material. Compressive forces are obtained upon increasing the temperature and the magnitude of these forces can be tuned by proper selection of the metal type and sealing material, based

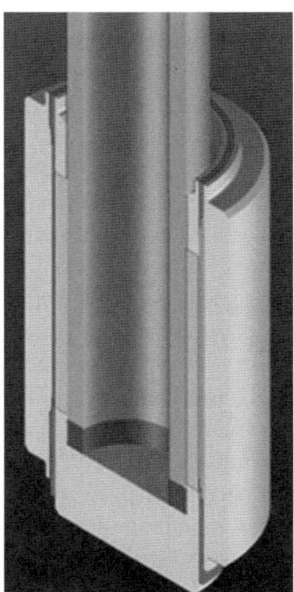

Fig. 2.12 Principle of compression sealing of tubular membranes

on their TEC's. This tunable principle is beneficial since the perovskite membranes will likely have moderate strength compared to steel.

Other sealing material options can be found in ceramic glues with tunable TEC, SCF-melt or SCF/binder systems. Last two options have the advantage of a proper TEC but are less preferred from an economical point of view since manifolding will be more complicated.

Module Concepts

The design of an adequate air separation unit highly depends on the type of membrane configuration chosen. Several conceptual options are available: hollow fibers [35], multi-channel monoliths [36], single tube [37], and tube-and-plate [38] configurations. Each membrane configuration has its advantages and disadvantages when taking reachable surface area, sealing technology and possible use of a sweep gas into account. Hollow fibers provide the highest membrane surface area but are relatively fragile and difficult to seal. Sealing of multichannel monoliths, especially when small channels are used for maximizing the amount of surface area, is also difficult. Single tube configurations, on the other hand, are less difficult to seal while up-scaling principles to one meter length has already been proven for other tubular membrane systems.

The use of a sweep gas is, in contrast to the other configurations, very complicated in the case of a tube-plate configuration since the space in between the plates can be poorly reached.

A conceptual design study of a full-scale plant with an oxygen production capacity of 3×10^4 Nm^3/hr or 9,000 tonnes per day, values comparable to a normal sized cryogenic distillation plant, was carried out. In this numerical study, the merits of these membrane geometries were compared in terms of packing density, expressed as surface area to module volume ratio (m^2/m^3) manufacturability and so on, leaving influences of pressure drops and presuming a constant oxygen flux. The maximum allowable gas velocity (25 m/s) has also been taken into account here since higher gas velocities would result in resonances that might result in damaging of the membranes, especially for fragile hollow fibers. The study reveals that the tubular membrane with an outer diameter of ~20 mm leads to the optimal configuration [39] despite the fact that hollow fibers provide more membrane surface area. The maximum allowable gas velocity restricts the membrane length severely. With a decrease of diameter of single-hole tubes, the maximum allowable length when reaching the maximum gas velocity also decreases rapidly. This argument is especially valid for thin hollow fibers and results in a high amount of small separation units that are needlessly costly. Tubular configurations with dimensions of ~20 mm diameter allow membrane lengths exceeding 2.5 meter. Since the cost price of the stainless steel vessel will make up ~50% of the total cost price of the module, minimized use of steel is to be preferred.

Dimensional limitations contribute to the design of an air separation unit. Only parts of the volume of the module can be used because of the arrangement of the membranes. In this respect, hexagonal packed volumes are most desirable since they allow for the highest module space-filling. In case of monoliths, the same argument is valid, but additionally, a choice can be made with respect to the outer monolith dimensions. Minimization of the lost space in between the monoliths by the use of square monoliths rather then tubular one's is to be preferred but, as can be seen from Fig. 2.13, lost space at the boundary of module vessel and monolith will still exist. Square monoliths will only increase the m^2/m^3 ratio with $\sim$10%.

Additionally, space is required for manifolding and heat insulation while space is also required at the top and bottom of the module so as to gain maximal profit of the available membrane surface area, Fig. 2.14. A module vessel consisting of an insulated double wall has the advantage of minimization of expensive heat- and oxygen-resistant stainless steel, used for relatively thin inner walls, and therefore reduction of the total costs of the module. Accommodation of the pressure inside the module is reached by a thick cold external wall that can be manufactured from cheap steel.

Based on the tubular configuration, and presuming a total module diameter of about 2 meter including the space required for insulation, the module of Fig. 2.15 can be designed with a height of about 4.5 meter which contains $\sim$160 m^2 of membrane surface area. In order to meet an oxygen production of 30.000 Nm^3/hr, 32 of these module units will be needed.

For reaching the same oxygen production, single-hole tubes of 10 mm diameter would increase the amount of units to 41, comparable to multi-channel monoliths (39 units). The extreme restriction in membrane length of hollow fiber systems would result in $\sim$1,800 modules. The difference in expansion behavior of vessel wall and membrane tubes can be compensated by mounting the tubes in a

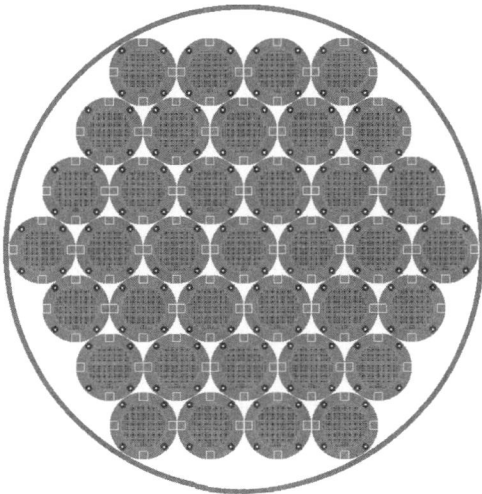

Fig. 2.13 Hexagonal packing of multi-channel tubular monoliths

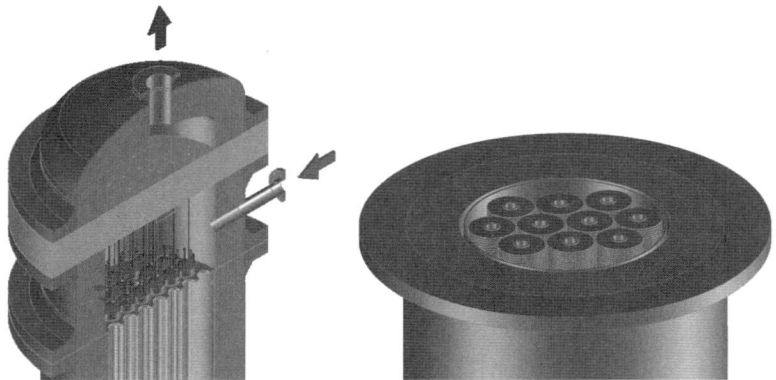

Fig. 2.14 Schematic view of the insulated double wall vessels filled with tubes (*left*) and tube-and-plate assemblies (*right*), showing the space lost due to the manifolding and insulation

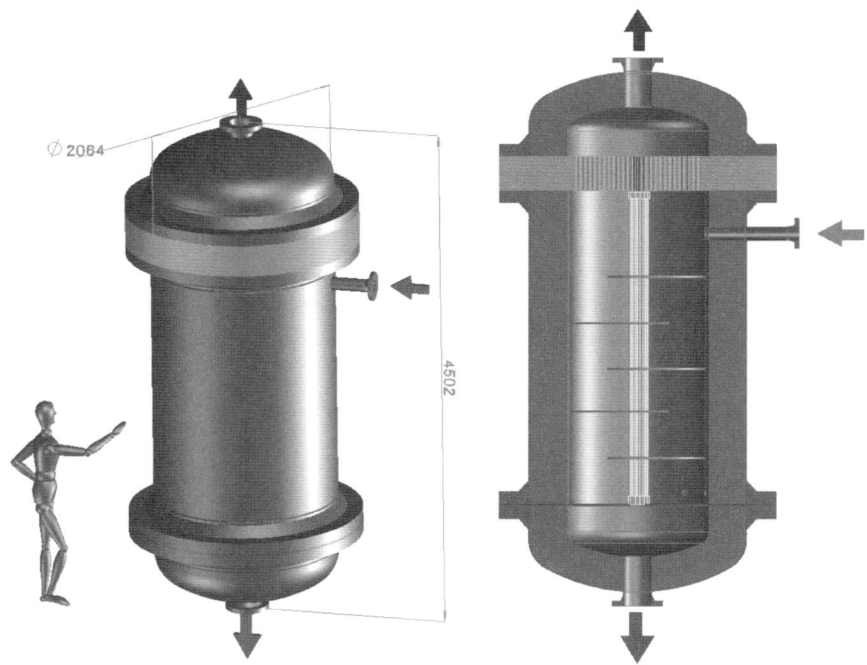

Fig. 2.15 Representation of a full scale module containing about 160 m² of membrane area

"pipe-plate" at the top side of the vessel and a "centre-plate" at the bottom of the vessel, keeping the membrane tubes centered while maintaining the possibility of moving, see Fig. 2.16. In order to make use of a sweep gas, sweep gas supply tubes with a diameter much smaller than the membrane tube diameter are connected to the upper plate and stuck through each membrane tube over nearly the full length of the membranes.

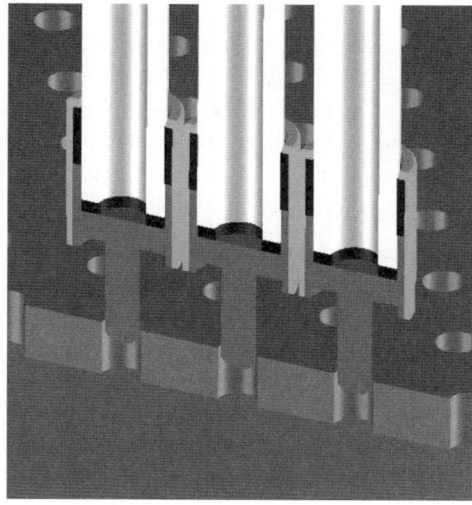

Fig. 2.16 The pipe-plate from where the membrane tubes are suspended (*left*) and the centre-plate at the bottom of the module (*right*)

This reasoning leading to the best option being tubular membranes in a module does not necessarily apply for membrane reactors in which (partial) combustion takes place. The necessity of catalysts also sets extreme demands as to bed permeability.

Concluding Remarks

From the discussion in this chapter, the conclusion can be drawn that the actual implementation of ITM technology cannot be expected in the short term. We believe that there is a number of technology challenges that still needs to be met. However, the small amount of public information from two large leading consortia in the US, led by Air Products and Praxair, hampers a thorough and reliable assessment of the current state-of-the–art ITM technology. Our vision on the viability of high-temperature ITM technology can thus be unnecessarily pessimistic. From our viewpoint, we can point at the following barriers that have to be removed before this technology will be successfully applied

Economic Viability

This is one of the main issues for the implementation of any breakthrough technology. We believe that external factors like energy price, CO_2 penalty and legislation will play a decisive role in this respect. Our economic calculations are still not exact enough for the determination of the maximum allowable membrane in module price. Significantly, increased membrane fluxes will lead to lower surface areas and smaller number of modules. However, we do not anticipate a flux increase of a factor 2–3 is likely with any material that is stable under the given conditions. However, the inclusion of heat integration options in our assessments may result in a much more optimistic view. Those options are certainly available when an ITM air separation unit is combined with a partial oxidation reactor, for example, in coal gasification, natural gas reforming, and iron ore reduction.

Technological Viability

One of the first assumptions in our economic considerations was that all the technological challenges were met. This remains, at the moment of writing, uncertain. In the above discussion, a number of issues have passed, such as sealing technology, thermodynamic phase stability of the cubic perovskite versus brownmillerite, and kinetic demixing. In relation to this, we have not mentioned creep resistance as yet. For SCF, this was shown to be very low [40], which is likely to be the case for BSCF as well.

Acceptability

Before this technology can be applied on a full-scale as envisaged in our conceptual module design, the end-users should gain sufficient trust in long term performance and reliability of this membrane solution. We believe that to achieve this, first, a number of smaller on-site production facilities must be set up. Only through wide publication of the information obtained in those demonstration plants can significant industrial application be expected.

Oxygen Production or Partial Oxidation?

At this moment, it is uncertain which of the two major possible applications will be first on the market. The production of oxygen puts a much smaller demand on the material properties. The change of oxygen partial pressure will be limited to about one order of magnitude. The amount of impurities that may cause membrane degradation in the feed is likely to be limited to H_2O and CO_2. When a membrane reactor is considered, the change in oxygen partial pressure over the membrane is expected to be much larger; up to 20 orders of magnitude can be envisaged. This will put significant constraints on the membrane materials as they have to be stable under a wide range of reducing conditions. However, thanks to the large driving force, the demands on the intrinsic transport properties of the materials can be much relaxed. As a result, materials with a larger creep resistance can be chosen. In this case of partial oxidation, catalyzed chemical reactions take place that are exothermic. Precautions to prevent the occurrence of hot spots and runaways have to be in place.

Finally

Despite the fact that the current class of Co/Fe perovskites has been under investigation for over 20 years [10], there are still a large number of uncertainties in getting this technology towards implementation. A joint research agenda combining the worldwide activities directed at achieving this implementation is likely to be required. The involvement of public and private funding appears to be also prerequisite.

Acknowledgments The work presented in this article has been financed in part by the Dutch ministry of economic affairs through SenterNovem agency in several projects including Captech (EOSLT04003), and two projects directed at the energy efficient production of oxygen (EDI03201 and EDI02106).

References

1. IPCC, 2007: Summary for policymakers. In: Metz, B, Davidson OR, Bosch PR, Dave R, Meyer LA, editors. Climate change 2007: mitigation. Contribution of Working Group III to the Fourth Assessment, Report of the Intergovernmental Panel on Climate Change. Cambridge, United Kingdom and New York: Cambridge University Press; 2007.
2. Department of Energy, Office of Clean Coal, Strategic Plan: Moving America Towards an Affordable "Zero" Emissions Coal Energy Option. 2006.
3. Mattisson T, Garciá-Labiano F, Kronberger B, Lyngfelt A, Adánez J, Hofbauer H. Chemical-looping combustion using Syngas as fuel. Int J Greenhouse Gas Control. 2007;1:158.
4. Lyngfelt A, Kronberger B, Adánez J, Morin JX, Hurst P. The GRACE Project. Development of oxygen carriers particles for chemical-looping combustion. Design and operation of a 10kW chemical-looping combustor. 2004.
5. Acharya D, Krishnamurthy KR, Leison M, MacAdam, Sethi VK, Anheden M, et al. Development of a high temperature oxygen generation process and its application to oxycombustion power plants with carbon dioxide capture. 2005.
6. Ruddlesden SN, Popper P. Acta Crystallogr. 1958;11:541.
7. Cava RJ, van Dover RB, Battlog B, Rietman EA. Bulk superconductivity at $36\,K$ in La1.8Sr0.2CuO4. Phys Rev Lett. 1987;58:408.
8. Fu WT, IJdo DJW. Crystal structure of superconducting $BaPb_{0.7Bi0.15Sb0.15O3}$. Solid State Commun. 2001;118:291.
9. Reinen D, Wegwerth J. Local and cooperative bonding effects in K2NiF4-type copper superconductors. Structural, spectroscopic and magnetic investigations on mixed crystals La1 + xSr1-xGa1-xCuxO4. Physica C. 2007;183:261.
10. Teraoka Y, Zhang HM, Furukawa S, Yamazoe N Oxygen permeation through perovskite-type oxides. Chem Lett. 1985;14:1743.
11. Kilner JA, Shaw CKM. Mass transport in La2 Ni1-xCoxO4 + d oxides with the K2NiF4 structure. Solid State Ionics. 2002;523:154–5.

12. Bredesen R, Sogge J. A technical and economic assessment of membrane reactors for hydrogen and Syngas production. Seminar on the Ecological Applications of Innovative Membrane Technology in the Chemical Industry, 1996.

13. Bouwmeester HJM Dense ceramic membranes for methane conversion. Catal Today. 2003;82:141.

14. Kiesewetter J. Personal Commiunication, 2007.

15. Peters MS, Timmerhaus KD. Plant design and economics for chemical engineers. 4th ed. New York: McGraw-Hill; 2007. p. 1991.

16. Eklund HR, Sundkvist SG. A Development project of a future NGCC power process with CO2-capture. 2004.

17. Sundkvist SG, Klang A, Sjödin M, Wilhelmsen K, Asen K, Tintinelli A, McCahey S. AZEP Gas Turbine Combined cycle power plants – thermal optimisation and Lca analysis. 2004.

18. Yantovski E, Gorski J, Smyth B, Elshoften J. Zero emission fuel-fired power plants with ion transport membrane. 2003.

19. Switzer L, Sirman J, Rosen L, Thompson D, Howard H, Bool L. In: Thomas DC, Benson MS, editors. Cost and feasibility study on the Praxair advanced boiler for the CO2 capture Project's Refinery Scenario. Vol. 1. Elsevier Publishers; 2005. p. 561.

20. Sundkvist SG, Klang A, Thorshaug NP. AZEP – development of an integrated air separation membrane – gas turbine. 2001.

21. Bredesen R, Jordal K, Bolland O. High-temperature membranes in power generation with CO2 capture. Chem Eng Process. 2004;43:1129.

22. Maurstad O, Bredesen R, Bolland O, Kvamsdal HM, Schell M. SOFC and gas turbine power systems – evaluation of configurations for CO2 capture. 2004.

23. Lowe C, Andersen H. CO2 Capture project-pre-combustion technology overview. 2004.

24. Vente JF, Haije WG, Rak ZS. Performance of functional perovskite membranes for oxygen production. J Membr Sci. 2006;276:178.

25. McIntosh S, Vente JF, Haije WG, Blank DHA, Bouwmeester HJM. Phase stability and oxygen non-stoichiometry of $SrCo0.8Fe0.2O3-d$ measured by in-situ neutron diffraction. Solid State Ionics. 2006;177:833.

26. Kruidhof H, Bouwmeester HJM, van Doorn RHE, Burggraaf AJ. Influence of order-disorder transitions on oxygen permeability through selected nonstoichiometric perovskite-type oxides. Solid State Ionics. 1993;63–65:816.

27. S. McIntosh JF, Vente WG, Haije DHA, Blank, Bouwmeester HJM. Oxygen stoichiometry and chemical expansion of $Ba0.5Sr0.5Co0.8Fe0.2O3-d$ measured by in-situ neutron diffraction. Chem Mater. 2006;18:2187.

28. van Doorn RHE, Bouwmeester HJM, Burggraaf AJ. Kinetic decomposition of $La0.3Sr0.7CoO3-d$ perovskite membranes during oxygen permeation. Solid State Ionics. 1998;111:263.

29. Ullmann H, Trofimenko N, Naoumidis A, Stover D. Ionic/electronic mixed condution relations in perovskite type oxydes by defect structure. J Eur Ceram Soc. 1999;19:791.

30. Deganello F, Liotta LF, Longo A, Casaletto MP, Scopelletti M. Cerium effect on the phase structure, phase stability and redox properties fo ce-doped strontium ferrates. Solid State Chem. 2006;179:3406.

31. Paulsen JM. Thermodynamics, Oxygen stoichiometric effects and transport porperties of ceramic materials in the sytem Sr-Ce-M-O (M = Co, Fe). 1997.

32. Zhu X, Wang HH, Yang W. Structural stability and oxygen permeability of cerium lightly doped $BaFeO3-d$ ceramic membranes. Solid State Ionics. 2006;177:2917.

33. Faaland S, Einarsrud MA, Grande T. Reactions between calcium and strontium substituted lanthanum cobaltite ceramic membranes and calcium silicate sealing materials. Chem Mater. 2001;13:723.

34. Rusting FT, de Jong G, Pex PPAC, Peters JAJ. Sealing socket and method for arranging a sealing socket to a tube. WO 01/63162 A1. 2001.

35. Liu S, Gavalas GR. Oxygen selective ceramic hollow fiber membranes. J Membr Sci. 2005;246:103.

36. Bruun T. Design issues for high temperature ceramic membrane reactors. 6th International Conference on Catalysis in Membrane Reactors, 2004.
37. van Hassel BA, Prasad R, Chen J, Lane J. Ion-transport membrane assembly incorporating internal support. US 6,565,632, 2001.
38. Armstrong PA, Bennett DL, Foster EPT, van Stein EE. Ceramic membrane development for oxygen supply to gasification applications. Gasification Technol. 2002.
39. Vente JF, Haije WG, IJpelaan R, Rusting FT. On the full-scale module design of an air separation unit using mixed ionic electronic conducting membranes. J Membr Sci. 2006;278:66.
40. Majkic G, Wheeler L, Salama K. Creep of polycrystalline $SrCo0.8Fe0.2O3-d$. Acta Mater. 2000;48:1907.

Chapter 3
Oxygen-Ion Transport Membrane and Its Applications in Selective Oxidation of Light Alkanes

Weishen Yang and Rui Cai

This article provides a brief survey of oxygen-ion transport membrane materials and their applications as membrane reactors in the selective oxidation of light alkanes.

The development of *oxygen-ion transport membranes* (OITM) has attracted great interests for their ability to supply high-purity oxygen and their potential applications for many industrial processes and advanced energy technologies. Compared to cryogenic distillation of air, the conventional technology of oxygen production, membrane separation process will cut the cost by nearly one-third [1]. One of the most promising applications of OITM is as a membrane reactor for selective oxidation of light alkanes, in which oxygen (or different oxygen species) can be controlled when supplied from the oxygen-rich side to the reaction side.

The first section of this chapter gives a brief survey of major membrane concepts and different membrane reactor configurations. Membrane materials are discussed in the second section. The third section will present the recent development of OITM reactors for selective oxidation of light alkanes.

Membrane Concepts and Reactor Configurations

Oxygen-ion transport membranes are made from materials that possess high oxygen ion conductivity at proper temperatures, such as oxygen-ion *solid electrolyte* (SE) or *mixed ionic and electronic conducting* (MIEC) materials. In the case of using solid electrolyte membrane (Fig. 3.1a), only oxygen ions can permeate through the membrane barrier. The SE membrane is sandwiched between two porous electrodes and electrons flow through the external circuit. Under this condition, the driving force of oxygen permeation is electric potential gradient. In the latter case, both electrons and oxygen ions can simultaneously transfer through the membrane but in the opposite direction in the same phase (Fig. 3.1b) or different phases (Fig. 3.1c).

W. Yang (✉)
Satate Key Lab of Catalysis, Dalian Institute of Chemical Physics, Chinese Academy of Sciences, Dalian 116023, China,
e-mail: yangws@dicp.ac.cn

A.C. Bose (ed.), *Inorganic Membranes for Energy and Environmental Applications*, DOI 10.1007/978-0-387-34526-0_3, © Springer Science+Business Media, LLC 2009

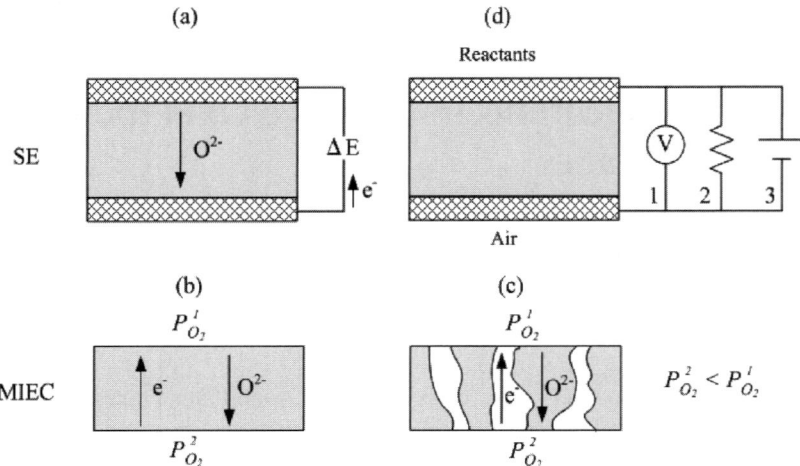

Fig. 3.1 The schematic diagram of oxygen-ion transport membrane. (**a**) SE; (**b**) single-phase MIECM; (**c**) dual-phase MIECM; (**d**) operation modes of SE: 1. Open circuit, 2. Fuel cell mode, 3. Electrolytic cell mode

Oxygen partial pressure between the two sides of the membrane is the driving force of oxygen permeation. Figure 3.1c shows a dual-phase MIEC membrane, which can be visualized as a dispersion of a continuous electronic conducting phase into an SE matrix. The electronic conducting phase is usually made from precious metal or metal oxides.

When the solid electrolytes are used as membranes, there are three different operation modes, as shown in Fig. 3.1d. Mode 1 is under open circuit operation, in which no net current passes through the membrane. The reactor in this mode often serves as a sensor or an in situ characterization technique for catalytic gas-solid reactions under work conditions, named *solid electrolyte potentiometry* (SEP) [2]. Mode 2 is the Galvanic cell (i.e. fuel cell) mode and mode 3 is the Volta cell (i.e. electrolytic cell) mode. Both modes are under closed-circuit operation. Mode 3 is also called electrochemical oxygen pump. The oxygen pumping rate normally obeys Faraday's law, that is, current I corresponds to I/4F mol oxygen per second transport through the SE. Interestingly, in some cases where oxygen and other reactants were co-fed to the reaction side and air was supplied to the other side, when applying a small current or potential, the reaction rates are dramatically promoted or restrained, deviating from Faraday's law. This is the famous phenomena called *non-faradaic electrochemical modification of catalytic activity* (NEMCA). This has been explained successfully with the catalyst work function change caused by the change of electrode overpotential, as outlined in a recent book by Vayenas and co-workers [3]. In this chapter, we will focus on the *Oxygen-Ion Transport Membrane Reactors* (OITMRs) as oxygen distributors for selective oxidation, as the open circuit operation mode and NEMCA phenomena are not of concern here.

The two most common membrane geometries are the flat plate and the tube. Single flat plate membranes are usually used in laboratory scale investigations due to their ease of fabrication. Tubular membranes are more and more popular due to their much larger ratio of the membrane surface area to the equipment volume than flat plate membranes [4]. OITM reactor configurations with multi-planar or multi-tubular structures are required for commercial use.

Membrane Materials

As we mentioned in Sec. 1, OITMs are made from two different types of membranes, SE or MIEC.

The most frequently used SE is cubic fluorite *yttria stabilized zirconia* (YSZ). More than 90% of the literatures of SE membrane reactors for selective oxidation are based on this kind of materials. The substitution of lower valence cations (e.g., Ca^{2+}, Y^{3+}, Sc^{3+}, Yb^{3+}, etc.) introduces oxygen vacancies and, in turn, oxygen-ion conductivity [5]. Doped CeO_2 has the same structure and much higher ion conductivity than YSZ at 600–800°C, while Ce^{4+} is easy to be reduced to Ce^{3+} and the material presents electronic conductivity at low oxygen partial pressures [6]. δ-Bi_2O_3, which also has fluorite-related structure, has intrinsic high oxygen-ion conductivity at 729–825°C without substitution of other cations. The δ-phase can be stabilized to lower temperatures by cation substitution for Bi in Bi_2O_3, such as W^{6+}, Y^{3+}, Gd^{3+}, Er^{3+} and the like. The conductivity of $Bi_{1.5}Y_{0.5}O_3$ reaches 10^{-2} S.cm^{-1} at 500°C. Unfortunately, the materials based on Bi_2O_3 suffer easy reduction at low oxygen partial pressures [7]. Pyrochlores structure is derived from an oxygen-deficient fluorite structure. $Gd_2(Ti_{1-x}Zr_x)O_7$ is an oxygen-ion conductor when x > 0.4, but conductivity of 10^{-2} S.cm^{-1} at 1,000°C is not competitive with other oxygen-ion SE materials [8].

Perovskite (ABO$_3$) and related structure SE materials show promising future for their higher conductivity than YSZ and good stability. Ishihara et al. [9] first reported a pure oxygen-ion conductor $La_{0.9}Sr_{0.1}Ga_{0.8}Mg_{0.2}O_{2.85}$ (LSGM) with oxygen-deficient, cubic perovskite structure. The oxygen-ion conductivity is more than 10^{-2} S.cm^{-1} at 600°C and almost independent of oxygen partial pressure from 10^{-20} to 1 atm. Brownmillerite (A$_2$B$_2$O$_5$) is a deriving structure from perovskite [10]. Brownmillerite $Ba_2In_2O_5$ presents higher oxygen-ion conductivity than YSZ at temperatures more than 930°C [11].

Abraham et al. [12] found that the tetragonal γ-phase (stable at 400–500°C) of bismuth vanadate ($Bi_4V_2O_{11}$) exhibits high oxygen ionic conductivity at moderate temperatures. This phase can be stabilized to room temperature by partially substituting other elements for vanadium that was identified as BIMEVOX. $Bi_4Cu_{0.2}V_{1.8}O_{11-\delta}$ exhibits excellent low temperature oxygen-ion conductivity, 10^{-2} S.cm^{-1} at 350°C [13, 14]. However, high-chemical reactivity, low-mechanical strength and a high-thermal-expansion coefficient limit its further application.

The search for new oxygen-ion conductors never stops. For example, $La_{2-x}GeO_{5-\delta}$ [15] and $La_2Mo_2O_9$ [16], neither of which belongs to the structures mentioned above, were reported as alternative SE material in recent years. Further information on oxygen-ion SE materials can be found in a recent review [8].

For MIEC, most efforts were focused on perovskite structure materials after Teraoka [17] firstly reported high oxygen permeability of $La_{1-x}Sr_xCo_{1-y}Fe_yO_{3-\delta}$ perovskite membranes. Differing from perovskite solid electrolytes, a demand for sufficient electronic conductivity as well as matching oxygen-ion conductivity should be met. Electronic conductivity is introduced by substitution of multivalent cations for B site cations of perovskite, which can be explained by Zerner's *double exchange* mechanism [18] or a small polaron mechanism with a thermally activated mobility [19]. The oxygen permeation flux of perovskite-related materials in the literatures is listed in Table 3.1. The stability of MIEC materials under severe conditions is also of great importance. Perovskite $SrCo_{0.8}Fe_{0.2}O_{3-\delta}$ is thermodynamically stable only at high oxygen partial pressures (> 0.1 atm) and high temperatures (> 790°C) [20]. Its perovskite phase with disordered oxygen vacancies was transformed to brownmillerite phase in which oxygen vacancies are ordered when the temperature is lower than 790°C [21]. Proper substitutions of metal ions for A or B site cations of $SrCo_{0.8}Fe_{0.2}O_{3-\delta}$ may improve the phase stability. For example, partial substitution of La, Ba for Sr or Ti, Zr for Co significantly improves the stability, but at the cost of oxygen permeability. Doping less reducible ions for Co, such as Ga^{3+}, Ti^{4+}, Zr^{4+}, is an effective method to improve the stability of perovskite MIEC materials in reducing atmosphere at high temperatures. $BaZr_{0.2}Co_{0.4}Fe_{0.4}O_{3-\delta}$, as an example, can withstand syngas production from methane at 850°C for more than 2200 h [22]. A series of new cobalt-free perovskite membrane materials $BaCe_xFe_{1-x}O_{3-\delta}$ was reported recently in our group [23]. $BaCe_{0.15}Fe_{0.85}O_{3-\delta}$ exhibits the highest oxygen permeation flux among this series, which can compare to what $La_{0.6}Sr_{0.4}Co_{0.2}Fe_{0.8}O_{3-\delta}$ exhibits at the same temperature.

Alternative MIEC materials with other structures were also developed. A brownmillerite-derived cobalt-free MIEC, $La_{2-x}Sr_xGa_{2-y}Fe_yO_{5+\delta}$, was reported to be stable for more than 1,000 h for syngas production from methane [29]. $Sr_4Fe_6O_{13}$ adopts an orthorhombic intergrowth structure composed of alternating blocks of $(Fe_4O_5)^{2+}$ and perovskite-derived $(Sr_4Fe_2O_8)^{2-}$ [30]. The tubular membrane of $Sr_4Fe_6O_{13}$ also ran more than 1,000 h for oxygen permeation [31]. Manthiram et al. [32] reported another kind of materials with perovskite-related intergrowth

Table 3.1 Oxygen permeation fluxes of some MIEC membrane materials

Materials	O_2 flux (ml cm^{-2}min^{-1})	Temp. (°C)	Thickness (mm)	Ref.
$La_{0.6}Sr_{0.4}Co_{0.8}Fe_{0.2}O_{3-\delta}$	0.62	850	2.0	[17]
$La_{0.2}Ba_{0.8}Co_{0.2}Fe_{0.8}O_{3-\delta}$	0.4	900	2.3 − 3.1	[24]
$La_{0.8}Sr_{0.2}Ga_{0.6}Fe_{0.4}O_{3-\delta}$	0.85	900	0.5	[25]
$SrCo_{0.8}Fe_{0.2}O_{3-\delta}$	0.5	830	1.0	[26]
$Ba_{0.5}Sr_{0.5}Co_{0.8}Fe_{0.2}O_{3-\delta}$	1.3	900	1.5	[27]
$BaZr_{0.2}Co_{0.4}Fe_{0.4}O_{3-\delta}$	0.7	900	1.0	[28]
$BaCe_{0.15}Fe_{0.85}O_{3-\delta}$	0.42	900	1.0	[23]

structure, *Ruddlesden–Popper* (R–P) series $A_{n+1}B_nO_{3n+1}$, in which a number of perovskite blocks having corner-shared BO_6 octahedra alter with AO rock salt layers along the c-axis. The oxygen permeability of these series materials is lower than that of perovskite-type counterparts. Additionally, R–P series materials with larger oxygen vacancies along c-axis, such as $Sr_3FeCoO_{7-\delta}$, are extremely sensitive to atmospheric moisture [33].

Introduction of electronic conductivity to fluorite solid electrolyte was also attempted by dissolution of oxides having multivalent cations [34, 35]. However, the electronic conductivity is often orders of magnitude less than those of perovskite based MIEC materials.

Dual-phase composite materials, which are made from oxygen ionic conducting phase and electronic conducting phase, were also suggested as MIEC membrane materials. Stabilized bismuth oxide or zirconia mixed with noble metals forms a kind of dual-phase membrane, where the noble metal (Pd, Pt, Au, Ag, etc.) phase should exceed 30 vol. percent in order to form a continuous electronic conducting phase [36]. The high material cost and low oxygen permeability hinder this type of composite membrane for practical applications. Perovskite MIEC also shows high electronic conductivity and can be used as the electronic conducting phase. A dual-phase $La_{0.15}Sr_{0.85}Ga_{0.3}Fe_{0.7}O_{3-\delta}$ (LSGF)-$Ba_{0.5}Sr_{0.5}Co_{0.8}Fe_{0.2}O_{3-d}$ (BSCF) composite membrane presents a structure defined by the closed packing of LSGF grains with a three-dimensional thin BSCF film running between the boundaries of the connected LSGF grains [37]. The film phase is percolated at a volume percent as low as 7.2%. The oxygen permeation flux of LSGF-BSCF membrane ($\sim$0.45 ml/cm^2.min) is nine times as high as that of the LSGF membrane ($\sim$0.05 ml/cm^2.min) at 915°C. H_2 reduction experiment also shows that LSGF-BSCF membrane has a good stability in a H_2 containing atmosphere.

For practical applications, OITM materials used in membrane reactors must meet a number of requirements, as shown in Fig. 3.2. These include (1) the materials must be stable for long-term operation under strongly reducing atmosphere, such as syngas (CO$+$H$_2$); (2) the materials must have considerable high oxygen permeability under the operation conditions; (3) the materials must have enough mechanical strength for constructing the membrane reactor; (4) the oxygen permeability of the

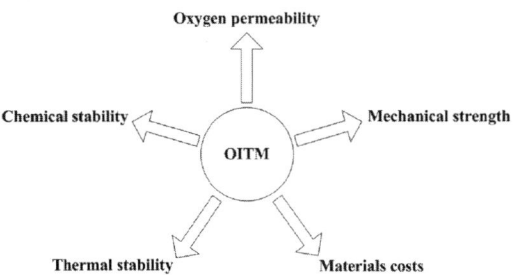

Fig. 3.2 Criteria for oxygen permeable membrane

membrane materials should avoid a decline with time; and (5) the materials should be cheap enough for large-scale industrial applications.

OITM Reactors for Selective Oxidation of Light Alkanes

Alkanes are among the most chemically stable of all organic molecules with their abundant reserves, such as in natural gas, petroleum, coal and synthetic fuels [38]. They are mostly burned as fuels for heat and power generation but their potential as feedstock for more useful chemicals is of great practical importance. Selective oxidation of light alkanes is one of the most attractive and challenging route and has been intensively studied. The major problem holding back the large-scale use of selective oxidation process is the low selectivity to object products because products are usually more active than feed alkanes, resulting in complete oxidation. The effective method to increase selectivity to object products can be accomplished by controlling the oxygen concentration and/or oxygen species. It is convenient to be achieved by using a membrane reactor as an oxygen distributor. Furthermore, using OITM also provides different types of active species (e.g. O^{2-}, O^{-}, O_2^{-}) at the membrane surface where the catalyst and reactants are located, which may also enhance product selectivity. In the past 20 years, many efforts have been focused on the selective oxidation of light alkanes by using OITM reactors. Several books and review articles that covers partly or wholly this field have been published recently [39–45]. Here, most attention is paid to the work published after 2000 in this field.

Oxidative Coupling of Methane

Oxidative coupling of methane (OCM) to ethylene and ethane is one of the most desirable approaches to the direct conversion of methane. Although many efforts have been devoted to this reaction on novel catalysts in the past two decades, the per-pass C_2 yield reported was limited to about 25% except for some occasional reports. Researchers realized that the design of new reactors should also be considered. OITM reactors show promising application on OCM not only because of the advantages of membrane reactors we mentioned above but also the intrinsic catalytic activity of membrane materials for OCM [46].

Based on the work of Lin [47] and Yang [48], the reaction mechanism of OCM in MIECM reactors is illustrated in Fig. 3.3. The ideal condition for the OCM is that all permeated lattice oxygen is consumed on the membrane surface and gaseous oxygen is absent in the reaction side.

Both disk [49] and tubular [48] type membrane reactor based on BSCF material were investigated in the application for OCM. Between 800 and 900°C, methane conversion was about 2–14%, which was apparently higher than 3.25–0.65% conversion obtained in disk type reactors. On the other hand, the corresponding C_2 ($C_2H_4 + C_2H_6$) selectivity was 38–52%, which was lower than 48.4–72.8% C_2

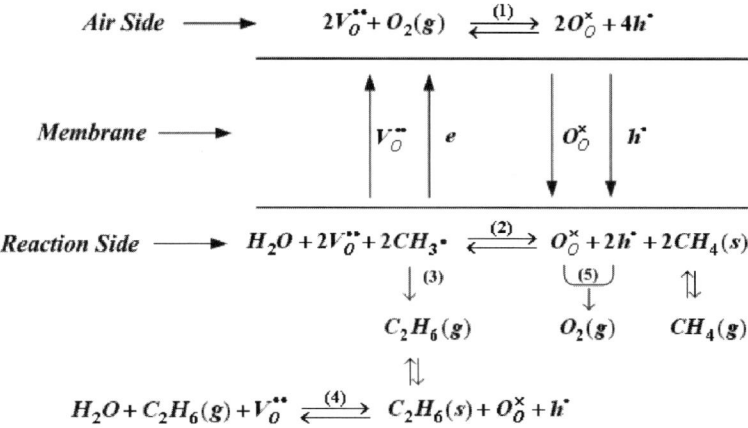

Fig. 3.3 The schematic diagram of mechanism of OCM in MIECM reactor

selectivity obtained in the disk type reactor. Compared to the packed-bed reactor using BSCF as catalyst, the C_2 selectivity increased by 20%. When active OCM catalyst (La-Sr/CaO) was added to the tubular BSCF membrane reactor, both C_2 selectivity and CH_4 conversion were improved, and about 13–15% C_2 yield was obtained. Interestingly, the C_2H_4/C_2H_6 ratio reached 12, which was much higher than that obtained in a packed-reactor under the same conditions. Actually, C_2H_4 is most desired product in OCM reactions.

Lu et al. [50] investigated the OCM in a tubular membrane reactor, which consisted of two layers, with $SrFeCo_{0.5}O_3$ as the MIECM and $BaCe_{0.6}Sm_{0.4}O_3$ as the secondary layer deposited inside the $SrFeCo_{0.5}O_3$ tube to minimize the effect of the total oxidation catalytic activity of $SrFeCo_{0.5}O_3$. La/MgO as OCM catalyst was added inside the reactor. C_2 product yields of up to 7% were reported.

Recently, good results obtained in single-layer MIECM reactors in the absence of additional catalyst were also reported. C_2 yield of 16.5% with a C_2 selectivity of 62.5% was obtained in a single-layer $BaCe_{0.8}Gd_{0.2}O_3$ tubular membrane reactor [50]. Lin et al. [51, 52] studied the catalytic performance of a fluorite-structured yttria-doped bismuth disk type membrane for OCM reaction. The reported C_2 selectivity and yield were in the range of 20–90% and 16–4% respectively in the membrane reactor of 25 mol percent yttria-doped bismuth oxide (BY25). When samarium was doped in the BY25 lattice, the oxygen permeability and catalytic performance were improved. In $Bi_{1.5}Y_{0.3}Sm_{0.2}O_{3-\delta}$ (BYS) disk-type membrane reactor, the highest C_2 yield achieved is 17% with a C_2 selectivity of about 80%. More promising results were reported in a BYS tubular membrane reactor. As high as 35% C_2 yield was achieved with a C_2 selectivity of 54% at 900°C [51–55]. These were the best results of C_2 yield obtained for oxidative coupling of methane and higher than the computed upper bound on OCM yield of 28% under conventional, packed-bed, continuous-feed operation [56].

Although the first study of OCM in OITM reactors for oxygen separation was in a YSZ solid electrolyte membrane reactor with Ag and Ag/Bi_2O_3 as electrodes [57], only a few articles in this field were reported after 2000. A kind of *temperature-programmed desorption* (TPD) technique was used in SE membrane reactors to investigate the oxygen species for the combustion reaction and the coupling reaction separately [58].

Partial Oxidation of Methane

During the last decades, partial oxidation of methane to syngas has received much attention and many investigations have been devoted to these studies over many different active catalysts with a result of high CH_4 conversion, high CO selectivity and proper H_2/CO ratio of 2 for the downstream process. However it is a high-energy and capital-intensive process because the downstream process requires the syngas stream to be free from nitrogen, and consequently pure oxygen is needed. Using OITM reactors can combine air separation and POM into a single unit operation resulting in saving significant cost, which is expected to be cut by 25–35% from the present method [59]. The hot-spot problem of the conventional co-fed reactor can also be avoided by the separate feeding of methane and oxygen. In this process, the OITM acts mainly as an oxygen separator and distributor, while its catalytic properties are less important than that in the OCM process since the active catalyst for POM is present.

From the point of view of the practical application in the partial oxidation of methane to syngas, the membrane materials (including the electrode material in SE membrane reactors) must be chemically and mechanically stable under severe operation conditions. One side of the membrane is exposed to an oxidized atmosphere (usually air) and the other side is exposed to a very reductive gas mixture (CH_4, CO, H_2, CO_2, H_2O and so on). Therefore, it is crucial that the membrane materials should have a stable lattice structure under a wide range of oxygen partial pressure and high resistance to the highly reductive atmosphere besides possessing the characteristics with the desired oxygen permeability.

$SrCo_{0.8}Fe_{0.2}O_{3-\delta}$ shows the highest oxygen permeability in the $(La,Sr)(Co,Fe)O_x$ series. However, Pei et al. [60] found two types of structure failure occurred during the POM process. The first type, occurring shortly after the beginning of reaction, is the consequence of the lattice mismatch caused by the oxygen pressure gradient between the two sides of membrane. The other one is the result of a chemical decomposition in the severe operation atmosphere, which occurred after days of reaction. The failure of structure crack was also found in other $(La,Sr)(Co,Fe)O_x$ series materials used as membranes for POM reaction. Diethelm et al. [59] developed a bi-layered membrane, fabricated by depositing a thin film of $(Sm_2O_3)_{0.1}(CeO_2)_{0.9}$ (SDC) on a $La_{0.6}Sr_{0.4}Co_{0.8}Fe_{0.2}O_{3-\delta}$ substrate via pulsed laser deposition, for POM reaction. They compared the stability of the bi-layered membrane with the single layer $La_{0.6}Sr_{0.4}Co_{0.8}Fe_{0.2}O_{3-\delta}$ membrane under POM operation conditions. In case of the single layer membrane, rapid deterioration of the oxygen transport properties was seen within 1 h due to phase decomposition, whereas, in the case of the bi-layer membrane coated SDC layer acted as a protective layer in the reaction side, the reaction ran more than 50 h with no obvious decay. Another interesting result also found that the methane conversion rate and CO formation rate showed a periodic oscillatory behavior in SDC-coated membrane reactors due to a redox cycle of the SDC layer.

Tsai et al. [61, 62] used $La_{0.2}Ba_{0.8}Co_{0.8}Fe_{0.2}O_{3-\delta}$ disk membrane successfully in POM reaction for 850 h at 850°C with 5% Ni/Al_2O_3 packed on the top of the membrane. After the induction period of 500 h, CH_4 conversion reached its steady state of about 80% due to the gradual increment of oxygen permeation rate. Diethelm et al. [59] used both planar and tubular $La_{0.6}Ca_{0.4}Co_{0.25}Fe_{0.75}O_{3-\delta}$ membrane reactor for POM with an unspecified POM catalyst. For the planar reactor, the CO selectivity and CH_4 conversion were 99 and 75% respectively at 918°C with pure methane. The tubular reactor was operated stably to over 1,400 h with > 95% CH_4 conversion and > 90% CO selectivity at 900°C.

BSCF membrane reactors, with disk [49] or tubular [63] shape, were also successfully used for POM reaction for more than 500 h by packing $LiNaNiO/\gamma - Al_2O_3$ catalyst. In disk type BSCF membrane, a CH_4 conversion of 97–98%, CO selectivity of 95–97% and an oxygen permeation flux of around 11.2–11.8 ml/cm^2.min were achieved during the 500 h run. H_2-TPR results demonstrated that BSCFO was unstable under reducing atmosphere. Fortunately, the material was found to have excellent phase reversibility. The high oxygen permeability of BSCF membrane can stabilize or re-oxidize the surface of the membrane exposed to syngas atmosphere during the POM process. The oxygen permeation flux of the tubular BSCF membrane was higher than 8.0 ml/cm^2.min under POM conditions, which was 8 times the value obtained under air/He experiment. The selectivity to CO and the conversion of methane was 94% and higher than 95% respectively during the 500 h run. It was also found that the mechanism of POM in membrane reactor was another stabilization factor for BSCF membrane used for POM besides the quite good structure reversibility of BSCF. POM in disk BSCF membrane reactors at high pressures (2–10 atm) was investigated recently for further large-scale commercial use [64]. The syngas production rate was 79 ml/cm^2.min (SPT) and oxygen permeation flux was 15.5 ml/cm^2.min (SPT) at 850°C and 5 atm.

The disk type membrane reactor made of $BaCo_{0.4}Fe_{0.4}Zr_{0.2}O_{3-\delta}$ in syngas production experiments at 850°C can be operated steadily for more than 2,200 h with a high catalytic performance of 96–98% CH_4 conversion, 98–99% CO selectivity and an oxygen permeation rate of 5.4–5.8 ml/cm^2.min. Moreover, a short induction period of 2 h was also obtained.

$La_{0.3}Sr_{1.7}Ga_{0.6}Fe_{1.4}O_{5+\delta}$ [24] and $SrFeCo_{0.5}O_x$ [65] were also successfully used as MIECM for POM more than 1,000 h. A membrane reactor based on a brownmillerite structure materials could be continuously operated for over one year under syngas atmosphere at 900°C [66]. The syngas production rate was 60 ml/cm^2.min, and equivalent oxygen permeation flux was 10–12 ml/cm^2.min. The composition of the membrane was not specified in the literature.

Kharton [67] studied the POM reaction on $Ce_{0.8}Gd_{0.2}O_{2-\delta}/Pt$ anodes in the reactor made of YSZ and compared the performance with those in the MIECM reactors made of $La_{0.3}Sr_{0.7}Co_{0.8}Ga_{0.2}O_{3-\delta}$ and $La_2Ni_{0.9}Co_{0.1}O_{4+\delta}$ respectively. The reactors with either MIECM or SE membranes showed similar tendency and high CO_2 yields, which suggests the heavy dependence of the complete methane oxidation on the interface between an oxygen ion-conducting membrane and gas phase, thus making it necessary to incorporate reforming catalysts in the reactors. The solid electrolyte membrane reactor, Ni catalyst-Au anode|YSZ|Ag had been employed for partial oxidation of methane [68]. At 900°C, both CH_4 conversion and syngas selectivity were more than 90% after 10 h that was required for reaching a steady state.

In solid electrolyte membrane reactors, it is attractive to produce electricity as well as syngas during POM process. A large power density and a high CH_4 conversion were attained by using $La_{0.8}Sr_{0.2}Ga_{0.8}Mg_{0.115}Co_{0.085}O_3$ or $La_{0.8}Sr_{0.2}Ga_{0.8}Mg_{0.115}Co_{0.085}O_3$ as electrolyte and Ni and $La_{0.6}Sr_{0.4}CoO_3$ as the catalysts for POM [69].

Selective Oxidation of Other Light Alkanes

There are only a few studies of oxidation of alkanes other than methane because OITM reactors are mostly used at high temperatures, typically more than 700°C [40].

Oxidative dehydrogenation of ethane (ODE) is one route of its direct conversion to ethylene. Per-pass ethylene yield of 67% with ethylene selectivity of 80% was achieved in a BSCF disk-type membrane reactor, while only 53.7% ethylene selectivity was obtained in a conventional fixed-bed reactor under the same reaction conditions with BSFC pellets as catalyst [70]. It was also clarified that the high ethylene selectivity in MIECM resulted from the continuous supply of lattice oxygen [71]. Lin et al. [72] studied ODE in a tubular membrane reactor made of $Bi_{1.5}Y_{0.3}Sm_{0.2}O_3$ (BYS). At 875°C per-pass ethylene yield was 56% with an ethylene selectivity of 80%.

Partial oxidation of ethane to syngas was also studied in a BSFC tubular reactor with $LiNaNiO/\gamma - Al_2O_3$ as the catalyst [73]. The reactor was operated steadily for 100 h at 875°C with ~100% ethane conversion and higher than 91% CO selectivity. Hamakawa [74] studied the selective oxidation of ethane to acetaldehyde in Au|YSZ|Ag solid electrolyte membrane reactor at 475°C. When oxygen was pumped to the Au electrode, the selectivities to acetaldehyde and CO_2 were 45% and 55% respectively.

The *oxidative dehydrogenation of propane* (ODP) to propylene was studied in BSCF tubular membrane reactor at 700°C and 750°C [75] and in a Pt $|Bi_4Cu_{0.2}V_{1.8}O_{11-\delta}|$ Pt solid electrolyte membrane reactor with $V_2O_5/TiO_2 - ZrO_2$ catalyst within the temperature range of 350–450°C [76]. A solid electrolyte membrane reactor, using YSZ as a solid electrolyte and gold and silver as the anode and cathode respectively, has been employed for the selective partial oxidation of C_2–C_4 hydrocarbons at 500°C [77]. MoO_3 or V_2O_5 was simultaneously deposited as the catalyst film on the Au anode. When oxygen was pumped to the anode catalyst film through the YSZ, ethane and propane were selectively dehydrogenated to ethene and propene respectively, while isobutane gave a small amount of methacrolein.

Conclusion

The present paper summarizes the recent development of oxygen-ion transport membrane and its applications in selective oxidation of light alkanes. Although OITMRs show a lot of advantages, there is not yet an easy way to go for their successful commercial applications. For OCM, the crucial step is the proper selection of a membrane material with good intrinsic OCM catalytic properties as well as the appropriate oxygen permeation rate or the modification of membrane surface with good OCM catalysts. For POM, the material must survive under severe reaction conditions. The match of membrane material and the additional POM catalyst is also needed to be considered. For other alkane conversion except methane, it is necessary to develop a membrane material with sufficient oxygen permeation at low temperatures ($< 500°C$) to meet the need for selective oxidation of alkanes.

Acknowledgments The authors gratefully acknowledge financial supports from the National Nature Science Foundation of China (Grant No.50332040) and the Ministry of Science and Technology, China (Grant No. 2005CB221404). The authors would also like to thank the colleagues in the research group for their work summarized in this chapter.

References

1. DOE Begins Research Effort to Revolutionize Oxygen Production for Future Energy, Industrial Processes. DOE – Fossil Energy Techline, 7 Oct 1998.
2. Estenfelder M, Hahn T, Lintz HG. Catal Rev Sci Eng. 2004;46:1–29.
3. Vayenas CG, Bebelis S, Pliangos C, Brosda S, Tsiplakides D. Electrochemical activation of catalysis : promotion, electrochemical promotion, and metal-support interactions. New York: Kluwer Academic Publishers; 2001.
4. Dixon AG. In: Spivey JJ, editor. Catalysis. Vol. 14. London: Royal Society of Chemistry; 1999. pp. 40–92.
5. Såomiya S, Yamamoto N, Yanagida H. Science and technology of Zirconia III; Westerville, OH: American Ceramic Society; 1988.
6. Mogensen M, Sammes NM, Tompsett GA. Solid State Ionics. 2000;129:63–94.
7. Sammes NM, Tompsett GA, Nafe H, Aldinger F. J Eur Ceram Soc. 1999;19:1801–26.
8. Goodenough JB. Annu Rev Mater Res. 2003;33:91–128.
9. Ishihara T, Matsuda H, Takita Y. J Am Chem Soc. 1994;116:3801–3.
10. Anderson MT, Vaughey JT, Poeppelmeier KR. Chem Mater. 1993;5:151–65.
11. Goodenough JB, Ruiz-Diaz JE, Zhen YS. Solid State Ionics. 1990;44:21–31.
12. Abraham F, Debreuillegresse MF, Mairesse G, Nowogrocki G. Solid State Ionics. 1988;28:529–32.
13. Abraham F, Boivin JC, Mairesse G, Nowogrocki G. Solid State Ionics. 1990;40–1:934–7.

14. Kendall KR, Navas C, Thomas JK, zurLoye HC. Chem Mater. 1996;8:642–9.
15. Ishihara T, Arikawa H, Akbay T, Nishiguchi H, Takita Y. J Am Chem Soc. 2001;123:203–9.
16. Lacorre P, Goutenoire F, Bohnke O, Retoux R, Laligant Y. Nature 2000;404:856–8.
17. Teraoka Y, Zhang HM, Furukawa S, Yamazoe N. Chem Lett. 1985;1743–6.
18. Zener C. Phys Rev. 1951;82:403–5.
19. Anderson HU. Solid State Ionics. 1992;52:33–41.
20. Xu SJ, Thomson WJ. Ind Eng Chem Res. 1998;37:1290–9.
21. Kruidhof H, Bouwmeester HJM, Vondoorn RHE, Burggraaf AJ. Solid State Ionics. 1993; 63–5:816–22.
22. Tong JH, Yang WS, Cai R, Zhu BC, Lin LW. Catal Lett. 2002;78:129.
23. Zhu XF, Wang HH, Yang WS. Chem Commun. 2004;1130.
24. Stevenson JW, Armstrong TR, Carneim RD, Pederson LR, Weber WJ. J Electrochem Soc. 1996;143:2722–9.
25. Ishihara T, Yamada T, Arikawa H, Nishiguch H, Takita Y. Solid State Ionics. 2000;135:631–6.
26. Qiu L, Lee TH, Liu LM, Yang YL, Jacobson A. J. Solid State Ionics. 1995;76:321–9.
27. Shao ZP, Yang WS, Cong Y, Dong H, Tong JH, Xiong GX. J Membr Sci. 2000;172:177–88.
28. Tong JH, Yang WS, Zhu BC, Cai R. J Membr Sci. 2002;203:175.
29. Schwartz M, White JH, Sammelis AF. U.S. Patent 6,033,632, 2000.
30. Ma B, Victory N, Balachandran U, Mitchell B, Richardson J. J Am Ceram Soc. 2002;85:2641–5.
31. Balachandran U, Dusek J, Mieville R, Poeppel R, Kleefisch M, Pei S, et al. Appl Catal A. 1995;133:19–29.
32. Manthiram A, Prado F, Armstrong T. Solid State Ionics. 2002;152:647–55.
33. Breard Y, Michel C, Hervieu M, Studer F, Maignan A, Raveau BB. Chem Mater. 2002;14:3128–35.
34. Arashi H, Naito H. Solid State Ionics. 1992;53–6:431–5.
35. Nigara Y, Kosaka Y, Kawamura K, Mizusaki J, Ishigame M. Solid State Ionics. 1996; 86–8:739–44.
36. Chen CS. Ph.D. thesis, University of Twente, Enschede, Netherlands, 1994.
37. Wang HH, Yang WS, Cong Y, Zhu XF, Lin YS. J Membr Sci. 2003;224:107–15.
38. Akhmedov VM, Al-Khowaiter SH. Catal Rev Sci Eng. 2002;44:455–98.
39. Bouwmeester HJM. Catal Today. 2003;82:141–50.
40. Gellings PJ, Bouwmeester HJM. The CRC handbook of solid state electrochemistry. Boca Raton, FL: CRC; 1997. p. 630.
41. Gellings PJ, Bouwmeester HJM. Catal Today. 2000;58:1–53.
42. Miachon S, Dalmon J. Top Catal. 2004;29:59–65.
43. Sanchez Marcano JG, Tsotsis TT. Catalytic membranes and membrane reactors. Weinheim, Germany: Wiley-VCH; 2002.
44. Stoukides M. Catal Rev Sci Eng. 2000;42:1–70.
45. Thursfield A, Metcalfe I. J Mater Chem. 2004;14:2475–85.
46. Liu SM, Tan XY, Li K, Hughes R. Catal Rev Sci Eng. 2001;43:147–98.
47. Wang W, Lin YS. J Membr Sci. 1995;103:219–33.
48. Wang H, Cong Y, Yang W. Catal Today. 2005;104:160–7.
49. Shao Z, Dong H, Xiong G, Gong Y, Yang W. J Membr Sci. 2001;183:181–92.
50. Lu Y, Dixon A, Moser W, Ma Y, Balachandran U. Catal Today. 2000;56:297–305.
51. Zeng Y, Lin Y. AIChE J. 2001;47:436–44.
52. Zeng Y, Lin Y. J Catal. 2000;193:58–64.
53. Akin F, Lin Y. AIChE J. 2002;48:2298–306.
54. Akin F, Lin Y. Catal Lett. 2002;78:239–42.
55. Akin F, Lin Y, Zeng Y. Ind Eng Chem Res. 2001;40:5908–16.
56. Su Y, Ying J, Green W. J Catal. 2003;218:321–33.
57. Otsuka K, Yokoyama S, Morikawa A. Chem Lett. 1985;319–22.
58. Kiatkittipong W, Tagawa T, Goto S, Assabumrungrat S, Praserthdam P. Solid State Ionics. 2004;166:127–36.

59. Diethelm S, Sfeir J, Clemens F, Van Herle J, Favrat D. J. Solid State Electrochem. 2004;8:611–7.
60. Pei S, Kleefisch MS, Kobylinski TP, Faber J, Udovich CA, Zhangmccoy V, et al. Catal Lett. 1995;30:201–12.
61. Tsai CY, Dixon AG, Ma YH, Moser WR, Pascucci MR. J Am Ceram Soc. 1998;81:1437–44.
62. Tsai CY, Dixon AG, Moser WR, Ma YH. AIChE J. 1997;43:2741–50.
63. Wang HH, Cong Y, Yang WS. Catal Today. 2003;82:157–66.
64. Lu H, Tong J, Cong Y, Yang W. Catal Today. 2005;104:154.
65. Balachandran U, Dusek JT, Maiya PS, Ma B, Mieville RL, Kleefisch MS, et al. Catal Today. 1997;36:265–72.
66. Sammells AF, Schwartz M, Mackay RA, Barton TF, Peterson DR. Catal Today. 2000;56:325–8.
67. Yaremchenko AA, Valente AA, Kharton VV, Tsipis EV, Frade JR, Naumovich EN, et al. Catal Lett. 2003;91:169–74.
68. Takehira K, Shimomura J, Hamakawa S, Shishido T, Kawabata T, Takaki K. Appl Catal B. 2005;55:93–103.
69. Ishihara T, Takita Y. Catal Surv Japan. 2000;4:125–33.
70. Wang HH, Cong Y, Yang WS. Catal Lett. 2002;84:101–6.
71. Wang HH, Cong Y, Yang WS. Chem Commun. 2002;1468–9.
72. Akin FT, Lin YS. J Membr Sci. 2002;209:457–67.
73. Wang HH, Cong Y, Yang WS. J Membr Sci. 2002;209:143–52.
74. Hamakawa S, Sato K, Hayakawa T, York APE, Tsunoda T, Suzuki K, et al. J Electrochem Soc. 1997;144:1–5.
75. Wang HH, Cong Y, Zhu XF, Yang WS. React Kinet Catal Lett. 2003;79:351–6.
76. Cai R, Tong JH, Ji BF, Yang WS, Douvartzides S, Tsiakaras P. Ionics. 2005;11:184–8.
77. Takehira K, Sakai N, Shimomura J, Kajioka H, Hamakawa S, Shishido T, et al. Appl Catal A. 2004;277:209–17.

Chapter 4
Ceramic Membrane Devices for Ultra-High Purity Hydrogen Production: Mixed Conducting Membrane Development

S. Elangovan, B. Nair, T. Small, B. Heck, I. Bay, M. Timper, J. Hartvigsen, and M. Wilson

The cost of hydrogen will be a major factor in establishing a commercially viable hydrogen economy. Economic projections suggest that a cost of $1.50/kg of ultra-high purity (UHP) hydrogen is needed, compared with current costs of $5–6/kg. One of the few ways in which such stringent cost targets can be met is to extract UHP hydrogen from a process where it is a byproduct or from gasification products using cost-effective fuels such as biomass. Current membrane technologies for hydrogen separation are incompatible with the high-temperature, high-pressure environment in these chemical processes and Integrated Gasification Combined Cycle (IGCC) systems. A ceramic composite membrane has been developed with high proton conductivity and exceptional stability in high-pressure syngas environment. Initial results using a 35-μm thick membrane have shown a concentration driven hydrogen permeation flux of over 20 cc/cm^2-min ($\sim$40 scfh/ft^2). Thinner membranes of $\sim$10 μm thickness are projected to achieve a flux of 140 scfh/ft^2.

Introduction

The commercial viability of hydrogen economy depends on the cost of high purity hydrogen. Economic projections suggest that a cost of $1.50/kg of UHP hydrogen is needed, compared with the current costs of $5–6/kg. Such stringent cost targets can be met by cost-effectively extracting UHP hydrogen from a process where hydrogen is a byproduct or from gasification products using cost-effective fuels such as biomass. Examples of select commercial processes include alkene synthesis, syngas generation and IGCC systems. Current membrane technologies for hydrogen separation have two primary drawbacks: low hydrogen flux and chemical incompatibility with the high-temperature, high-pressure environment in these chemical processes and IGCC systems. Thus, there is a critical need for the development of a robust and efficient membrane system that can overcome these problems. An overview of present membrane technology is given below.

S. Elangovan (✉)
Ceramatec, Inc., 2425 South 900 West, Salt Lake City, UT 84119-1517, USA
e-mail: elango@ceramatec.com

A.C. Bose (ed.), *Inorganic Membranes for Energy and Environmental Applications*, DOI 10.1007/978-0-387-34526-0_4, © Springer Science+Business Media, LLC 2009

Hydrogen Membrane Technology Overview

A variety of metallic, ceramic and polymer membranes have been used for H_2 separation from gas streams. The most common metallic membrane materials are palladium (Pd) and palladium alloys [1]. However, these materials are unsuitable for H_2 separation from raw syngas due to the fact that they are poisoned by hydrocarbons at concentrations as low as 0.5 part per million (ppm). Furthermore, oxygen concentrations higher than 50 ppm can lead to catalytic oxidation of hydrogen to form water in the presence of Pd, resulting in localized hot spots and premature failure of these membranes. A number of organic membranes (Nafion, for example) have also been identified as protonic conductors, but these are limited to lower temperature applications (less than 150°C) and even at those temperatures they are severely degraded by CO gas [2]. Ceramic membranes are the only alternatives available for operation under conditions similar to the high-temperature, high-pressure environment expected in IGCC systems.

There are two main categories of ceramic hydrogen separation membranes, namely dense membranes and porous membranes. An overview of membrane technologies for H_2 separation was given in a recent IEA publication [3]. The dense membranes are proton conducting membranes that selectively transport H^+ ions at high temperatures (typically >800°C) under driving forces such as a pressure differential or an applied voltage. The main families of ceramic proton conductors are perovskites, pyrochlores and acidic phosphates. Porous ceramic membranes, on the other hand, separate hydrogen from gas mixtures by a mechanism of pressure-driven transport of hydrogen molecules through networked pores that have characteristic dimensions of a few nanometers (e.g. zeolites, mesoporous phosphates or oxides). The very small size of the hydrogen molecule allows selective transport of H_2 over other heavier and larger molecules such as CO, CO_2, O_2 and N_2. Membranes fabricated from both dense and porous ceramic membranes have their respective advantages and disadvantages for hydrogen separation. Dense proton-exchange membranes offer the possibility of recovering a very high purity hydrogen stream as the proton-transport mechanism allows for very high selectivity over any other gas species, but they are generally practical only at temperatures as high as 900°C. Porous ceramics are limited to lower temperature operation (typically 300°C) due to the instability of zeolites in H_2O containing atmospheres, although efforts are currently underway to enhance flux by developing porous membranes that can function at temperatures as high as 600°C. The devices based on porous membranes however do not generate high purity hydrogen, and it is clear that to meet the requirements of automotive PEM fuel cells (<15 ppm CO), a membrane separation process is required.

One category of dense proton conducting membranes that has received considerable attention in the preceding decade is proton conducting perovskite type oxide ceramics [4–6]. The stoichiometric chemical composition of perovskites is represented as ABO_3, where A is a divalent ion (A^{2+}) such as calcium, magnesium, barium or strontium and B is a tetravalent ion (B^{4+}) such as cerium or zirconium. Although simple perovskites such as barium cerate ($BaCeO_3$) and strontium cerate

(SrCeO$_3$) have some proton conductivity, it is now well known that doping these perovskites at the B sites can significantly enhance their protonic conductivities. The most common examples are structures of the form AB$_x$D$_{1-x}$O$_{3-\delta}$, formed by doping ABO$_3$ with trivalent ions such as yttrium (Y^{3+}) and ytterbium (Yb^{3+}). For example, BaCe$_{0.95}$Y$_{0.05}$O$_{3-\delta}$ has been shown to have a proton conductivity as high as 1.27×10^{-2} S/cm at 800°C in a H$_2$/H$_2$O atmosphere [7]. The authors of this work concluded that while the protonic conductivity was very good, the electronic conductivity of these materials is poor and needs to be improved in order for these materials to be used as mixed conducting membranes. The addition of other cations such as neodymium (Nd), europium (Eu) and gadolinium (Gd) has been shown to further increase the hydrogen conductivity [8–10].

Although alternative dopants could also increase the electronic conductivity, the increase in electronic conductivity is insufficient to allow the membrane to function effectively as a pressure-driven hydrogen separation membrane. Alternatively, if a two-phase composite can be fabricated wherein an electronically conducting phase and a protonic conducting phase form interpenetrating networks within a dense ceramic, one can independently control the fluxes of protons and electrons. Ambipolar conductivity of 8×10^{-3} S/cm has been reported for a cermet composition at Eltron Research Inc., with corresponding hydrogen separation rate of 0.3 cc/cm^2/min for 1 mm thick membrane at 850°C [11]. They also reported that by using a cermet with graded ceramic to metal composition, a four-fold increase in hydrogen flux could be obtained compared to a membrane with no surface metal layer [12]. Argonne National Laboratory (ANL) has also developed dense ceramic/metal composite mixed conducting membranes for hydrogen separation [13, 14]. The ANL group has reported hydrogen fluxes as high as 20 cc/cm^2/min when the metallic phase is also a hydrogen conductor. The composite membrane had a ceramic matrix of either Al$_2$O$_3$ or yttria-stabilized zirconia. In spite of recent progress, the selection of an appropriate metallic second phase with thermochemical and thermomechanical stability is also a concern. Most metals are embrittled by H$_2$ or corroded by trace components in syngas at elevated temperatures [15]. Relatively inert metals, such as platinum and gold, have compatibility issues with the protonic conducting phase or an inert ceramic matrix due to thermal expansion mismatch. Therefore, while the use of cermets for pressure-driven H$_2$ separation remains an interesting possibility if additional technological improvements over the state of the art can be achieved, there is a critical need for the identification and development of alternative systems having both adequate thermochemical and thermomechanical properties, and sufficient hydrogen flux to facilitate commercial applications. Table 4.1 summarizes the various approaches for the use of ceramic membranes in hydrogen separation.

In addition to the challenges mentioned above, the thermochemical stability of the perovskite membranes in syngas environment is a major hurdle that needs to be overcome for successful implementation of membrane technology for hydrogen separation. Both SrCeO$_3$ and BaCeO$_3$ based compositions are shown to be unstable in the presence of CO$_2$ and H$_2$O [16–18]. It was shown that replacing a fraction of Ce in the perovskite with Zr also provided improved stability [19]. However this

Table 4.1 Options for ceramic membranes for H_2 separation

Membrane concept	Benefits	Drawbacks/challenges
Porous ceramic (zeolites)	High H_2 flux	Poor selectivity
		Low purity hydrogen
		Requires significant cooling of IGCC exhaust gas
Dense ceramic (doped $BaCeO_3$, $SrCeO_3$)	High purity H_2	Inadequate electronic conductivity to function as pressure driven membrane
	Temperature compatibility with IGCC exhaust	Requires electrical driving force, thus cost and efficiency penalty
Dense ceramic/metal composites (Perovskite/metal composite)	Very high flux if hydrogen permeable metal second phase used	Limited high temperature stability of metal
		High propensity for coking in hydrocarbon environment
		Thermal mismatch
		Poor corrosion resistance and mechanical properties of the metallic phase

substitution of Zr^{4+} ions at the B-site results in a substantial decrease in protonic conductivity. In general, dopants that increase stability have been found to lower the ionic conductivity of $BaCeO_3$ [20]. The use of pyrochlores, of the form $A_2B_2O_7$, has been studied as potential hydrogen separation membranes [21]. Without the presence of Ba or Sr cations in the structure, they have also been shown to possess good stability [22].

Ceramatec Membrane Concept

In a pressure driven system, both hydrogen ions and electrons generated by dissociation of H_2 molecules at the high-pressure surface must be transported through the membrane to recombine at the low-pressure surface. Since these two are parallel kinetic processes, the overall kinetics are limited by the slowest process. In the case of a conventional mixed conducting single phase membrane, the paths for proton conduction and electron conduction are the same. The proton flux (j_{H+}) through a membrane where the primary charge carrying species are H^+ and e^- can be shown [4] as

$$j_{H^+} = \left| -\frac{kT}{2e^2} \int_i^a (\sigma t_{H^+} t_{e^-}) d \ln p_{H2} \right| \qquad (4.1)$$

where k is the Boltzman constant, T is the absolute temperature in Kelvin, e is the magnitude of the electronic charge, σ is the total conductivity, p_{H2} is the partial pressure of hydrogen and t_{H+} and t_{e-} are the transference numbers of H^+ ions and electrons through the membrane. The hydrogen flux thus depends on having high conductivity for both species. The electronic conductivity of perovskites is very low and therefore is usually the limiting factor to these materials being used effectively as pressure-driven hydrogen separation membranes. For example, the electronic conductivity of $SrCe_{1-x}Y_xO_{3-\delta}$ is shown to be 2–3 orders of magnitude lower than its protonic conductivity at 800°C [23].

Our approach is to separate the conduction paths for H^+ ions and electrons through the incorporation of a ceramic second phase. This approach essentially eliminates the combined dependence of hydrogen flux on electronic and proton conductivities. The approach is to short-circuit the electron flow-paths so that the overall flux is limited only by the proton conductivity. A similar mixed conducting requirement exists for electrodes in high-temperature proton conducting fuel cells, and some work has been carried out to develop mixed conductors as electrodes [24].

In addition to being a good electronic conductor at these temperatures, the material chosen as the second phase should also possess good thermomechanical and thermochemical stability (with ambient gases and the proton conducting perovskite) at testing conditions involving temperatures as high as 900°C and high pressures of up to 5–10 atm. Furthermore, the material must be relatively cost-effective and easily processed. A ceramic-ceramic composite material has been developed, in which independent migration of proton and electron species occurs through an interpenetrating network of proton and electron conducting ceramic phases.

Advantages of All-Ceramic Composite Membrane

The primary benefit in selecting a ceramic second phase as the electronic conductor arises from the thermochemical compatibility of the two phases. By appropriate selection of compositions, the thermal expansion coefficient of the two phases can be matched to allow for thermal cycling of the membrane device. Unlike a cermet composite, the all-ceramic composite is not subjected to dewetting of the second phase during fabrication and device operation. In addition to the lower cost of the second phase material, traditional low cost ceramic processing techniques such as tape casting and air-sintering can be used to fabricate the composite membrane devices. Thus the hydrogen membrane devices fabricated using ceramic composite materials are expected to be of low cost. It is estimated that the total cost (including materials and processing) of a supported membrane structure with an active area of over $400\,cm^2$, as installed in a stack, would be less than \$14.

Materials Selection

The cerate based perovskite materials such as $SrCeO_3$ and $BaCeO_3$ have been extensively characterized for hydrogen conductivity. Thus they make good candidates as the primary ion conducting phase. By themselves these perovskite phases exhibit very low electronic conduction, and thus require an addition of an electronically conductive second phase. However, the selection of a second ceramic phase that is an electronic conductor in reducing atmosphere is more challenging. In addition, it would be beneficial if the second phase helps overcome the stability issues that have been hindering the potential use of cerate materials in hydrogen separation application. One of the fundamental limitations encountered in the use of cerate materials is their inherent instability in CO_2 and H_2O containing atmospheres. The cerate perovskite materials tend to dissociate according to

$$ACeO_3 + CO_2 = ACO_3 + CeO_2 \tag{4.2}$$

and

$$ACeO_3 + H_2O = A(OH)_2 + CeO_2 \tag{4.3}$$

where A is Ba or Sr. One approach to reduce the tendency for these materials to react with CO_2 or H_2O is to add the reaction product CeO_2 to the perovskite phase. An intimate and well-dispersed mixture of the two phases is less likely to react with CO_2 or H_2O than the baseline perovskite.

Doped ceria compositions have been evaluated as a solid electrolyte material for oxygen separation [25–27] and solid oxide fuel cell [28–32] applications due to their high oxygen ion conductivity. Although they function well as an oxygen conductor in air for oxygen separation, their applicability in fuel cell application is compromised by their mixed conducting property in the low oxygen partial pressures (pO_2) of fuels. In fact, the electronic conduction of ceria compositions is an order of magnitude higher in hydrogen than their oxygen ion conductivity in air [33]. The mixed conductivity in ceria results from the oxygen non-stoichiometry in low pO_2. In reducing atmospheres the ceria phase reduces according to

$$2CeO_2 = Ce_2O_3 + 1/2O_2 \tag{4.4}$$

Using Kröger-Vink notation, it can be written as

$$4O_O + 2Ce_{Ce} = 2Ce'_{Ce} + 3O_O + V_O^{\bullet\bullet} + 1/2O_2 \tag{4.5}$$

and

$$Ce'_{Ce} = Ce_{Ce} + e' \tag{4.6}$$

Although the mixed conduction is detrimental to fuel cell application, the high electronic conductivity of doped ceria in low pO_2 conditions can be exploited in hydrogen separation membrane technology. The chemical, thermal expansion and processing compatibilities of the two oxides in the composite make perovskite-ceria combination suitable for the mixed conducting membrane application.

Membrane Evaluation

A series of perovskite compositions were synthesized using oxides and carbonates of the cations by conventional ceramic process. The synthesized powders were characterized using powder x-ray diffraction technique to ensure phase purity. Conductivity measurements were made in H_2-H_2O atmosphere to determine proton conductity. As the perovskite compositions are inherently mixed conducting, the transference numbers for proton and electron conduction were also determined by varying the partial pressures of hydrogen and steam across the membrane.

The composite material was synthesized by mixing the perovskite and doped ceria in the desired volume ratio, using conventional ball milling process. In general, the dopant for ceria was kept the same as the B-site dopant of the perovskite. Discs were pressed and sintered for evaluating the mixed conducting properties of the membrane. Stability of the proton conductor as well as the composite was studied by exposing the material to syngas compositions at ambient and up to 2.7 atm pressure at 900°C for 4 hours. Stability evaluation in CO_2 containing oxidizing atmosphere was also performed. The materials were analyzed using x-ray diffractography. Permeation experiments were done by measuring concentration driven hydrogen flux through the membrane. Pressure driven permeation tests were also performed at lower temperatures due to concerns over seal leaks at higher temperatures.

Conductivity Measurement

A variety of dopants were evaluated to study their effects on the conductivity of the perovskite material. The measured protonic conductivity of the doped $BaCeO_3$ perovskite is compared to standard Y-doped $BaCeO_3$ and $SrCeO_3$ compositions. As can be seen in Fig. 4.1, the new composition exhibits a conductivity nearly three times that of baseline perovskites.

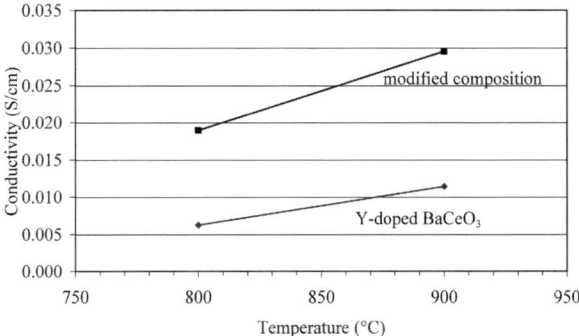

Fig. 4.1 Comparison of proton conductivity of modified perovskite to baseline compositions

Mixed Conduction in Ceramic Composite

The selection of ceramic second phase compositions was based on its high electronic conductivity such that the overall hydrogen permeation is not limited by the
electronic transport through the membrane. In order to verify the electronic conductivity of the second phase, a composite disc was made using a well-known proton
conducting ceramic, $SrCeO_3$. The overall transference numbers of the proton conductor and the composite were measured using $H_2 - H_2O$/air concentration cells.
The comparison of the two materials is shown in Fig. 4.2. The open circuit potential across each type of membrane is shown as a function of cell temperature. The
calculated transference number, based on theoretical values of open circuit potential
for the gas composition used is also shown (open symbols). The baseline perovskite
clearly shows a proton transference number $t_{H+} \sim 0.8$ to 1, while the composite
membrane exhibits a proton transference number $t_{H+} \sim 0.6$–0.8 confirming the
enhanced mixed conducting behavior for the composite. When the membrane faces
reducing atmosphere on both sides, as would be in the intended application conditions, the transference number is expected to be even lower thereby eliminating the
electronic conduction as the limiting step that was common in all ceramic single
compositions that have been evaluated previously.

Stability Evaluation

The stability of the ceramic composite was compared to baseline perovskite using
Thermogravimetry and powder x-ray diffraction techniques. Figure 4.3 shows the
comparison of weight gain measurements of baseline perovskite and the composite
material in CO_2 + air mixture. The baseline material continues to gain weight during
the heat-up and retains much of the gained weight during cool down resulting in a
net weight gain of about 3.5%. The composite membrane in comparison does not
show any measurable gain in weight and in fact does not show a net weight change
after cool down. A similar weight gain measurement of the composite was made in

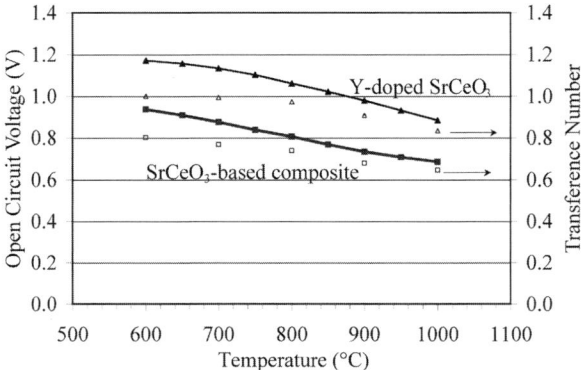

Fig. 4.2 Evaluation of mixed conduction in composite membrane

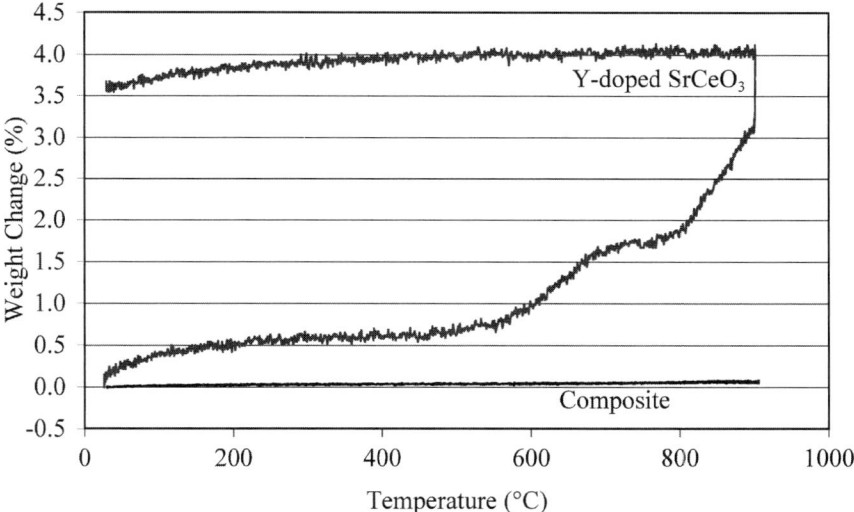

Fig. 4.3 Thermogravimetry comparison of baseline perovskite and the composite membrane in CO_2-containing oxidizing atmosphere

syngas atmosphere. No weight gain was noted even in the presence of CO_2 and H_2O in reducing atmosphere. The results are shown in Fig. 4.4.

Two compositions of the composite materials were evaluated for stability in syngas. Sintered pellets of the composite were crushed to powder form to increase the surface area of the material. The powdered composites were exposed to syngas at 900°C for 4 hours. A baseline perovskite composition was also exposed to similar conditions. X-ray diffraction analysis of the materials is shown in Fig. 4.5. As can be seen, while the baseline perovskite readily formed a carbonate phase, the composite materials showed significantly improved stability in syngas.

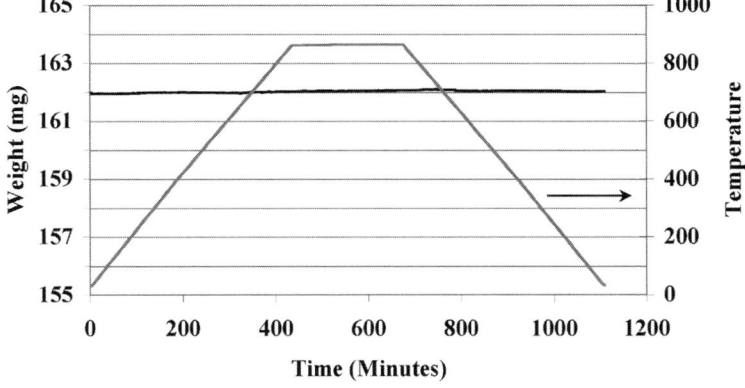

Fig. 4.4 Thermogravimetry of the composite material in syngas atmosphere

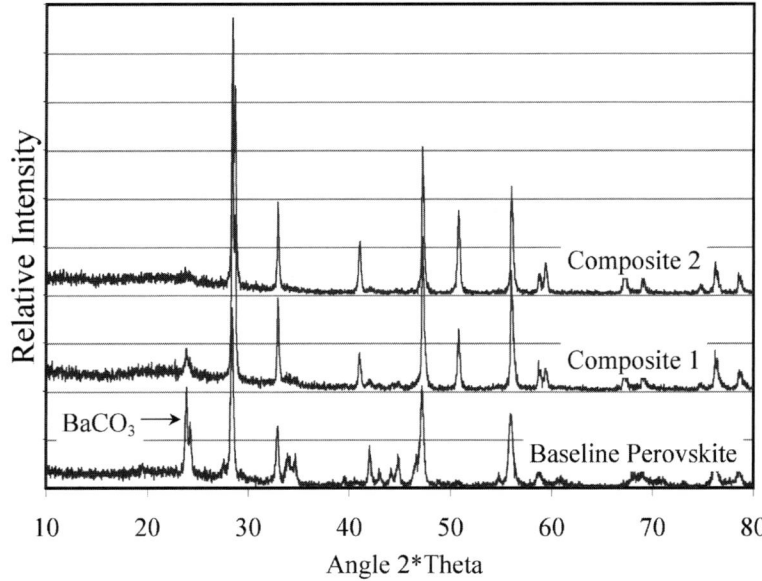

Fig. 4.5 Comparison of XRD pattern indicating improved stability of composite material

Thin Membrane Fabrication

The hydrogen flux through the composite membrane is inversely proportional to its thickness assuming fast surface kinetics. It is anticipated that a membrane thickness in the range of 10–30 μm may be needed in order to achieve the high hydrogen flux required for commercial devices. In order to improve the reliability of the thin membranes, they are typically supported on a porous substrate. A variety of deposition techniques has been reported in the literature for fabricating thin ceramic membranes. They range from physical and chemical vapor deposition to spin coating of polymeric precursors. The selection of an appropriate technique depends on materials compatibility, process yield, and cost. Tape cast processing has been successfully adapted for fabrication of thin planar ceramics by the electronics industry with proven commercial success. Thus, conventional, low cost tape cast processing was evaluated for membrane fabrication.

Dense, composite membranes that are 35 μm in thickness were successfully fabricated. A cross section of a supported, thin membrane structure is shown in Fig. 4.6.

Membrane Scale Up

Functional hydrogen separation devices will consist of multiple layers of the membrane with each layer having an active area as large as practical. Prototype membranes with a sintered dimensions of ∼ 4.5 inch× ∼ 4.5 inch were fabricated

Fig. 4.6 Thin, supported composite membrane

using tape cast and lamination technique. These membranes had a dense layer with a thickness of about 50 µm with porous support layers. Photographs of a typical sub-scale coupon that was used for permeation tests and a sintered full size laminate are shown in Fig. 4.7.

H_2 *Permeability Measurements*

Hydrogen separation experiments were carried out from hydrogen/nitrogen mixtures that clearly showed that partial pressure/concentration driven H_2 separation can be achieved through our membranes. The experiments were performed using 500 µm thick composite disks and supported 35 µm thick membranes similar to the ones shown in Fig. 4.6. The active surface area of these membranes typically ranged from 1 to 1.5 cm^2. The feed side gas was at ambient pressure and the product side was swept with nitrogen at a known flow-rate. The gas mixture from the product

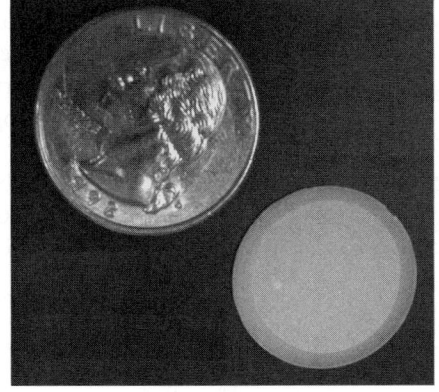

Sub-scale Membrane   Full Size Membrane

Fig. 4.7 Photographs of sintered composite membranes

side was then passed through a pre-calibrated yttria-stabilized zirconia tube and the open circuit potential with respect to ambient air was determined from which the hydrogen concentration at the product side was determined. The flux measurement method is estimated to have an error of ±20% and is intended only as a preliminary demonstration. The absence of any leaks was verified by demonstrating that when the flow-rate was doubled, the flux did not change substantially within the limits of the measurement, which showed that the H_2 permeation is partial pressure/concentration driven. Although the maximum flux obtained through the thick membrane was relatively low ($\sim 1.4 \pm 0.3$ cc/cm^2/min) as expected, the maximum flux obtained through the 35 μm membranes was $\sim 20 \pm 4$ cc/cm^2/min, or ~ 40 scfh/ft^2. These results, shown in Fig. 4.8, are close to commercial targets for membrane separation (>60 scfh/ft^2) and by reducing the membrane thickness further to 10–20 μm, and providing a pressure differential across the membrane it is expected that commercially viable fluxes can be achieved.

In order to verify the effect of pressure on hydrogen permeation, measurements were made using composite membranes of about 35–50 μm at the University of Tennessee, Knoxville, TN [34]. These measurements were made at relatively low temperatures of 600° and 700°C due to limitations of the experimental set up. A mixture of H_2, CO, CO_2 and CH_4 was used as the feed gas at the desired H_2 partial pressure. Nitrogen sweep gas was used on the permeate side. Both the feed and permeate compositions were monitored using a gas chromotagraph. The results showed that only hydrogen was permeating through the composite membrane. The measured hydrogen fluxes are shown in Fig. 4.9. The results clearly demonstrate the pressure dependence of hydrogen flux.

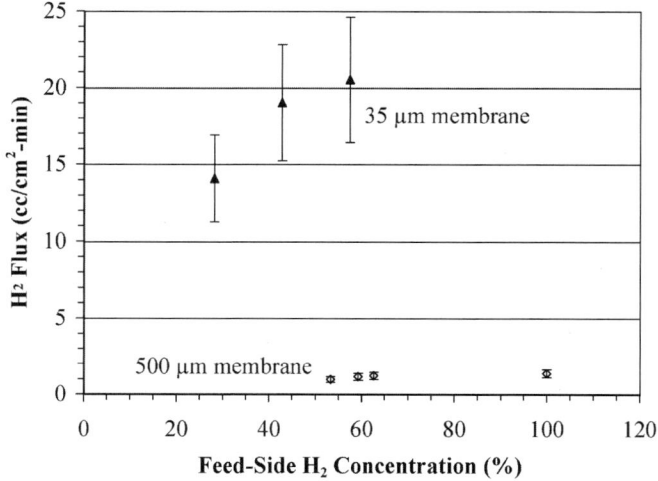

Fig. 4.8 Hydrogen flux obtained through thick and thin composite membranes

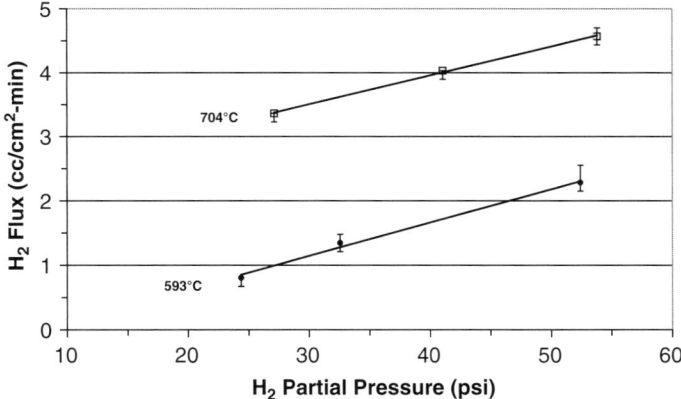

Fig. 4.9 Pressure-driven hydrogen flux

Summary

Pressure-driven membrane separation devices are ideally suited for hydrogen extraction from syngas produced in IGCC. The major technology hurdles in membrane development are related to the selection of materials that are compatible with IGCC operating conditions. We have identified a set of proton and electron conducting ceramic materials that are well suited for the fabrication of an all-ceramic composite membrane that addresses many of the known technology hurdles such as stability, manufacturability, and cost, encountered by present cermet membranes.

It has been experimentally demonstrated that the ceramic – ceramic composite material is a proton-electron mixed conductor selective to hydrogen transport and is stable in syngas atmosphere. Both concentration-driven and pressure-driven hydrogen permeation test results demonstrate the viability of the composite membranes for separating high purity hydrogen from high pressure hydrogen-rich gas streams. Integration of composite membrane modules in the high pressure, high temperature syngas streams for IGCC systems provides an opportunity of economic production of UHP hydrogen.

References

1. Mordkovich VZ., Biachtock YK, Sosna MH. Large-Scale production of hydrogen from gas mixtures; a use of ultra-thin palladium alloy membranes. Int. J. Hydrogen Energy. 1993;19(7):539–44.
2. Fielding HC. Fluoropolymer membranes. Prog Rubber Plastics Technol. 1993;9(4):253–270.
3. Collot A. Prospects for hydrogen from coal. 2003. IEA Clean Coal Centre; London, UK ISBN 92-9029-393-4.
4. Norby T, Larring Y. mixed hydrogen ion-electronic conductors for hydrogen permeable membranes. Solid State Ionics. 2000;136–137:139–148.

5. Norby T. Solid-state protonic conductors: principles, properties, progress and prospects. Solid State Ionics. 1999;125:1–11.

6. Sata N, Yugami H, Akiyama Y, Sone H, Kitamura N, Hattori T, Ishigame M. Proton conduction in mixed perovskite type oxides. Solid State Ionics. 1999;125:383–387.

7. Guan J, Dorris SE, Balachandran U, Liu M. Transport properties of $BaCe0.95Y0.05O3-\alpha$ mixed conductors for hydrogen separation. Solid State Ionics. 1997;100:45–52.

8. Shima D, Haile SM. The influence of cation non-stoichiometry on the properties of undoped and Gadolinia-doped barium cerate. Solid State Ionics. 1997;97(1–4):443–55.

9. Song SJ, Wachsman ED, Dorris SE, Balachandran U. Electrical properties of p-type electronic defects in the protonic conductor $SrCe_{0.95}Eu_{0.05}O_{3-\delta}$. J Electrochem Soc. 2003;150(6): A790–A795.

10. Stevenson DA, Jiang N, Buchanan RM, Henn FEG. Characterization of Gd, Yb and Nd doped barium cerates as proton conductors. Solid State Ionics. 1993;62(3–4):279–85.

11. Roark SE, Mackay R, Sammels AF. Hydrogen separation membranes for vision 21 fossil fuel plants. Vision 21 Program Review Meeting, 2001 Conference Proceedings, Morgantown, WV, DOE-NETL Publications, 2001.

12. Sammels AF, Barton TF, Mundschau MV. Catalytic membrane reactors for gas, liquid and solid reforming to syngas. 225th American Chemical Society National Meeting – Fuel Division Preprint, New Orleans, LA, 2003.

13. Guan J, Dorris SE, Balachandran U, Liu M. Development of mixed-conducting ceramic membranes for hydrogen separation. Ceram Trans. 1998;92:1–12.

14. Ceramic membrane development. Argonne National Laboratory. http://www.et.anl.gov/sections/ceramics/research/ceram_mem.html

15. Siriwardane RV, Poston JA, Jr., Fisher EP, Lee TH, Dorris SE, Balachandran U. Characterization of ceramic hydrogen separation membranes with varying nickel concentrations. Appl Surf Sci. 2000;167(1):34–50.

16. Taniguchi N, Nishimura C, Kato J. Endurance against moisture for protonic conductors of perovskite-type ceramics and preparation of practical conductors. Solid State Ionics. 2001;145:349.

17. Bhide SV, Virkar AV. Stability of AB'1/2B"1/2O3-type mixed Perovskite proton conductors. J Electrochem Soc. 1999;146(12):4386.

18. Tanner CW, Virkar, AV. Instability of BaCeO3 in H2O-containing Atmospheres. J Electrochem. Soc. 1996;143(4):1386.

19. Carneim RD, Armstrong TR. Chemical stability of Barium cerates-based high temperature proton conductors. 225th American Chemical Society National Meeting – Fuel Division Preprint, New Orleans, LA, 2003.

20. Matsumoto H, Kawasaki Y, Ito N, Enoki M, Ishihara T. Relation between electrical conductivity and chemical stability of BaCeO3-based proton conductors with different trivalent dopants. Electrochem. Solid-State Lett. 2007;10(4):B77.

21. Omataz T, Otsuka-Yao-Matsuo S. Electrical properties of proton-conducting Ca^{2+} Doped $La_2Zr_2O_7$ with a pyrochlore-type structure. J Electrochem Soc. 2001;148(6):E252.

22. Gade S, Schaller R, Berland B, Schwartz M. Novel composite membranes for hydrogen separation in gasification processes in vision 21 Plants. 19th Annual Pittsburgh Coal Conference Proceedings, 2002.

23. Phillips RJ, Bonanos N, Poulsen FW, Ahlgren EO. Structural and electrical characterization of $SrCe_{1-x}Y_xO_{3-\delta}$. Solid State Ionics. 1999;125:389–95.

24. Rauch WL, Liu M. $BaCuGd_2O_5 - BaCeO_3$ composite cathodes for barium cerate-based electrolytes. J Electrochem. Soc. 1997;144(11):4049–54.

25. Arai H, Kunisaki T, Shimizu Y, Seiyama T. Solid State Ionics. 1986;20:241.

26. Yahiro H, Eguchi Y, Arai H. Solid State Ionics. 1986;21:37.

27. Yahiro H, Eguchi Y, Eguchi K, Arai H. J App Electrochem. 1988;18:527.

28. Choudhury N, Patterson J. J Electrochem Soc. 1971;118(9):1398–1403.

29. Tannhauser D, J Electrochem Soc. 1978;125(8):1277–1282.

30. Riess I. J Electrochem Soc. 1981;128(10):2077.

31. Gödickemeier M, Sasaki K, Gauckler L. In: Dokiya M, Yamomoto O, Tagawa H, Singhal SC, editors. Solid oxide fuel cells IV. The Electrochemical Society, Pennington, NJ 1995.
32. Milliken C, Elangovan S, Hartvigsen J, Khandkar A. Ceria Electrolyte for Solid Oxide Fuel Cell Applications. Report TR-109199, Electric Power Research Institute, Palo Alto, CA, Nov. 1997.
33. Tuller H, Nowick A. J Electrochem Soc., 1979:126(2):209–217.
34. Singh BK. Bench scale evaluation of dense ceramic membranes for production of high purity hydrogen from gasification, M.S. Thesis, The University of Tennessee, Knoxville, TN, Dec 2005.

Chapter 5
Decomposition of Yttrium-Doped Barium Cerate in Carbon Dioxide

Theodore M. Besmann, Robert D. Carneim, and Timothy R. Armstrong

Doped barium cerate and related perovskite ceramics currently dominate the high-temperature proton conductor field. Unfortunately, these materials have very stringent environmental limitations necessitating the costly and complex conditioning or cleaning of the application feed-gas. This work shows through both experimental results and thermodynamic modeling that $Ba(Ce, Y)O_{3-\delta}$ is ill-suited for use in environments containing CO_2. Despite efforts to improve the stability of $Ba(Ce, Y)O_{3-\delta}$ through partial Zr_{Ce} substitution and adjusting the A-site to B-site cation ratio, the perovskite phase is shown to be unstable in the presence of CO_2 at temperatures below ca. $1,000°C$.

Introduction

Proton conducting ion transport membranes are required to extract absolutely pure hydrogen from mixed gas streams in the processing of fossil fuels and other petroleum and petrochemical processes. Ceramic materials are required to withstand these high-temperature processes. Yttrium-doped barium cerate ($BaCe_{0.80}Y_{0.20}O_{3-\delta}$) is currently one of the best materials when considering only proton conductivity. However this material is very sensitive to the carbon dioxide (CO_2) and sulfur (e.g., H_2S) contamination found in the fossil fuel process stream. Barium cerate will decompose into barium carbonate ($BaCO_3$) and ceria (CeO_2) in the presence of CO_2; and CeO_2 will react with H_2S to produce a sulfided phase, particularly in a highly reducing fuel environment. Two approaches for improving the stability of barium cerate are examined: manipulation of the A- to B-site (in this case, Ba and Ce/Y/Zr, respectively) stoichiometry and partial substitution of Ce with Zr. Individually these techniques have been previously shown to enhance stability [1–4]. In the present work, the effect of applying both modifications simultaneously is explored. High-temperature x-ray diffraction (HTXRD) and

T.M. Besmann (✉)
Materials Science and Technology Division, Oak Ridge National Laboratory, Oak Ridge,
TN 37831, USA
e-mail: besmanntm@ornl.gov

A.C. Bose (ed.), *Inorganic Membranes for Energy and Environmental Applications*,
DOI 10.1007/978-0-387-34526-0_5, © Springer Science+Business Media, LLC 2009

simultaneous thermogravimetric and differential thermal analysis (TGA/DTA) were used to determine decomposition temperatures, extent of decomposition, and reaction kinetics of various yttrium-doped barium cerate compositions in CO_2-contaminated environments.

Experimental Procedure

Sample Preparation

Small batches ($\approx$ 10 g) of powders were prepared by the glycine-nitrate process (GNP) [5]. Appropriate amounts of metal salts (barium-, cerium-, yttrium-, and zirconyl nitrates, ammonium cerium [IV] nitrate, and barium acetate were [Alfa Aesar, Ward Hill, MA]) and glycine were dissolved in deionized water. The solution was heated on a hot plate in a stainless steel beaker until a sufficient amount of solvent had evaporated allowing the precursor to ignite. The resulting ash was sieved (100–mesh) and calcined for 20 min to 30 min at 1,200°C to produce the desired single-phase perovskite. Additional material was obtained commercially (Praxair). The following compositions were synthesized:

BCY—: Four compositions with varying barium stoichiometry: $BaCe_{0.80}Y_{0.20}$ $O_{3-\delta}$ with x = 1.00, 0.98, 0.95, and 0.90; designated BCY108, BCY88, BCY58, and BCY08, respectively.

BCZ—: Three barium-deficient, non-acceptor-doped zirconium-substituted compositions: $Ba_{0.98}Ce_{1-y}Zr_yO_{3-\delta}$ with y = 0.15, 0.10, and 0.05; designated BCZ885, BCZ89, and BCZ895, respectively.

BCZY15: One barium-deficient, acceptor-doped, zirconium-substituted composition: $Ba_{0.98}Ce_{0.65}Zr_{0.15}Y_{0.20}O_{3-\delta}$; designated BCZY15.

The following compositions were purchased:

BCY—p: Two compositions with varying barium stoichiometry: $Ba_xCe_{0.80}Y_{0.20}$ $O_{3-\delta}$ with x = 1.00 and 0.98; designated BCY108p and BCY88p, respectively.

High-Temperature X-Ray Diffraction

Initial phase identification of the calcined powders was carried out using room temperature powder x-ray diffraction (XRD) using Ni-filtered Cu Kα x-rays (Scintag PAD V diffractometer). HTXRD was conducted with the same radiation in a Phillips X'Pert diffractometer with a platinum heater strip/sample stage in a hermetically sealed chamber (HTK 16 High Temperature Camera). Resistance to degradation due to the presence of CO_2 was quantified by determining the temperatures at which $BaCO_3$ formed during heating and cooling of the single-phase perovskite in flowing CO_2 (p_{CO2} = p = 101 kPa [1 atm]) using the following procedure (Fig. 5.1):

HTXRD: Heat sample in N_2 to t = 1,400°C and cool to 200°C; perform scan; change sample environment to flowing CO_2 and hold 20 min; perform scan; raise temperature by 50°C and hold 5 min; perform scan; repeat preceding two steps until

Fig. 5.1 Schematic of temperature profile and gas flow schedule used for the high -temperature x-ray diffraction procedure

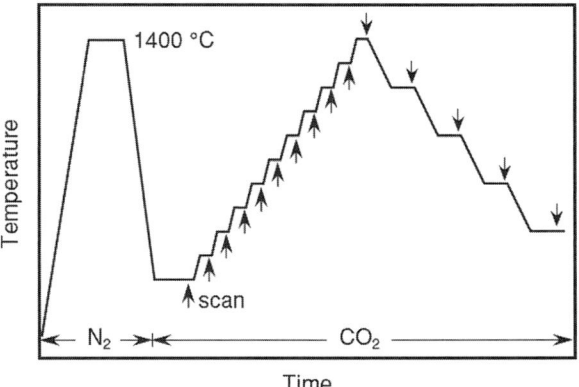

$t = 1,400°C$; lower temperature by $100°C$ and hold $10\,min$; perform scan; repeat preceding two steps until $t = 500°C$.

Measurements were taken over the range from $2\theta = 20°$ to $2\theta = 44°$. Temperature was measured and controlled based on a type-S thermocouple welded to the bottom of the platinum strip. At sufficiently high temperatures ($t > 800°C$), more accurate sample temperature measurements were possible through the use of an optical pyrometer (Optical Pyrometer, Leeds & Northrup Co., Philadelphia, PA) trained on the sample surface through a quartz window in the sample chamber.

Data obtained by the HTXRD procedure are presented by plotting the logarithms of the peak ratios (areas under peaks attributable to only the perovskite phase vs. those attributable to only non-perovskite phases; background intensity and noise ensure finite results), $\lg(r_{p/c})$ vs. temperature during heating and cooling (Fig. 5.2).

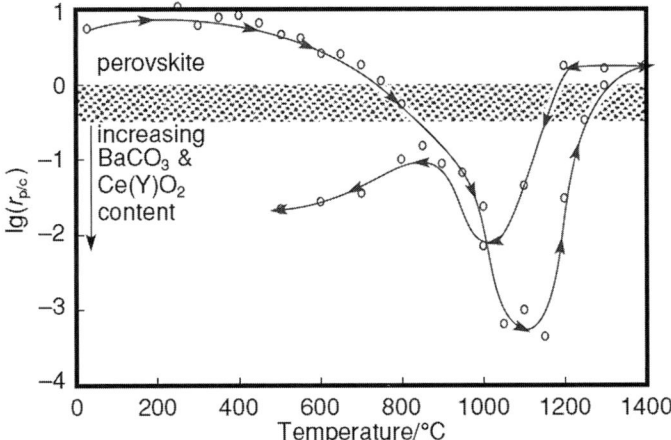

Fig. 5.2 Example data set obtained by HTXRD showing raw data points from the individual scans, a freehand curve fit of the data (arrows indicate increasing time), and phase fields: top section indicates no detected decomposition products, the shaded section indicates just detectable $BaCO_3$ and $Ce(Y)O_2$ peaks, and the bottom region indicates significant decomposition

Based on examination of the raw XRD data, a cut-off value of $\lg(r_{p/c}) = -0.25$ was chosen to define transition temperatures and $\lg(r_{p/c}) > 0$ indicates no x-ray observable $BaCO_3$ and $\lg(r_{p/c}) < -0.5$ indicates the presence of a significant amount of $BaCO_3$. The temperatures at which carbonate phase forms and disappears on heating, and forms again on cooling can be easily seen and compared.

An additional feature that stands out in this representation of the data is hysteresis in the high-temperature transition (i.e., the temperature at which $BaCO_3$ disappears on heating or appears on cooling). The high-temperature transition hysteresis (Δt_{hyst}) is defined as the difference between the transition temperatures measured during heating and cooling. This quantity may be used for comparing reaction kinetics between related materials undergoing the same reaction at comparable temperatures.

Thermal Analysis

Simultaneous TGA analysis and DTA were performed (T/A Instruments SDT 2960 or Stanton Redcroft STA780) on select compositions. Multiple scans were performed on certain compositions at heating rates between $\phi = 1.6°C \cdot min^{-1}$ and $\phi = 25°C \cdot min^{-1}$ in flowing CO_2 to allow estimation of the reaction activation energies using the Kissinger method [6, 7]. Additional experiments were carried out at $\phi = 1.6°C \cdot min^{-1}$ in mixtures of flowing CO_2 and Ar-4%H_2 to examine the effect of CO_2 in reducing environments. The two overall gas compositions used were 3.85% CO_2 + 3.85% H_2 + 92.3% Ar and 95% CO_2 + 0.2%H_2 + 4.8% Ar (Air Liquide, Oak Ridge, TN).

Modeling

A thermodynamic equilibrium analysis was carried out to predict the transformation temperatures for reaction of $BaCO_3$, CeO_2, and $YO_{1.5}$ to $BaCe_{0.8}Y_{0.2}O_{3-\delta}$. Thermodynamic values (298 K heats for formation, 298 K entropies, and heat capacities) were obtained from the SGTE database [8] and calculations using FactSage^{TM} computational thermodynamics software. Neither the SGTE database nor the literature contained values for the $BaCe_{1-z}Y_zO_{3-\delta}$, thus this phase was treated as an ideal solution of $BaCeO_3$ and $BaYO_{2.5}$. An ideal solution of the phases is a reasonable assumption given that $BaCeO_3$ and $BaCe_{0.8}Y_{0.2}O_{3-\delta}$ are both orthorhombic ($BaYO_{2.5}$ is not a reported phase). Data for these phases were obtained from the literature and estimates. The 298 K heat of formation and entropy of $BaCeO_3$ are -1690 kJ $\cdot$ mol^{-1} and 144.5 J $\cdot$ mol$^{-1} \cdot$ K^{-1}, respectively [9]. The heat capacity of $BaCeO_3$ over the temperature range of interest was taken as the sum of the BaO and CeO_2 heat capacities provided by the SGTE database.

The 298 K heat of formation and entropy of $BaYO_{2.5}$ were estimated by interpolation of the values for the phases formed from the unary oxides along the BaO–$YO_{1.5}$ join; these were -16 kJ $\cdot$ mol^{-1} and 2 J $\cdot$ mol$^{-1} \cdot$ K^{-1}, respectively [10].

These, together with the 298 K values for the unary oxides from the SGTE database, yielded the required 298 K enthalpy of formation and entropy: -1524.5 kJ $\cdot$ mol^{-1} and 122.82 J$\cdot$mol$^{-1}\cdot$K^{-1} respectively. Again, the heat capacity relations were found by summing the unary oxide values.

The phases CeO_2 and $YO_{1.5}$ are cubic and form extensive solid solutions, although with a significant miscibility gap [11]. For the purpose of the current calculations this was treated as a single ideal solid solution.

Results and Discussion

Barium Stoichiometry

High-Temperature X-ray Diffraction

Preliminary HTXRD experiments in CO_2 showed no differences in the behaviors of the three barium-deficient compositions (BCY88, BCY58 and BCY08), with $BaCO_3$ diffraction peaks becoming apparent at $t \leq 1,050°$C on cooling. This was deemed a slight improvement over the stoichiometric composition, BCY108, which presented $BaCO_3$ peaks already at $t \approx 1,100°$C on cooling. Initial results during heating showed no difference between the four compositions. Based on these preliminary results, any of the barium-deficient compositions would serve to realize improvement over the stoichiometric composition and stay well clear of a barium-excess situation that would result in severe degradation of electrical properties [1, 2]. Therefore, the least (non-zero) barium-deficient composition ($x = 0.98$) was chosen for more detailed study.

Measurements performed on BCY108 and BCY88 using the HTXRD procedure described above revealed details of the differences in behavior of the stoichiometric and barium-deficient compositions with respect to degradation in CO_2. The various transition temperatures are nearly identical for these two compositions (Fig. 5.3, Table 5.1). The consequential difference between these two compositions is the extent of transformation: the peak ratio ($r_{p/c}$) is nearly one order of magnitude larger for BCY88 than for BCY108. The temperature range in which $BaCO_3$ is present and the high-temperature transition hysteresis were similar for these two compositions.

Thermal Analysis

The transition temperatures of these compositions as determined by DTA are consistently lower than as determined by HTXRD (Table 5.1). There is good agreement in the DTA data between the synthesized and purchased powders (BCY88 and BCY88p, respectively), allowing comparisons between results of the two methods applied to different source powders. The DTA results, relative to the HTXRD results, are approximately 250°C to 300°C lower for the initial decomposition on heating and approximately 50°C to 150°C lower for the subsequent reactions (perovskite formation on heating and decomposition on cooling).

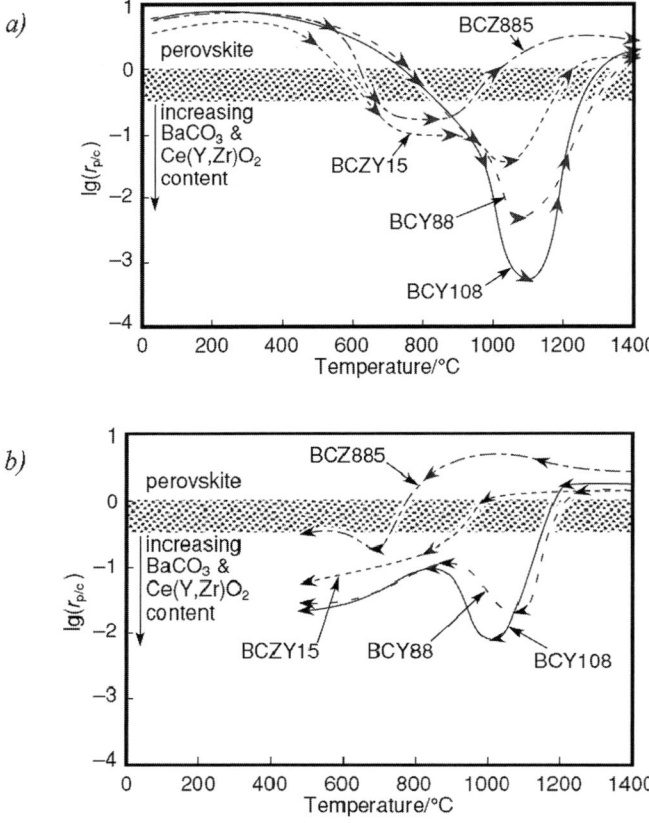

Fig. 5.3 High-temperature x-ray diffraction results for $BaCe_{0.8}Y_{0.2}O_{3-\delta}$, $Ba_{0.98}Ce_{0.8}Y_{0.2}O_{3-\delta}$, $Ba_{0.98}Ce_{0.85}Zr_{0.15}O_{3-\delta}$, and $Ba_{0.98}Ce_{0.65}Zr_{0.15}Y_{0.2}O_{3-\delta}$, during heating (a) and cooling (b)

The differences between the results obtained from the two methods arise due to a number of factors. The thermodynamic modeling results elucidate two features relevant at this point (Fig. 5.4a): $BaCO_3 + Ce(Y)O_2$ is the stable low-temperature phase compositions; and secondly, there is a three-phase region in which the perovskite and the two decomposition products are in equilibrium approximately 155°C wide. Therefore, at low temperature in CO_2, the perovskite is metastable and the decomposition reaction is determined by kinetics. Slight differences in the sample configuration and experimental conditions can greatly affect the observed transition temperature. Furthermore, the two methods detect different phenomena that are not necessarily dependent: the thermal analysis peaks reflect the addition of mass and the chemical reactions to incorporate that mass into the structure; this may all occur without immediately producing a crystalline phase, which is required for detection using the HTXRD method. Finally, the differences observed in the high-temperature reaction can further be due to the two methods being sensitive to the reaction at different points while traversing the three-phase region. HTXRD requires a certain

Table 5.1 Summary of results in CO_2 from HTXRD[a] and DTA[b] experiments

Composition	Method	Decomp. on heating/°C	Perovskite forms on heating/°C	Decomp. on cooling/°C	Hysteresis/°C	Max. $\lg(r_{p/c})$ decomp. peak deflection
BCY108	HTXRD	780	1280	1160	120	−3.25
BCY108p	DTA	486	1234	1056	178	n/a
BCY88	HTXRD	780	1330	1180	150	−2.35
BCY88	DTA	515[d]	≈ 1188[e]	1038	≈ 150	n/a
BCY88p	DTA	513	1239	1060	179	n/a
BCZ885	HTXRD	640	980	750	230	−0.80
BCZ885	DTA	550	1113	no rxn	n/a	n/a
BCZY15	HTXRD	600	1200	950	250	−1.45
BCZY15	DTA	539	1192	855	337	n/a

[a] Tabulated values were extracted from data as in Fig. 5.2 using $\lg(r_{p/c}) = -0.25$ cutoff value.

[b] $4°C \cdot min^{-1}$

[c] *N.B.*, $\lg(r_{p/c}) < -0.5 \equiv$ significant decomposition.

[d] Thermal analysis of the synthesized BCY88 powder shows a broad two-stage peak for this reaction; the reported value corresponds to the initial, but smaller peak.

[e] Thermal analysis of this material showed a doublet peak; the average temperature is reported.

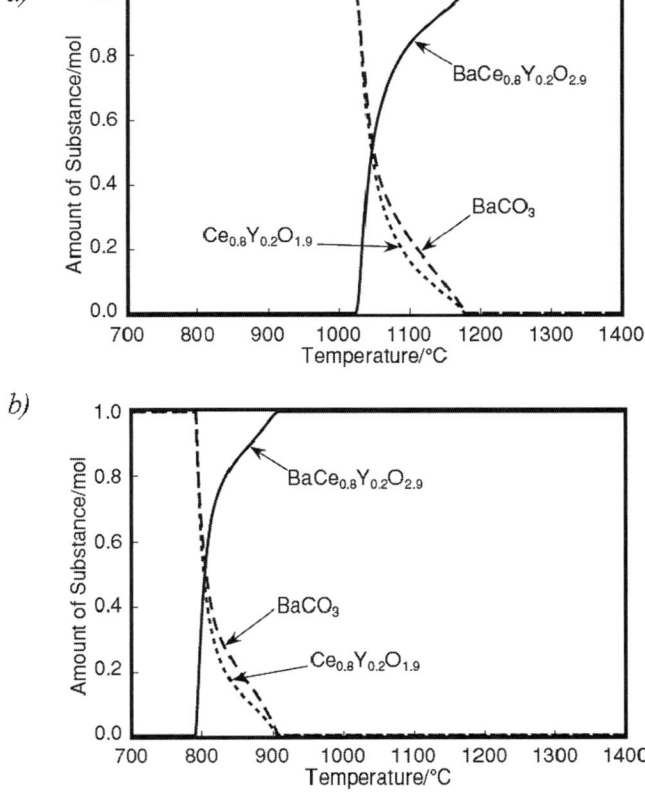

Fig. 5.4 Thermodynamic modeling results for $BaCe_{0.8}Y_{0.2}O_{3-\delta}$ in CO_2 and 95% CO_2 + 5% 96Ar • 4H$_2$ (calculation for these two environments produced indistinguishable plots) (a) and in 3.85% CO_2 + 96.15% 96Ar • 4H$_2$ (b)

amount of crystalline material to be present before it can be detected, while the DTA peak maximum corresponds to the maximum reaction rate; these two events do not necessarily occur simultaneously.

Environment

Table 5.2 summarizes the results obtained by DTA experiments and thermodynamic modeling for BCY108 in different atmospheres. Experiments in the 100% CO_2 and 95% CO_2 + 0.2% H_2 (balance Ar) environments resulted in essentially identical transition temperatures. Reducing the CO_2 concentration to 3.85% and including an equal amount of H_2 (balance Ar) resulted in a slightly higher transition temperature for the low-temperature reaction and significantly lower transition temperatures (i.e., during both heating and cooling) for the high-temperature reaction.

The temperature at which the metastable perovskite decomposes (i.e., the low temperature reaction during heating) increased a mere 7°C when the CO_2 concentration was reduced from 100% to 95%. Further reduction in CO_2 concentration to 3.85% resulted in a further reaction temperature increase of only 33°C. The high-temperature reaction was also not significantly affected by the reduction in CO_2 concentration from 100% to 95% and addition of 0.2% H_2. There was a significant response to the further reduction in CO_2 concentration to 3.85% and relative increase in H_2-content (H_2 : CO_2 increased from 1:475 to 1:1). The reaction temperature decreased by approximately 300°C.

Modeling

Equilibrium calculations were performed over the temperature range from 700°C to 1, 400°C for the initial composition 1 mol BaO, 0.8 mol CeO_2, and 0.2 mol $YO_{1.5}$. The above solution phases and all known condensed phases were included in the calculations at one bar total pressure and the three different gas compositions. The only stable phases computed to be present under all conditions were $BaCO_3$ and the two solution phases (Fig. 5.4). The results for 100% CO_2 and the high CO_2 : H_2 ratio were identical (Fig. 5.4a). From 700°C to 1, 025°C the $Ce_{0.8}Y_{0.2}O_{1.9}$ solution phase was the only stable condensed phase. Between 1, 025°C and 1, 180°C there is a continuous conversion of the $Ce_{0.8}Y_{0.2}O_{1.9}$ phase to the perovskite phase with concomitant compositional changes within the phases (i.e., the Ce:Y ratio is

Table 5.2 Summary of DTA[a] and modeling results of $BaCe_{0.8}Y_{0.2}O_{3-\delta}$ in different gas mixtures

Gas composition	Decomposition on heating/°C	Perovskite formation on heating/°C	Decomposition on cooling/°C	Equilibrium 3-phase field range/°C
CO_2	470	1225	1062	1025–1180
$95CO_2 + 0.2H_2 + 4.8Ar$	477	1206	1054	1025–1180
$3.85CO_2 + 3.85H_2 + 92.3Ar$	510	945	717	795–910

[a] $1.6°C \bullet min^{-1}$

not constant in each phase within the transformation temperature range). Above $1,180°C$ only the $BaCe_{0.8}Y_{0.2}O_{2.9}$ is stable. At the lower $CO_2 : H_2$ ratio the reaction forming the perovskite occurs at lower temperatures (Fig. 5.4b). The DTA results are in amazingly good agreement with the predicted phase diagrams produced by this relatively simple ideal solution model. The ranges defined by the hysteresis in the DTA data correlate closely with the ranges defined by the predicted 3-phase fields in the model (Table 5.2).

Zirconium Substitution

Undoped

Preliminary HTXRD results in CO_2 on the three $Ba_{0.98}Ce_{1-y}Zr_yO_{3-\delta}$ compositions showed $BaCO_3$ formation in the two lower Zr-content samples ($y = 0.05$ and 0.10) and no x-ray detectable decomposition products in the $y = 0.15$ sample (BCZ885). The two lower Zr-content compositions were abandoned and the detailed HTXRD analysis and DTA were performed on BCZ885. This composition had considerably improved chemical stability over the most similar non–zirconium-substituted composition, the acceptor-doped composition BCY88. The data just enters the $BaCO_3$ region of the $lg(r_{p/c})$ vs. t plot during both heating and cooling (Figs. 5.3a and 5.3b respectively), and the DTA/TGA experiments showed no decomposition on cooling whatsoever (Table 5.1). Furthermore, the temperature range in which $BaCO_3$ is present is smaller ($t_{range} \approx 340°C$ vs. $550°C$) and occurs at lower temperatures.

Yttrium-Doped

Zirconium substitution resulted in significantly improved chemical stability in CO_2; however, the material must be acceptor doped to provide any hope of practical proton conductivity. The composition BCZY15 incorporates the stability-enhancing modifications examined earlier, slight Ba deficiency and 0.15 ZrCe substitution, and reintroduces the standard 0.20 Y doping. The HTXRD analysis and DTA in CO_2 show that BCZY15 exhibits an intermediate behavior. On heating, BCYZ15 possesses characteristics of both the two simpler compositions (i.e., BCY88 and BCZ885): a minor excursion into $BaCO_3$ territory at $t \approx 600°C$ similar to BCZ885; then further $BaCO_3$ formation above $t \approx 1000°C$, approximately the temperature of maximum conversion to $BaCO_3$ of BCY88. The extent of conversion is also intermediate to the two simple compositions. On cooling, BCZY15 behaves slightly differently. Barium carbonate begins to form at a temperature approximately midway between the onset temperatures for BCY88 and BCZ885. But rather than distinct peaks in the $lg(r_{p/c})$ data, as BCY88 has at $t \approx 1,050°C$ and BCZ885 at $t \approx 700°C$—BCZY15 displays a gradual, monotonic decrease in $lg(r_{p/c})$.

   Since BCZY15 behaves approximately as a combination of the two component compositions, the magnitudes of the $BaCO_3$-presence temperature range and high-temperature transition hysteresis are approximately equal to the greater of the two

corresponding values for the simple compositions: $t_{range} \approx 600°C$, comparable to
the t_{range} value for BCY88; and $\Delta t_{hyst} \approx 250°C$, similar to the Δt_{hyst} value of
BCZ885. Furthermore, the high-temperature transition temperatures of BCZY15
fall between those of BCY88 and BCZ885 while the hysteresis is still comparable
to that of BCZ885.

Kinetics

Two methods of characterizing the reaction kinetics were employed. The hysteresis
in the high-temperature reaction temperature (Δt_{hyst}; i.e., the difference between
the transition temperatures observed during heating and cooling) provides a qualitative measure of reaction kinetics for similar materials at comparable temperatures. The change in the reaction temperatures during heating as a function of

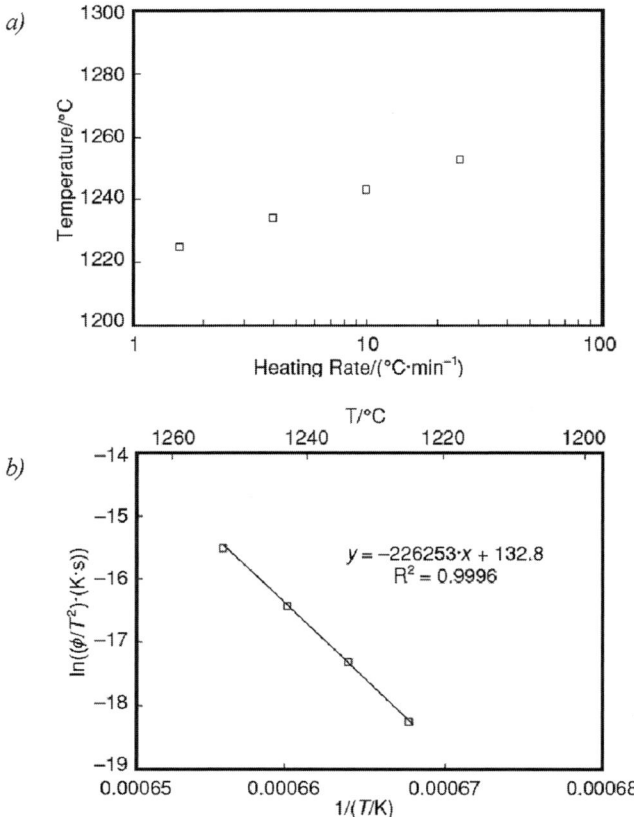

Fig. 5.5 Example data set used for calculation of reaction activation energies (BCY108p, perovskite formation on heating): change in DTA peak temperature with changing heating rate (a) and
the same data plotted in the form of the Kissinger equation (b) (the slope of the line is –E/R, where
E is the activation energy and R is the universal gas constant

Table 5.3 Reaction activation energies[a] in CO_2 on heating

Composition	Decomposition on heating/(kJ · mol⁻¹)	Perovskite formation on heating/(kJ · mol⁻¹)
BCY108p	242	1881
BCY88p	205	2142
BCY88	148[b]	1452[c]
BCZ885	163	4152
BCZY15	230	1735

[a] Calculated by the Kissinger equation using results from up to four heating rates. ($R^2 > 0.95$)
[b] Initial, but smaller peak used.
[c] Average temperature of doublet peak used.

heating rate allows the calculation of the reaction activation energies for both the low-temperature decomposition reaction of the metastable phase as well as the high-temperature reaction. Fig. 5.5a illustrates this effect for the high-temperature transition in BCY108p with the corresponding Kissinger analysis plot in Fig. 5.5b.

Considering first the compositions BCY88 and BCZ885, it is clear that the high-temperature transition occurs at significantly lower temperatures in BCZ885. The high-temperature transition hysteresis, however, is greater: $\Delta t_{hyst} \approx 230°C$ for BCZ885 (when this reaction occurs) vs. $\Delta t_{hyst} \approx 150 - 179°C$ for BCY88. Such an increase in this quantity indicates an increase in the energy barrier for this reaction. This is supported by the calculated activation energies (Table 5.3). The activation energy for this reaction in BCZ885 is approximately three times its activation energy in BCY88. (The difference in the activation energies calculated for the synthesized and purchased powders, BCY88 and BCY88p, respectively, is likely due to differences in powder characteristics such as surface area and amorphous content.)

The high-temperature transition temperatures of BCZY15 fall between those of BCY88 and BCZ885 while the hysteresis is still comparable to that of BCZ885. However, the calculated activation energy, though it does fall between the values for BCY88 and BCZ885, is comparable again to that for BCY88. This discrepancy indicates that the materials being compared are too dissimilar to allow estimation of the relative kinetics using the temperature hysteresis.

The calculated activation energies for the initial decomposition of the perovskite on heating were not influenced greatly by the compositional changes studied in this work. For this reaction, the highest value obtained for these materials was only approximately 1.6 times the lowest value (cf. the factor of 3 difference for the high-temperature reaction). Furthermore, the observed activation energies for the initial composition are relatively low due to the elevated internal energy inherent in metastable phases.

Summary and Conclusion

Two methods for improving the stability of yttrium-doped barium cerate in CO_2 have been investigated. Producing slightly barium-deficient compositions and partial substitution of Zr for Ce have been previously, and in this work, shown to increase the resistance to decomposition in the presence of CO_2. The effects of the individual modifications as well as their combined effects have been shown utilizing thermal analysis and high-temperature x-ray diffraction. Although definite improvements in CO_2-tolerance were observed, absolute immunity was not achieved at the levels of modification studied. Since it is known [8, 9] that significant amounts of Zr_{Ce} substitution or Ba-deficiency severely decrease conductivity, the prospect of producing a practical, durable high-temperature proton conductor in this system with these techniques is unlikely.

Acknowledgments This project was sponsored by the Office of Fossil Energy. The Oak Ridge National Laboratory is operated by UT-Battelle, LLC for the U.S. Department of Energy under Contract DE-AC05-00OR22725.

References

1. Shima D, Haile SM. Solid State Ionics. 1997;97:443–55.
2. Ma G, Shimura T, Iwahara H. Solid State Ionics. 1998;110:103–10.
3. Ryu KH, Haile SM. Solid State Ionics. 1999;125:355–67.
4. Taniguchi N, Nishimura C, Kato J. Solid State Ionics. 2001;145:349–55.
5. Chick LA, Pederson LR, Maupin GD, Bates JL, Thomas LE, Exarhos GJ. Mater Lett. 1990;44(1, 2):6–12.
6. Kissinger HE. J Res Natl Bur Stand. 1956;57(4):217–21.
7. Kissinger H. E Anal Chem. 1957;29(11):1702–6.
8. SGTE Pure Substance Database. 1996 Version, Scientific Group Thermodata Europe.
9. Cordfunke EHP, Booji AS, Huntelaar ME. J Chem Thermodyn. 1998;30:437–47.
10. Kale GM, Jacob KT. Solid State Ionics. 1989;34:247–52.
11. Longo V, Podda LJ. Mater Sci. 1981;16(3):839–41.

Chapter 6
Mixed-Conducting Perovskite Reactor for High-Temperature Applications: Control of Microstructure and Architecture

Grégory Etchegoyen, Thierry Chartier, Alain Wattiaux, and Pascal Del Gallo

High-temperature applications of perovskite-type membrane reactors require improved material performances and operational stability. The reactor microstructure and architecture controls were found to be crucial for thermo-mechanical integrity and oxygen permeation kinetics. A multilayer reactor was developed, using second-phase particles to control its microstructure and a co-sintering process to control its architecture.

Introduction

Mixed-conducting perovskite membranes were developed for several years now for potential industrial applications (syngas production, oxygen tonnage production, solid oxide fuel cell,...). As they operate at elevated temperatures and allow the separation of oxygen from air with high selectivity, these membranes could offer an interesting solution for the catalytic partial oxidation of natural gas thanks to their improved efficiency and lower cost [1–4]. The first applications of these catalytic membrane reactors (CMR) concern production of syngas or hydrogen from natural gas resources [5–9].

This technology is based on both the ionic and electronic conductivity of several oxides that allow oxygen transport through the bulk at temperature higher than 650°C by an oxygen-gradient driven mode. When a dense thin membrane is exposed to an oxygen pressure gradient between its opposite faces, oxygen is reduced on the oxygen-rich side, then ionic oxygen diffuses through the crystal lattice up to the permeate side where the reaction with methane occurs [10–13].

During the last two decades, an important research effort has been focused on materials providing high oxygen permeation rates. Perovskite-type oxides (ABO_3) have been largely studied since Teraoka [11] reported the mixed-conductivity of $La_{1-x}Sr_xCo_{1-y}Fe_yO_{3-\delta}$ systems. The partial substitution of A-sites trivalent cations by divalent cations is balanced by the creation of oxygen vacancies or change of

P. Del Gallo (✉)
Air Liquide, Centre de Recherche Claude-Delorme, Jouy-en-Josas, France
e-mail: Pascal.Del-Gallo@AirLiquide.com

A.C. Bose (ed.), *Inorganic Membranes for Energy and Environmental Applications*,
DOI 10.1007/978-0-387-34526-0_6, © Springer Science+Business Media, LLC 2009

B-site cation valence state to maintain charge neutrality. As a result, excellent ionic conductivity is ensured by high oxygen vacancy concentration and thus the electronic conductivity is achieved by the B-cation multivalence state.

More recently, research was focused mainly on material compositions with the objective to prioritize mechanical and chemical stability under industrial operating conditions than only on oxygen-flux performances. $La_{1-x}Sr_xFe_{1-y}Ga_yO_{3-\delta}$ systems were found to offer good chemical stability under a reducing environment, high oxygen-ionic conductivity and low dimensional variation when submitted to oxygen gradients [14–18].

An actual breakthrough is to design both the microstructure (grain size, porosity, ...) of the different system materials (support-membrane-catalyst) and the reactor architecture in order to obtain improved permeation properties and stability during the time on stream [19]. Considering all the parameters, a special consideration was given in this paper to the stack of the different materials at different length scale, from the microscopic-scale (microstructure) to the assembly of materials on the whole reactor thickness (architecture). An example of the architecture/microstructure concept was used in the Multi Electrode Assembly (MEA) approach.

Theoretical Models

The oxygen permeation flux through a dense membrane submitted to a chemical potential drop is a multistep process, including a sequence of reactions and different kinetics transports. As elementary steps are numerous and as their kinetics cannot be easily reached, it is common to consider the following three steps: gas-diffusion to surface, oxygen surface-exchange and oxygen-ionic diffusion in bulk [14, 20]. The overall flux kinetic is balanced by their respective resistive effect and thus permeation fluxes can be significantly improved by acting on the rate-limiting step [21–27].

Gas-diffusion to surfaces is mainly dependent on reactor configuration (geometry) and is to be taken under considerations when the membrane architecture includes a porous layer (grain size, porosity, interconnection shapes and dimensions, ...).

When oxygen-ionic diffusion is purely the limiting step, the one dimensional oxygen permeation flux can be expressed by Eq. 6.1 according to the Wagner theory:

$$jO_2 = \frac{RT}{16F^2L} \int_{lnpO_2''}^{lnpO_2'} \frac{\sigma i \cdot \sigma e}{\sigma i + \sigma e} \cdot dln(pO_2) \qquad (6.1)$$

where σi and σe are respectively the ionic and electronic conductivity, F the Faraday constant, L the membrane thickness and pO_2' and pO_2'' the oxygen rich-side and oxygen lean-side partial pressure respectively.

In some cases, the oxygen flux can be governed by the oxygen surface exchange reaction on the oxygen-rich side or permeate side (oxygen lean-side) of the membrane:

$$1/2O_2(g) + V_O^{\bullet\bullet} \Leftrightarrow O_O^x + 2h^\bullet \qquad (6.2)$$

where $V_O^{\bullet\bullet}$ and O_O^x denote respectively an oxygen vacancy and oxygen ion on an oxygen lattice site and $h^\bullet$ is an electronic hole. In a pure surface-exchange regime, the oxygen flux can be expressed as:

$$jO_2 = k.Cv.(pO_2)^{0.5} \qquad (6.3)$$

where k is a surface exchange coefficient, Cv the oxygen vacancy concentration and pO_2 the oxygen partial pressure. Thus for a given oxygen pressure gradient and temperature, the flux can be increased by reducing the membrane thickness L in Eq. 6.1 or by increasing the coefficient k in Eq. 6.3.

Our approach is then to define, to design and to develop an architecture and a microstructure for the reactor in order to optimize L and k parameters with respect to permeation performances, as well as thermo-mechanical and chemical integrity of the reactor.

Microstructure

Perovskite powders in the system $(LaSr)(FeGa)O_{3-\delta}$ were synthesized through a solid state route and shaped as described above [18, 28].

The grain size of the mixed-conducting material of dense membrane can have a great influence on:

- Oxygen permeation fluxes; oxygen exchange and transport models suffer from not considering that the membrane is a polycrystal material in which grain boundaries have different properties than bulk grain.
- Thermo-mechanical properties; the fracture and strength toughness of brittle ceramic materials can be significantly improved by controlling the microstructure in order to decrease the size of critical defects.
- Membrane selectivity; as the dense membrane thickness is typically about 100 μm, it is then obvious to keep a small grain size (about 1 μm) in order to prevent large defects and membrane leakage.

The sintering treatment of mixed-conducting materials generally leads to a coarse grain microstructure due to their high oxygen diffusion rate that enhances both densification and grain growth kinetics. The addition of dispersed secondary ceramic-phase in the membrane bulk makes it possible to maintain a small grain size during thermal treatment. These small particles presenting no chemical reaction with the matrix material enable to slow down and/or to delay grain growth during sintering by pinning grain boundaries. The pinned-grain size is balanced by the grain growth driving force and by its resisting force caused by the presence of stationary second phase particles at grain boundaries. Thus, the final microstructure of dense mixed-conducting materials can be controlled by the pinning particle volume fraction, grain size and morphology. Besides the smaller grain size impacting critical defects and increasing fracture resistance, the presence of well dispersed small particles is also expected to improve toughness [29–31].

The secondary phase particles have to be chosen to upgrade microstructure without altering oxygen permeation properties. A good candidate is magnesia that is well known for its good chemical compatibility with perovskite mixed-conducting materials when used as a porous support layer [24, 27]. Mixtures of mixed-conducting material (LaSr)(FeGa)O$_{3-\delta}$ and magnesia were referenced as LSFG, LSFG/2M, LSFG/5M and LSFG/10M for respectively 0, 2, 5 and 10 vol. of magnesia content.

Control of Microstructures with Magnesia Used as Pinning Particles

The preparation of LSFG/MgO composites has been described above [28, 32]. Characterization of shaped and sintered membranes was carried out by XRD, ICP, SEM and EPMA. The experimental procedure and equipment were reported elsewhere [28, 32]. Crystal structure of the LSFG perovskite phase was found to be rhombohedrally distorted. Incorporation of magnesia had no influence on the main phase composition (LSFG), of which chemical composition remained identical before and after sintering. ICP and EPMA data did not evidence gallium volatilization.

Small amounts of magnesia with a mean particle size of 0.5 μm, from 2 to 10vol%, significantly reduce grain size of sintered samples (Fig. 6.1). The perovskite material without magnesia addition presents a larger grain size (more than 4 μm) and a lower density than LSFG/MgO composites.

For a given sintering temperature and a second phase grain size (radius r), final grain size of the matrix (D) was found to depend on magnesia volume content (f) according to the Zener law [33].

$$\frac{D}{r} = a.f^{-b} \qquad (6.4)$$

where a and b are constants of the model. The grain size of LSFG matrix can be controlled by the amount and size of magnesia particles according to Eq. 6.4, with a = 1.64 and b = 0.41 (Fig. 6.2).

MgO addition has an additional beneficial effect by enlarging the sintering temperature range in which the material presents a high density. The microstructure of pure LSFG material was found to be very sensitive to sintering temperature and dwell time. The optimum sintering temperature is comprised between 1,250 and 1,300°C with a final relative density of 95%. Any further increase in temperature or dwell time leads to a decrease in density, that is, 93 to 87% for 1 h and 4 h dwell at

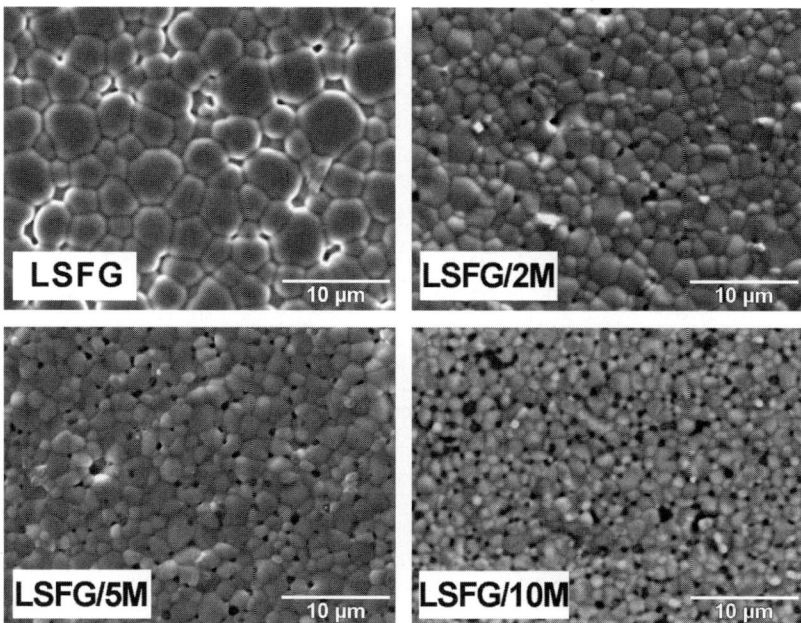

Fig. 6.1 SEM micrographs of $(LaSr)(FeGa)O_{3-\delta}$ containing Xvol% of MgO as pinning particles (*dark inclusions*); (LSFG) X = 0; (LSFG/2M) X = 2; (LSFG/5M) X = 5; (LSFG/10M) X = 10

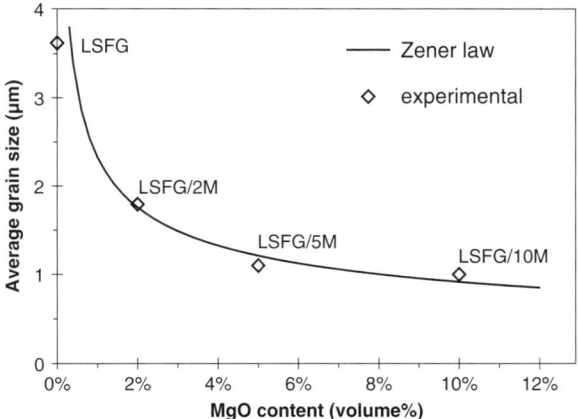

Fig. 6.2 Average grain size of sintered $(LaSr)(FeGa)O_{3-\delta}$ in function of MgO volume content

$1, 350°C$ respectively. The decrease in density associated with an increase in grain size (from 5 to 27 μm respectively) should likely affect membrane selectivity and thermo-mechanical properties. On the other hand, the presence of magnesia inclusions at grain boundaries decreases the grain boundary mobility and enables high density ($> 95\%$) and fine grain size until $1, 350°C$ (Fig. 6.3).

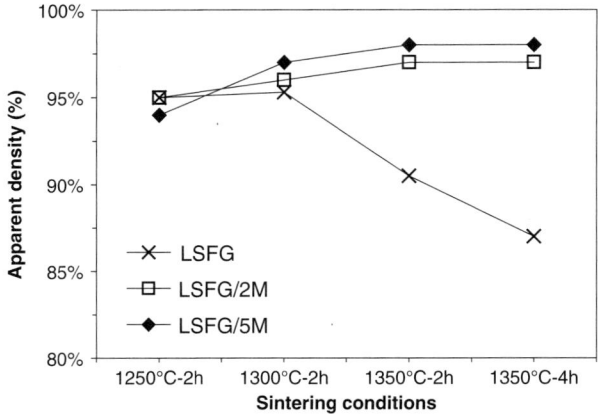

Fig. 6.3 Apparent density dependence on sintering conditions for $(LaSr)(FeGa)O_{3-\delta}$ based membrane containing 0vol% (LSFG), 2vol% (LSFG/2M) and 5vol% (LSFG/5M) of MgO

Microstructure Effects on Oxygen Permeation Rates

It was found that the microstructure control can be achieved without modification of its chemical nature, sintering temperature or powder synthesis protocol of the mixed-conducting phase. Thus any modification in oxygen permeation properties of the dense membrane can be completely attributed to microstructural changes. Figure 6.4 represents the Arrhenius plots of the flux through dense 1 mm thick membranes, sintered at 1, 300°C for 2 h in an air/argon gradient with different average grain sizes obtained with magnesia amounts of 0, 2 and 5vol%. The oxygen flux was found to increase significantly from 0 to 5vol% of magnesia, while the mean grain size was decreasing from 3.6 to 1.1 μm. The improved oxygen permeation flux is associated with a decrease in the flux activation energies that can be attributed to an evolution in the oxygen flux limiting step. The observed linear variation of the flux with $pO_2^{1/2}$ on air side (Fig. 6.5) and an argon flow on permeated side is in agreement with Eq. 6.3 and suggests a limiting oxygen surface exchange mechanism. The surface exchange coefficient k may be calculated from the line slopes using Cv values calculated from Mossbauer data on Fe^{4+}/Fe^{3+} ratio and thermogravimetric analysis (Cv increases from 3.09×10^{-3} to 3.28×10^{-3} mol.cm^{-3} between 800 and 900°C). The calculated values of k followed an Arrhenius law (Fig. 6.5) with an activation energy of the surface exchange reaction similar to the overall oxygen flux one determined in Fig. 6.4.

These results confirm that the flux in our system is limited only by the surface exchange reaction on the membrane air side. The decrease of the average grain size by adding small amounts of magnesia in the mixed-conducting phase results in the increase of the k-coefficient (Table 6.1) and consequently in the overall oxygen permeation kinetics. The effect of the microstructure was thought to be caused by the difference of grain boundary density on the surface of the membrane. Grain boundaries may offer a large number of catalytic sites for oxygen adsorption and charge transfer. As far as the microstructure of the material is concerned, it has been

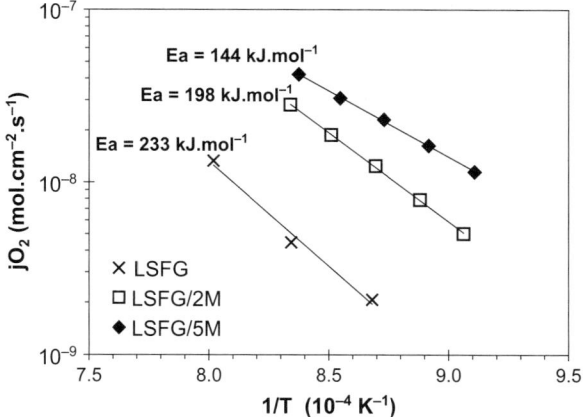

Fig. 6.4 Arrhenius plots of oxygen permeation fluxes through LSFG, LSFG/2M and LSFG/5M membranes (thickness=1 mm). Numbers in kJ/mol refer to the activation energies of oxygen permeation

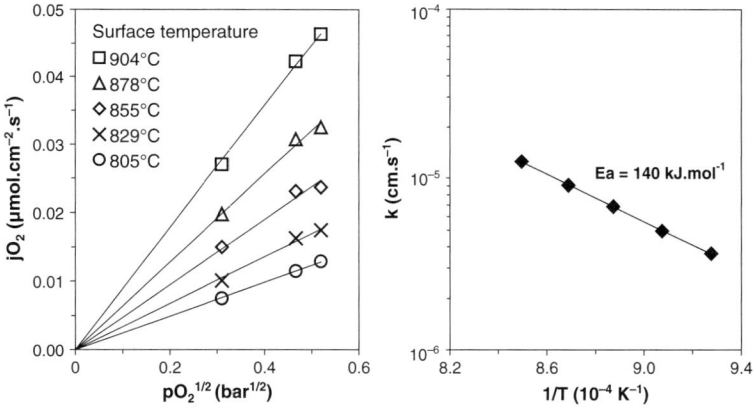

Fig. 6.5 Oxygen permeation flux through LSFG/5M membrane versus oxygen partial pressure on air side for different temperatures (*left*) and corresponding values of the surface exchange coefficient calculated with Eq. 6.3 (*right*)

demonstrated that the microstructure of membrane material has a great influence on the flux by acting on the surface exchange coefficient k.

Architecture

The term architecture refers to the specific spatial arrangement of different materials constituting the reactor. As the membrane placed in a catalytic reactor has to satisfy various functions such as oxygen separation from air, high oxygen permeation kinetics and catalytic reactions with methane, a reactor can be defined as a Functionally

Table 6.1 Surface exchange coefficient k of different membranes

Reference	MgO content	k at $pO_2 = 0.21$ atm and T $= 850°C$
LSFG	0 vol%	0.5×10^{-6} cm.s^{-1}
LSFG/2M	2 vol%	3.5×10^{-6} cm.s^{-1}
LSFG/5M	5 vol%	6.7×10^{-6} cm.s^{-1}

NOTE: k values were estimated from oxygen permeation fluxes versus $pO_2^{1/2}$.

Gradient Material (FGM) presenting discrete properties variation on its thickness. It is then necessary to identify the requirements of each layer in order to define an architecture in agreement with membrane functionalities:

- The dense layer has to present high oxygen conductivity and small dimensional variation caused by oxygen activity gradient, to be achieved by the material choice. On the other hand, the dense layer has to be thin (about 100 μm), of small grain size and without leakage.
- The porous layer is required to ensure the thermo-mechanical integrity of the thin dense layer. This support has (i) to allow gas species diffusion, (ii) to present good chemical and thermal compatibility with the dense layer, (iii) to have mixed-conducting properties in order to shift oxygen exchanges at gas/solid interfaces inside its whole thickness, and (iv) to be of low cost for industrial applications.
- The catalytic porous layer role is to promote the methane partial oxidation by offering numerous and dispersed catalytic sites.

The main challenge for a multilayer reactor concept is to provide the lowest number of manufacturing steps and to assure a perfect compatibility between each layer in operating conditions. We have used perovskite materials with similar compositions as membrane, support and catalytic layers. In addition, the multilayer reactor has been sintered in a single stage (co-sintering) to benefit from simultaneous shrinkages that reduces stresses, deformations or cracks occurrences. Reactors with different architectures were developed by stacking films obtained by tape-casting [19, 34].

Porous Layer Elaboration

Porous supports can be obtained by adding to mixed-conducting materials powder, a fugitive organic material that will burn out, leaving large connected pores in the membrane. A porous support was obtained by using 40vol% corn-starch particles per mineral part. Large connected pores, with size ranging from 8 to 20 μm, were obtained. The bulk surrounding the pores is dense with small grain size of about 1 μm. Pore size and pore volume would likely ensure easy gas diffusion through open porosity, although the narrow pore interconnections of about 2 μm could be

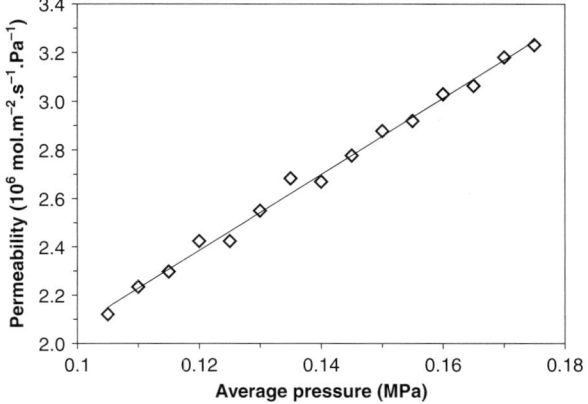

Fig. 6.6 Gas permeability versus average pressure of the porous support

resistant to the gas flow. Gas permeability (argon, room temperature) of a 1.2 mm thick porous support is presented in Fig. 6.6.

The linear pressure dependence of the gas permeability in function of the mean pressure suggests a viscous flow regime. Therefore, the permeability of the support can be improved by increasing gas pressure. In working conditions, the porous support is exposed to air pressure at 3 bars (0.3 MPa) that enables to reach a sufficient permeability of 2.9×10^{-6} mol.m^{-2}.s^{-1}.Pa^{-1}.

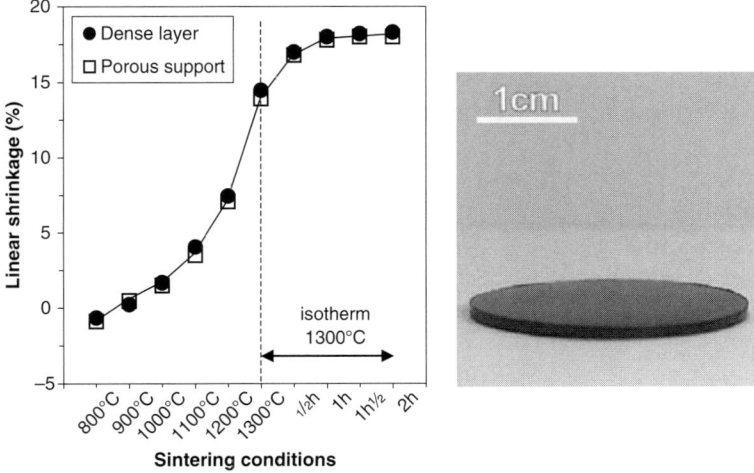

Fig. 6.7 Linear shrinkages of the dense layer and the porous support (*left*) and photograph of the resulting co-sintered asymmetric membrane (*right*)

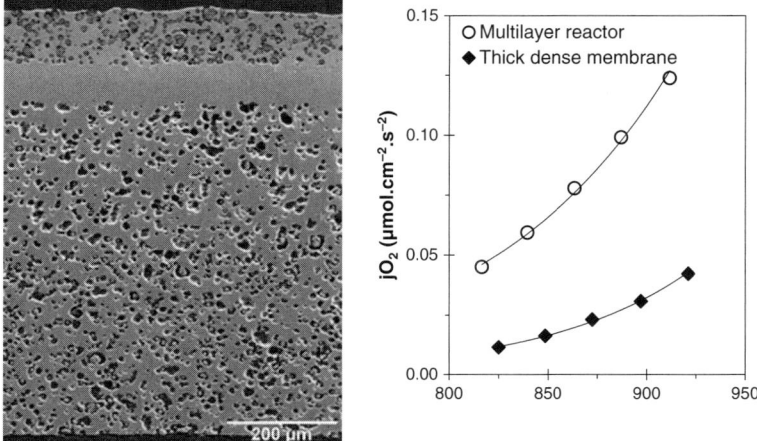

Fig. 6.8 SEM micrograph of a multilayer perovskite reactor (*left*) and its oxygen permeation flux with temperature compared with a thick dense membrane (*right*)

Multilayer Membrane

As dense and porous layers are co-sintered, they must present similar shrinkages and thermal expansion coefficients. Because the mineral skeleton density and sintering reactivity is unchanged, the shrinkage of the porous layer, the porosity of which is obtained with a fugitive material, was found to be equal to the dense layer one (Fig. 6.7). Similar thermal expansion coefficients can be achieved by choosing perovskite material compositions for both the membrane and the porous support. Consequently, a multilayer reactor composed of a thin dense layer, a thick porous layer and a thin catalytic porous layer can be developed using a co-sintering process.

A SEM micrograph of a co-sintered multilayer reactor is presented in Fig. 6.8; no delamination or interfacial reaction can be observed on the membrane. These co-sintered multilayer reactors, only based on perovskite materials, present higher oxygen flux than dense 1 mm thick membranes in an air/argon gradient.

Conclusion

Our research is focused on the control of both the microstructure and the architecture of perovskite-type mixed-conducting reactors for high temperature applications, especially for oxygen separation from air and syngas production.

The grain size of perovskite materials such as $(LaSr)(FeGa)O_{3-\delta}$ was adjusted by using magnesia second phase particles. The microstructures of perovskite/MgO composites were less sensitive to sintering conditions and high densities, and small grain sizes were obtained in a larger temperature range than samples without magnesia. This behavior is beneficial due to the decrease of the risk of leakage and of membrane failures. On the other hand, the surface exchange coefficient is improved by grain size reduction of the membrane and the overall oxygen permeation flux is promoted.

Improved performances were obtained with a multilayer reactor composed of a thick porous support, a thin dense layer and a thin porous catalytic layer. Perovskite materials were used for all three layers. Reactor development includes the use of a fugitive material to perform controlled-porosity in the support layer and co-sintering to avoid cracks or deformations in the sintered membranes.

Acknowledgments The authors wish to acknowledge Air Liquide and CNRS for supporting this research.

References

1. Armor JNJ. Membr Sci. 1998;147:217–33.
2. Sammells AF, Schwartz M, Mackay RA, Barton TF, Peterson DR. Catal. Today. 2000;56: 325–8.
3. Wilhelm DJ, Simbeck DR, Karp AD, Dickenson RL. Fuel Process Technol. 2001;71:139–48.
4. Balachandran U, Dusek JT, Sweeney SM, Poeppel RB, Mieville RL, Maiya PS, et al. Am Ceram Soc Bull. 1995;74:71–5.
5. Hazbun EA. US Patent 4,827,071, 1989.
6. Thorogood RM, Srinivasan R, Yee TF, Drake MP. US Patent 5,240,480, 1993.
7. Gottzmann CF, Prasad R, Schwartz JM. Eur Patent 0,962,422,B1, 1999.
8. Schwartz M, White JH, Sammells AF. US Patent 6,033,632, 2000.
9. Mazanec TJ, Cable TL. US Patent 5,648,304, 1997.
10. Tsai CY, Dixon AG, Moser WR, Ma YH. AIChE J. 1997;43:2741–50.
11. Teraoka Y, Zhang HM, Furukawa S, Yamazoe N. Chem Lett. 1985;11:1743–46.
12. Gellings PJ, Bouwmeester HJM. Catal Today. 1992;12:1–105.
13. Bouwmeester HJM. Catal Today. 2003;82:141–50.
14. Kim S, Wang S, Chen X, Yang YL, Wu N, Ignatiev A, et al. Electrochem Soc. 2000;147:2398–406.
15. Ishihara T, Tsuruta Y, Todaka T, Nishiguchi H, Takita Y. Solid State Ionics. 2002; 152–153:709–14.
16. Kharton VV, Yaremchenko AA, Patrakeev MV, Naumovich EN, Marques FMBJ. Eur Ceram Soc. 2003;23:1417–26.
17. Ritchie JT, Richardson JT, Luss D. AIChE J. 2001;47:2092–101.
18. Etchegoyen G, Chartier T, Del-Gallo P. French Patent FR2857355, 2003.
19. Etchegoyen G, Chartier T, Del-Gallo P. French Patent, 2004.
20. Deng H, Zhou M, Abeles B. Solid State Ionics. 1995;80:213–22.
21. Teraoka Y, Fukuda T, Miura N, Yamazoe NJ. Ceram Soc Jpn Int Ed. 1989;97:523–9.
22. Chen CH, Bouwmeester HJM, van Doorn RHE, Kruidhof H, Burggraaf AJ. Solid State Ionics. 1997;98:7–13.
23. Jin W, Li S, Huang P, Xu N, Shi JJ. Membr Sci. 2001;185:237–43.
24. Hong L, Chen X, Cao ZJ. Eur Ceram Soc. 2001;21:2207–15.
25. Kharton VV, Kovalevsky AV, Yaremchenko AA, Figueiredo FM, Naumovich EN, Shaulo AL, et al. Membr Sci. 2002;195:277–87.
26. Lee S, Lee KS, Woo SK, Kim JW, Ishihara T, Kim DK. Solid State Ionics. 2003;158:287–96.
27. Middleton H, Diethelm S, Ihringer R, Larrain D, Sfeir J, Van Herle JJ. Eur Ceram Soc. 2004;24:1083–6.
28. Etchegoyen G, Chartier T, Julian A, Del-Gallo PJ. Membr Sci. 2006;268:86–95.
29. Hen CC, Prasad R, Gottzmann CF. Eur Patent 0,850,679,A2, 1998.

30. van Calcar P, Mackay RA, Sammells AF. US Patent 6,471,921,B1, 2002.
31. Li D, Liu W, Zhang H, Jiang G, Chen C. Mater Lett. 2004;58:1561–4.
32. Etchegoyen G, Chartier T, Del-Gallo P. French Patent FR0350802, 2003.
33. Zener C, Smith CS. Trans AIME. 1948;175:15–51.
34. Chartier T, Guillotin F. French Patent FR 2,817,860, 2000.

Chapter 7
Mixed Protonic-Electronic Conducting Membrane for Hydrogen Production from Solid Fuels

Shain J. Doong, Francis Lau, and Estela Ong

Gas Technology Institute is developing a novel concept that incorporates a hydrogen-selective membrane closely coupled with a gasifier for direct extraction of hydrogen from coal/biomass. Mixed protonic-electronic conducting ceramic materials of perovskite type have been identified as good candidate membranes for the high temperature hydrogen separation application in the gasification system. In this article, hydrogen permeation rates for the candidate perovskite membranes at temperatures of up to 950°C and pressures of up to 12 bar were measured. A rigorous model for hydrogen permeation through mixed proton-electron conducting ceramic membranes was also developed based on non-equilibrium thermodynamics. The effect of pressure on the hydrogen flux and the proton/electron conductivity was elucidated from both experimental data and simulation results.

Introduction

Gasification combines carbonaceous materials, steam and oxygen to produce synthesis gas or syngas, mainly hydrogen and carbon monoxide. After cooling and cleaning, the syngas is shifted using water-gas shift reactor technology to generate additional hydrogen and convert carbon monoxide to carbon dioxide. Hydrogen is subsequently separated from the gas stream, typically through the use of Pressure Swing Adsorption (PSA) technology. Depending on the feedstock price, the cost of producing hydrogen from this conventional gasification process is currently not competitive to steam reforming of natural gas [1]. Gasification, shift reaction, and PSA operation are all relatively mature technologies. Continuing improving on these technologies may result in only incremental reduction in the overall cost of hydrogen from coal/biomass. To achieve significant reduction of costs, novel, advanced and breakthrough technologies must be developed.

One of the active research areas in reducing the hydrogen cost from the coal/biomass gasification processes is the development of high temperature

S.J. Doong (✉)
UOP, 25 E. Algonquin Rd. Des Plaines, IL 60016, USA
e-mail: Shain.Doong@uop.com

A.C. Bose (ed.), *Inorganic Membranes for Energy and Environmental Applications*, 107
DOI 10.1007/978-0-387-34526-0_7, © Springer Science+Business Media, LLC 2009

membranes that can be designed to separate hydrogen from the syngas. This type of membrane system is primarily targeted as a membrane reactor for the water-gas-shift reaction to convert syngas to hydrogen. Although pure hydrogen is generated directly from the membrane shift reactor, the remaining gas containing mostly CO_2 and some CO and H_2 is sent to a gas turbine to combust with oxygen for power generation. The working temperature for the membrane shift reactor is in the range of 200–500°C. The best candidate materials are palladium (Pd) and its alloys or microporous ceramic membranes. Recent studies performed by Parsons [1] and Mitretek [2] showed that coal to hydrogen plants employing this type of membrane system could achieve a significant reduction of hydrogen cost, compared with the conventional hydrogen plant for coal gasification.

Advanced Membrane Reactor for Hydrogen from Coal/Biomass

GTI has developed another novel concept of membrane reactor by incorporating a hydrogen-selective membrane near or within a gasifier for direct extraction of hydrogen from the synthesis gases. As more than 50–60% of the final hydrogen product is generated in the gasification stage, there is a great potential of maximizing hydrogen production by separating hydrogen directly from the gasifier.

By configuring a hydrogen-selective membrane with a gasification reactor in a closely coupled way, both gasification reactions and hydrogen separation can be accomplished simultaneously in one processing unit thus increasing hydrogen production and simplifying process operation. Figure 7.1 shows a simplified process diagram for the novel membrane gasification reactor in comparison with the conventional gasification process for hydrogen production from coal (or biomass).

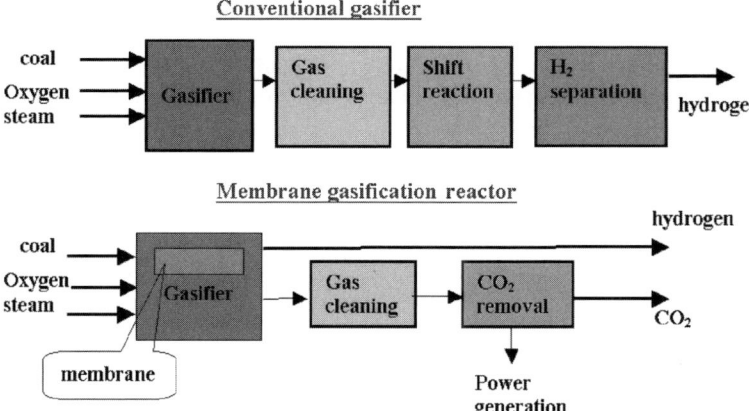

Fig. 7.1 Hydrogen production from coal gasification based on the conventional gasifier and the novel membrane gasification reactor concept

Under ideal conditions in which carbon in the solid feed is completely converted in a gasifier, the gasification reactions can be characterized by the following chemical reactions [3]:

$$CH_4 + H_2O = CO + 3H_2 \tag{7.1}$$

$$CO_2 + CH_4 = 2CO + 2H_2 \tag{7.2}$$

$$CO + H_2O = CO_2 + H_2 \tag{7.3}$$

If hydrogen is removed while it is being produced in the gasifier, the equilibrium will be shifted toward the right hand sides of the above three reactions. As a result, more H_2 and CO will be produced and less CH_4 will be present in the product gas. If additional steam is added to the gasifier, CO will also be shifted (Eq. 7.3) to produce hydrogen, potentially without the use of catalysts. Furthermore, because no other hydrogen purification unit such as PSA is needed in the membrane reactor process, the typical 20% loss of the hydrogen product in the PSA tail gas is completely eliminated. Thermodynamic and preliminary modeling results show that hydrogen production efficiency using the novel membrane gasification reactor concept can be increased by about 30–50% versus the conventional gasification process [4].

Mixed Protonic-Electronic Conducting (MPEC) Membrane

Due to the high temperature operation in the gasifier, only inorganic materials can be considered for this application. Dense ceramic membranes of perovskite type, which exhibit mixed protonic-electronic conductivity at high temperatures, represent one group of promising inorganic membranes for use in the high temperature membrane reactors. Under a gradient of chemical potential or partial pressure of hydrogen across the membrane, only hydrogen can permeate through the membrane (see Fig. 7.2). When a H_2-containing gas mixture is introduced to one side of the membrane, the H_2 dissociates into proton (H^+) and electron (e^+) on the surface. The

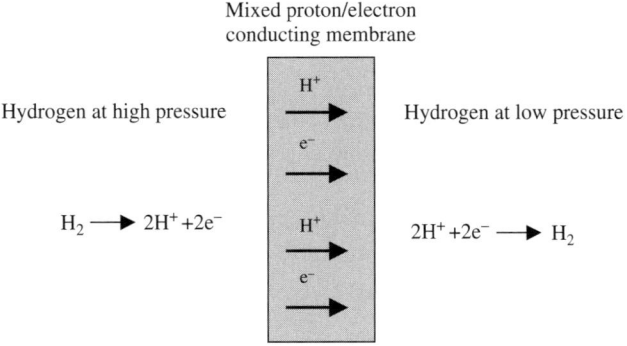

Fig. 7.2 Mixed proton/electron conducting membrane

dissociated species are transported through the membrane to its opposite side where the species recombine to H_2 molecule. Iwahara and coworkers [5] first discovered mixed proton/electron conduction in $SrCeO_3$ based perovskite type ceramics in high temperatures ($> 500°C$) in a hydrogen-containing atmosphere. Several research teams have also investigated the perovskite materials for high temperature hydrogen separation [6–12]. However, the perovskite membranes may not be suitable for low temperature ($< 500°C$) membrane shift reactor applications. The best working temperature range, $700–1,200°C$ of this type of material will be ideal for the membrane gasification reactor applications as proposed in Fig. 7.1.

Among the known MPEC materials, $BaCeO_3$ and $SrCeO_3$ perovskite oxides have the highest conductivity. To raise the conductivity, trivalent cation dopants such as Nd, Y, Eu, Gd, Sm, Tm are added to the compound to replace part of the Ce^{4+} and create oxygen vacancies, which are responsible for the transport of certain species through the lattice structure. This article discusses the fundamentals of the MPEC membrane materials in terms of the transport mechanism. The governing equations for modeling the hydrogen permeation and the simulation results are presented. The hydrogen fluxes for the Nd-doped $BaCeO_3$ membranes as measured from a high pressure permeation unit are also reported. Both the experimental data and the simulation results are used to elucidate the effect of pressure on the hydrogen flux and the proton/electron conductivity.

Transport Model for the MPEC Membrane

A rigorous model for hydrogen permeation through the MPCE ceramic membranes can be developed based on non-equilibrium thermodynamics. In a MPEC membrane, the driving forces for the transport of charged species come from both chemical and electrical potential gradients. The transport of four charged species, proton, oxygen vacancy, electron, and electron hole can be described by a combination of Fick's law and the equation for ion migration:

$$J_k = -\frac{\sigma_k}{z_k^2 F^2}\left(\frac{\partial \mu_k}{\partial x} + z_k F \frac{\partial \phi}{\partial x}\right) \tag{7.4}$$

where J_k is the flux of the species k, μ is the chemical potential, ϕ is the electrical potential, σ is the conductivity, z is the charge number of the species and F is the Faraday constant.

When no external current is imposed on the membrane, the net flux for each charged species is zero, that is

$$I = \sum_{k=1}^{n} I_k = \sum_{k=1}^{n} z_k F J_k = 0 \tag{7.5}$$

Combining Eqs. 7.4 and 7.5, a relationship between the electrical potential and the chemical potential can be obtained:

$$\frac{\partial \phi}{\partial x} = -\sum_{k=1}^{n} \frac{t_k}{z_k F} \frac{\partial \mu_k}{\partial x} \tag{7.6}$$

where t_k is the transport number of species k, which is a relative measure of conductivity of species k to the total conductivity.

$$t_k = \frac{\sigma_k}{\sum\limits_{i=1}^{n} \sigma_i} \tag{7.7}$$

The flux equation, Eq. 7.4, now becomes

$$J_k = -\frac{\sigma_k}{z_k^2 F^2} \left(\frac{\partial \mu_k}{\partial x} - z_k \sum_{i=1}^{n} \frac{t_i}{z_i} \frac{\partial \mu_i}{\partial x} \right) \tag{7.8}$$

Chemical potential μ is related to the chemical activity a_i by

$$\frac{\partial \mu_k}{\partial x} = RT \frac{\partial \ln a_k}{\partial x} \tag{7.9}$$

Under ideal conditions, activity a can be substituted with the concentration C. Furthermore, the conductivity of the defect species can be correlated with its concentration and diffusivity by the Nernst-Einstein equation:

$$\sigma_k = \frac{z_k^2 F^2}{RT} C_k D_k \tag{7.10}$$

Substituting Eqs. 7.9 and 7.10 into Eq. 7.8, the following equation can be obtained:

$$J_k = -C_k D_k \left[\frac{(1 - t_k)}{C_k} \frac{\partial C_k}{\partial x} - \sum_{\substack{i=1 \\ i \neq k}}^{n} \frac{z_k t_i}{z_i C_i} \frac{\partial C_i}{\partial x} \right] \tag{7.11}$$

Eq. 7.11 relates the flux of each species to the concentrations and the diffusivities of all the species inside the MPEC membrane.

In the proton-electron conductors, charged carriers are protons ($OH^\bullet$), vacancies ($V_{\ddot{O}}$), electrons (e^-), and electron holes ($h^\bullet$). The concentrations of the defect species in a typical proton conductor can be described by the following stoichiometric equation [22]:

$$1/2O_2 + V_{\ddot{O}} = O_O^x + 2h^{\bullet} \qquad K_1 = C_h^2/(C_V \, p_{O_2}^{1/2}) \qquad (7.12)$$

$$H_2 + 2O_O^x = 2OH^{\bullet} + 2e^- \qquad K_2 = (C_{OH}^2 C_e^2)/p_{H_2} \qquad (7.13)$$

$$H_2 + 1/2O_2 = H_2O \qquad K_3 = p_w/(p_{H_2} p_{O_2}^{1/2}) \qquad (7.14)$$

where O_O^x denotes the lattice oxygen. Eqs. 7.12–7.14 establish the relationships between the concentrations of the charged species inside the membrane to the gas partial pressures outside the membrane. The chemical potentials of each charged species can also be related to the chemical potentials of gases through the following equations corresponding to Eqs. 7.12–7.14:

$$1/2\mu_{O_2} + \mu_V = 2\mu_h \qquad (7.15)$$

$$\mu_{H_2} = 2\mu_{OH} + 2\mu_e \qquad (7.16)$$

$$\mu_{H_2} + 1/2\mu_{O_2} = \mu_w \qquad (7.17)$$

Also the electronic equilibrium requires

$$e^- + h^{\bullet} = nil \quad K_e = C_e C_h \qquad (7.18)$$

$$\mu_e + \mu_h = 0 \qquad (7.19)$$

The charged species concentrations can be converted to the gas partial pressures by Eqs. 7.12–7.14 and 7.18, and the concentration gradients can be converted to the gradients of the gas partial pressures by Eqs. 7.15–7.17 and 7.19. Eq. 7.11 for proton $OH^{\bullet}$ and vacancy $V_{\ddot{O}}$, will become

$$J_{OH} = -\frac{t_{OH}\sigma RT}{2F^2}\left[(t_h + t_e)\frac{\partial \ln p_{H_2}}{\partial x} + 4t_V\sigma\frac{\partial \ln p_w}{\partial x}\right] \qquad (7.20)$$

$$J_V = -t_V\sigma\frac{RT}{F^2}\left[(t_h + t_e)\frac{\partial \ln p_{H_2}}{\partial \ln x} - (t_h + t_e + t_{OH})\frac{\partial \ln p_w}{\partial x}\right] \qquad (7.21)$$

The hydrogen flux is related to the proton flux by

$$J_{H_2} = 2J_{OH} \qquad (7.22)$$

Essentially, a more general form of Wagner equation [8] was derived by including contributions from two other defect species, vacancy and electron hole. Furthermore, the conductivities are expressed in terms of species concentrations and diffusivities in the perovskite material according to Eq. 7.10. The Wagner equation can be obtained if the proton and the electron are assumed to be the dominating species and no steam is in the system. Thus with the aid of Eq. 7.22, Eq. 7.20 becomes

$$J_{H_2} = -\frac{RT}{4F^2L}\frac{(\sigma_{H^+})(\sigma_{el})}{\sigma_{H^+} + \sigma_{el}}(\ln(p_{H_2}^f) - \ln(p_{H_2}^p)) \qquad (7.23)$$

The term $\frac{(\sigma_{H^+})(\sigma_{el})}{\sigma_{H^+}+\sigma_{el}}$ in Eq. 7.23 is also called ambipolar conductivity.

Eqs. 7.20 and 7.21 can not be integrated directly because the transport numbers, t_i and the total conductivity σ are functions of the membrane position x. However at steady state, J_{OH} and J_V are constant and independent of the membrane positions. The above equations can be rearranged to give

$$\frac{RT}{F^2}\frac{\partial \ln p_{H_2}}{\partial x} = -\frac{4J_V}{\sigma(t_h + t_e)} - \frac{2J_{OH}(t_h + t_e + t_{OH})}{\sigma t_{OH}(t_h + t_e)} \qquad (7.24)$$

$$\frac{RT}{F^2}\frac{\partial \ln p_w}{\partial x} = \frac{J_V}{\sigma t_V} - \frac{2J_{OH}}{\sigma t_{OH}} \qquad (7.25)$$

Given the boundary conditions at both the feed side and the permeate side of the membrane, Eqs. 7.24 and 7.25 can be integrated with respect to x to obtain the profiles of hydrogen and water partial pressures across the membrane. The concentration profiles of the four defect species, proton (C_{OH}), vacancy (C_V), electron (C_e), and electron hole (C_h) are related to the gas partial pressure through Eqs. 7.12–7.14. The required parameters for the membrane material are equilibrium constants – K_1, K_2, K_3, and K_e – as well as the diffusivity data for the four defect species.

Experimental

Hydrogen Flux Measurement in High Pressure Permeation Unit

As coal gasification for hydrogen production occurs at temperatures above 900°C and pressures above 20 atm, it is critically important to evaluate the hydrogen flux of the candidate membrane materials under these operational conditions. To this end, a high pressure/high temperature permeation unit has been constructed. The unit is capable of operating at temperatures and pressures up to 1, 100°C and 60 atm respectively. The unit can allow screening and testing of the membrane materials at more realistic gasification temperature and pressure conditions. The permeation assembly consists of a permeation cell, a surrounding cylindrical heater, and an enclosing pressure vessel. A simplified schematic illustrating the concept of the permeation cell design is shown in Fig. 7.3. The membranes to be tested are in a disc form, with a diameter of about 2 cm. Hydrogen gas flows through the upper inner tube, and after contact with the membrane, it exits the system as a non-permeate gas diverted by an outer tube. An inert sweeping gas passing through the lower inner tube is used to sweep the hydrogen permeate from the membrane. Therefore, the pressures on both sides of the membrane can be adjusted to be equal, which would make the membrane sealing less difficult. A glass-based sealant material was used to seal the membrane along the edge to the metallic holding tube.

The hydrogen content of the permeate is analyzed by a GC to determine the hydrogen flux through the membrane. The inner tube, outer tube and the membrane holding tube are made of Inconel material for its good resistance to heat and easy

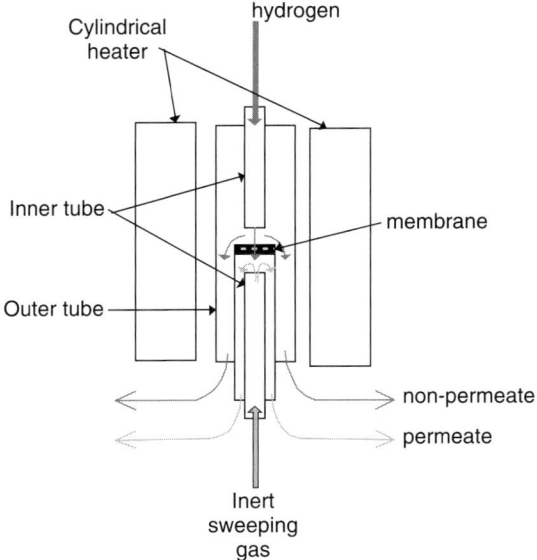

Fig. 7.3 High temperature/high pressure membrane permeation unit

machining and welding. The entire permeation cell assembly is heated by a cylindrical heater, which is enclosed in a pressure vessel purged with inert gas.

Before testing the membranes in the high pressure unit, helium was introduced to the feed side of the membrane while nitrogen was used in the permeate side as a sweeping gas to check the leakage across the membrane or the sealing material. Absence of helium in the permeate stream indicated good quality of the membrane and the seal. Pure hydrogen or hydrogen/helium mixture was used in the feed with flow rates generally in the order of 1,000 cc/min. The flow rates of sweeping nitrogen varied from 80 cc/min to about 380 cc/min to generate about 1% hydrogen compositions in the permeate stream. The data were obtained at pressures up to about 12 bar and temperatures up to 950°C.

Membrane Material Fabrication

Nd-doped $BaCeO_3$ was selected as the candidate membrane for testing because BCN ($BaCe_{0.9}Nd_{0.1}O_{3-x}$) was shown in the literature to be among the highest proton conductive materials of the perovskite [13]. Two BCN membranes, one unsupported and the other supported, were fabricated. The unsupported membranes (0.2 mm in thickness) was prepared by the tape casting method, followed by sintering at 1,450–1, 550°C for 2–3 hours. The supported membrane was prepared by a combination of the tape casting and the uniaxial pressing techniques. A thin (0.25 mm) membrane was first made by the tape casting process. Another thick (0.25–0.5 mm) membrane tape with 20vol.% of the organic pore former was then prepared as a membrane support. The two membrane tapes were pressed together to form a laminate. The laminate was then heated to 1,450–1,550°C to sinter and densify the thin membrane layer and create a porous support layer of about 0.33 mm with 31% porosity. The dense layer of the supported membrane sample was 0.2 mm.

Results and Discussion

Hydrogen Permeation Data for Perovskite Membrane

The hydrogen permeation testing results are shown in Figs. 7.4 and 7.5 for the unsupported and the supported BCN membranes respectively. For the supported membrane, the dense layer was facing the feed side in the experiment. The hydrogen flux for the unsupported membrane is slightly higher than the supported one probably due to the additional mass transfer resistance in the porous support layer. The hydrogen flux increases with the increasing hydrogen partial pressure in the feed and appears to reach a maximum at about six bar, after which the flux starts to drop. The reason why the flux decreases with the increasing hydrogen partial pressure will be discussed in the section of the modeling results below. The pressure probably affects the hydrogen flux through two mechanisms: (1) providing the driving force of the permeation by the hydrogen partial pressure difference across the membrane and (2) affecting the conductivity by the different proton and electron concentrations or proton diffusivities inside the perovskite membrane due to the different hydrogen pressures. This is the first time that the hydrogen permeation data at high pressures for the MPEC materials have been reported. Because of the higher operating pressures, the hydrogen flux generally is about one order of magnitude higher than those reported in the literature.

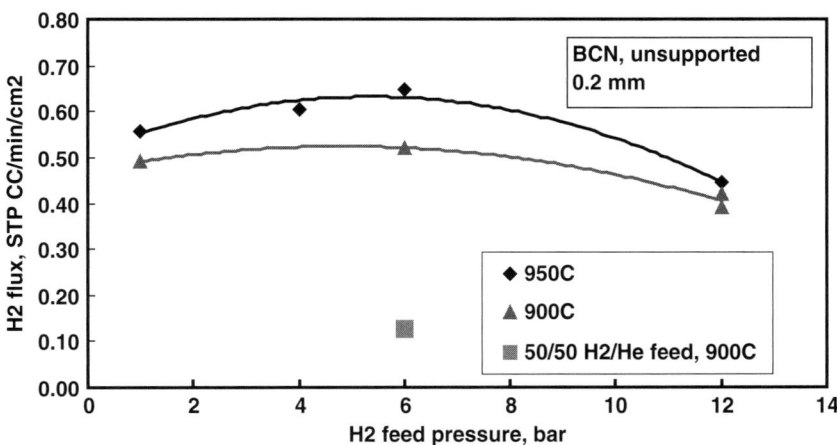

Fig. 7.4 Hydrogen fluxes measured from the high pressure permeation unit for the unsupported BCN membrane

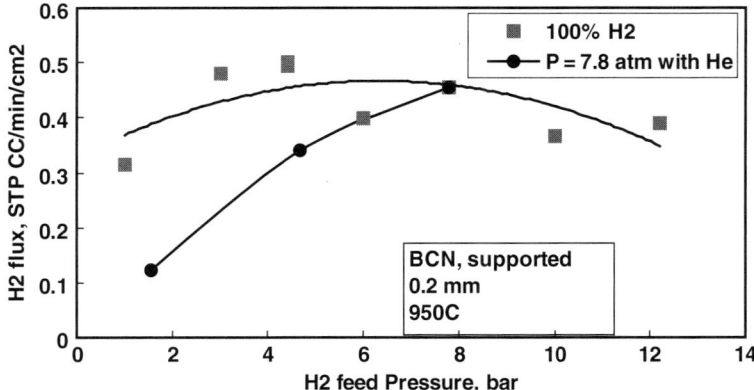

Fig. 7.5 Hydrogen fluxes measured from the high pressure permeation unit for the supported BCN membrane

Figure 7.5 also shows that the flux increases with the increasing hydrogen concentration in the feed side. This is simply due to the increasing hydrogen partial pressure difference across the membrane, which results in an increasing flux according to Eq. 7.23. The data were obtained at 7.8 bar with 20, 60 and 100% hydrogen in the feed side.

As expected, the hydrogen flux increases with the increasing temperature as shown in Fig. 7.6 for the unsupported BCN membrane. The calculated activation energy is about 11.8 Kcal/mole. An activation energy of 12 Kcal/mole for the proton conductivity of BCN material in the presence of steam was reported in the literature [14].

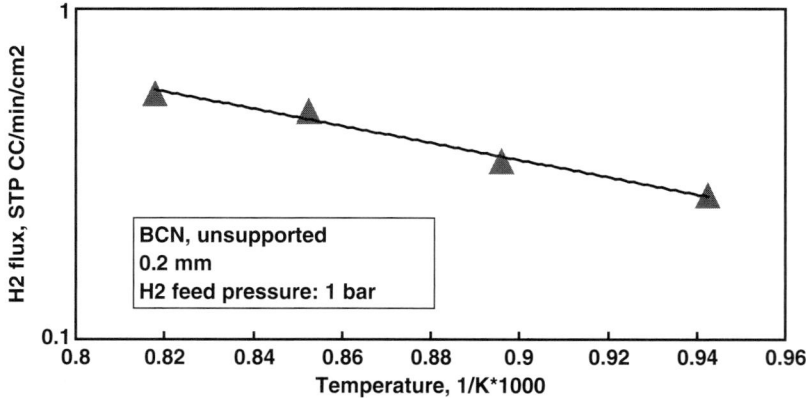

Fig. 7.6 Hydrogen permeation data for the unsupported BCN at different temperatures

Simulation Results for Hydrogen Transport in MPEC Membrane

Because the physical parameters such as the diffusivities and the equilibrium constants for the BCN membranes are not readily available in the literature, modeling analysis of hydrogen permeation through the MPEC membrane was carried out for the $SrCe_{0.95}Y_{0.05}O_{3-x}$ (SCY) perovskite membrane. The required physical parameters are taken from the literature [15–19] and are listed in Table 7.1.

The simulation results are first compared with the experimental hydrogen flux reported in [20] and are shown in Fig. 7.7. The feed was 4% hydrogen with the balance made of argon and the permeate side was maintained at 0.488% hydrogen with argon. The model appears to match the data quite well, considering the uncertainties of the parameters.

The concentration profiles of the 4 major defect species predicted from the model at 700°C are shown in Fig. 7.8. The feed is 4% hydrogen and the permeate is 0.488% hydrogen, the same conditions as in Fig. 7.7. Both the proton and the electron species dominate in the SCY membrane and the concentrations of the vacancy and the electron hole are very low. The results are reasonable because hydrogen permeation is mainly facilitated by both the proton and the electron while the vacancy and the electron hole are responsible for the oxygen transport. The proton and the electron concentrations decrease from the feed side to the permeate side as expected. Due to the low pressure operation, the proton concentrations generally are low,

Table 7.1 Equilibrium and diffusivity parameters used in the simulation

Equilibrium constant	Value, in mole/cc and atm	Diffusivity	Value, cm^2/sec
K_1 (Eq. 7.12)	5×10^{-6} [15]	Proton	$2.19 \times 10^{-3} \exp(-5339/T)$ [17]
K_2 (Eq. 7.13)	$4.12 \times 10^{-4} \exp(18884/T)$ [19]	Vacancy	$24.24 \exp(-23467/T)$ [18]
K_3 (Eq. 7.14)	$1.31 \times 10^{-3} \exp(29809/T)$ [16]	Electron hole	$49.38 \exp(-12589/T)$ [18]
K_e (Eq. 7.18)	$1.0 \times 10^{-6} \exp(-23188/T)$ [16]	Electron	$49.38 \exp(-12589/T)$ [18]

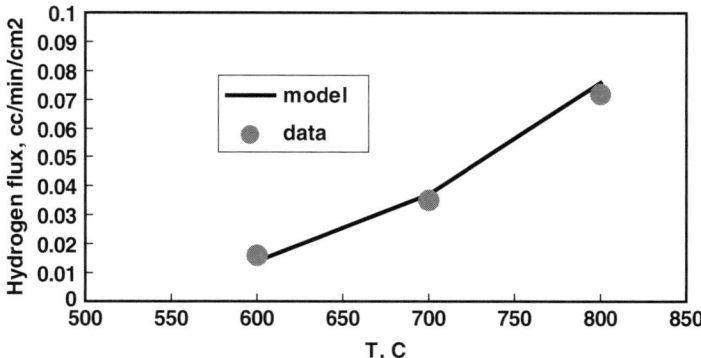

Fig. 7.7 Comparison of simulation results with the literature data [20]

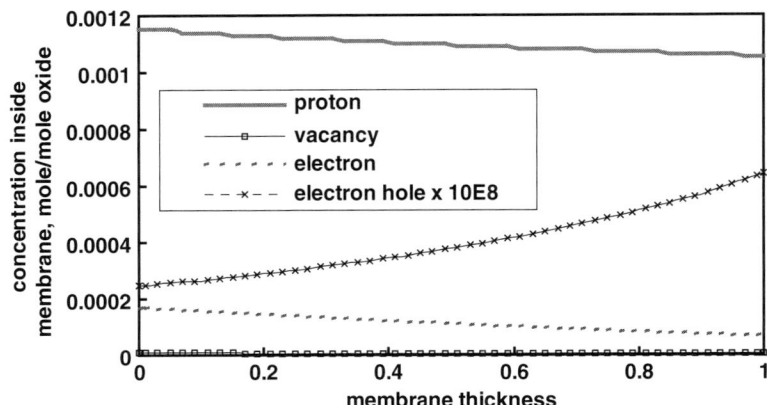

Fig. 7.8 Concentration profiles for the four defect species, proton, vacancy, electron and electron hole inside a SCY membrane

below 1.2×10^{-3} mole per mole of perovskite or 2.4×10^{-5} mole/cm^3, with a membrane density of 5.85 g/cm^3.

This model was applied to simulate the high pressure operating conditions for the SCY membrane. At 700°C, with pure hydrogen in the feed and the permeate stream kept at 1% hydrogen, the hydrogen fluxes at different hydrogen feed pressures are shown in Fig. 7.9. The fluxes continue increasing with the pressure without the maximum value as observed from the experimental data in Figs. 7.4 and 7.5. As mentioned, the hydrogen flux is dependent on $\ln(p_{H_2}^f) - \ln(p_{H_2}^p)$ according to Eq. 7.23. To eliminate the effect of $\ln(p_{H_2}^f) - \ln(p_{H_2}^p)$ on the hydrogen flux, the ambipolar conductivities versus the hydrogen partial pressures in the feed from the modeling results of the SCY membrane are plotted in Fig. 7.9. Likewise, the calculated ambipolar conductivities from the experimental data of the BCN membrane are shown in Fig. 7.10. Both Figs. 7.9 and 7.10 indicate that the hydrogen flux for the perovskite membrane cannot correlate with $\ln(p_{H_2}^f) - \ln(p_{H_2}^p)$ in a linear way as in Eq. 7.23. The conductivity of the proton/electron depends on the hydrogen partial pressure. The conductivity of the defect species such as proton or electron depends on the concentration and the diffusivity as expressed in Eq. 7.10. Moreover, the concentration of the proton inside the perovskite membrane is dependent on the hydrogen partial pressure according to Eqs. 7.12–7.14. Consequently, the conductivity of the proton/electron is pressure dependent. However the model cannot predict the trend of the data. The ambipolar conductivities from the model in Fig. 7.9 increase monotonically with the increasing hydrogen pressures, while the data in Fig. 7.10 shows a maximum conductivity.

One of the possible reasons that the model cannot predict the decreasing conductivity with the increasing hydrogen partial pressure in the high pressure region could be the inadequacy of Eq. 7.13. According to Eq. 7.13 the proton and the electron concentrations increase with the increasing hydrogen pressure. At an extremely

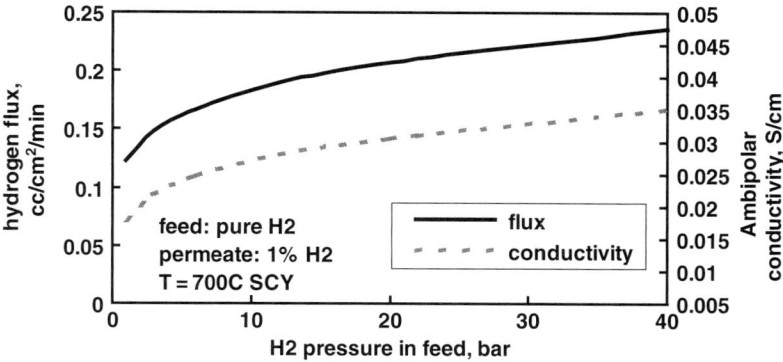

Fig. 7.9 Hydrogen flux and ambipolar conductivity at different hydrogen feed pressure from the modeling results

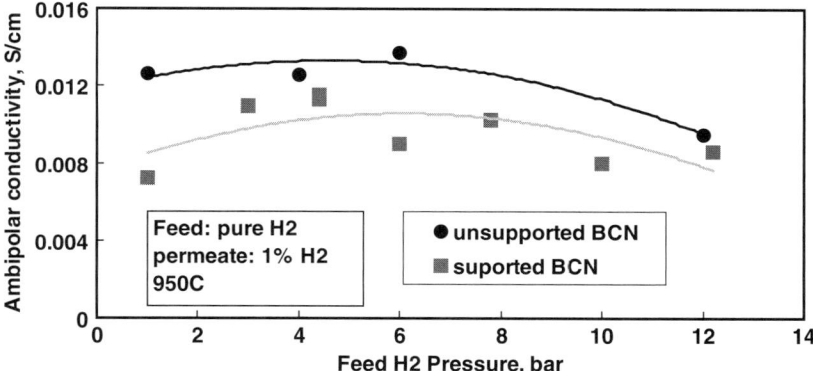

Fig. 7.10 The ambipolar conductivity as calculated from Eq. 7.26 based on the experimental data of Figs. 7.4 and 7.5

high hydrogen partial pressure, the proton and the electron concentrations would also reach infinitely high, which is unreasonable. Hydrogen concentrations in the SCY membrane in the presence of steam were measured by Yajima and Iwahara [21]. Their data showed that the proton concentrations or the water vapor solubility appeared to reach a saturation limit at the very high partial pressure of water. Assuming that the hydrogen solubility in the SCY membrane follows a Langmuir-type isotherm, as a first approximation, Eq. 7.13 can be rewritten as

$$K_2 = \frac{K_2'}{1 + V p_{H_2}} = (C_{OH}^2 C_e^2)/p_{H_2} \qquad (7.26)$$

Eq. 7.26 also implies that K_2 is a function of hydrogen pressure. At the low hydrogen pressure, $C_{OH}^2 C_e^2$ is a linear function of $K_2(= K_2')$, and at the very high pressure, it is $C_{OH}^2 C_e^2 = \frac{K_2'}{V}$. Using an arbitrary value of 10 for V, the effect of

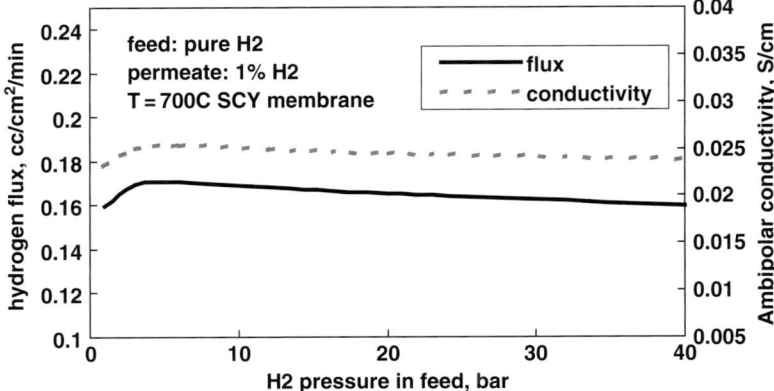

Fig. 7.11 Hydrogen flux and ambipolar conductivity at different hydrogen feed pressure from the modeling results based on Eq. 7.26

hydrogen partial pressure on the flux of the SCY membrane was calculated based on Eq. 7.26 instead of Eq. 7.13 in the model. The simulation results, along with the ambipolar conductivities are plotted in Fig. 7.11. In comparison with Fig. 7.9, a maximum value of hydrogen flux or conductivity can be seen, which indicates that the limited hydrogen solubility in the SCY membrane could be one reason that the ambipolar conductivity decreases with the increasing hydrogen pressure. It should be noted that the data for Figs. 7.4 and 7.5 and the simulation results for the Figs. 7.9 and 7.11 were obtained with pure hydrogen in the feed and 1% hydrogen in the permeate side. Therefore, the value of $\ln(p_{H_2}^f) - \ln(p_{H_2}^p)$ was constant and the flux only followed the change of the ambipolar conductivity. In general, the hydrogen flux increases with $\ln(p_{H_2}^f) - \ln(p_{H_2}^p)$ according to Eq. 7.23, except that the rate of the increase, or the ambipolar conductivity, will first increase then decrease with the increasing hydrogen pressure.

Conclusion

Mixed protonic and electronic conducting membranes are ideally suited for the high temperature membrane reactor applications for hydrogen production from coal/biomass gasification systems. We have measured hydrogen permeation rate for the supported and the unsupported BCN membranes at temperatures of up to 950°C and pressures of up to 12 bar in the high pressure permeation unit. The hydrogen flux increases with the increasing hydrogen pressure in the feed up to about 6 bar, after which the flux starts to decline. Modeling results seem to indicate that in the high pressure region (> 6 bar), the ambipolar conductivity could decrease with the increasing hydrogen partial pressure due to the finite solubility of hydrogen in the perovskite membrane. More research works are needed to improve the understanding of the fundamental transport mechanism for the MPEC membranes.

Acknowledgments The authors would like to acknowledge the support of the U.S. Department of Energy, under Award No. DE-FC26-03NT41851. However, any opinions, findings, conclusions, or recommendations expressed herein are those of the authors and do not necessarily reflect the views of the DOE.

References

1. Parsons infrastructure and technology group, inc. Hydrogen production facilities plant performance and cost comparisons. Final Report DOE Contract DE-AM26-99FT40465, March 2002.
2. Gary D, Mitretek GT. Hydrogen from coal. MTR 2002-31, DOE Contract DE-AM26-99FT40465, July 2002.
3. Probstein RF, Hicks RE. Synthetic fuels. New York: McGraw-Hill Inc. 1982.
4. Doong SJ, Ong E, Lau F, Bose AC, Carty R. Direct extraction of hydrogen from coal using a membrane reactor within a gasifier. Paper presented at 21st International Pittsburgh Coal Conference, Osaka, Japan, Sep 2004.
5. Iwahara H, Esaka T, Uchida H, Maeda N. Proton conduction in sintered oxides and its application to steam electrolysis for hydrogen production. Solid State Ionics. 1981;3/4:359.
6. Roark SE, Mackay R, Sammells AF. 19th Annual Pittsburgh Coal Conference, 2002;43–2.
7. Balachandran U, Lee TH, Wang S, Zuo C, Dorris SE. 20th Annual Pittsburgh Coal Conference, Pittsburgh, PA. 15–19 Sep 2003.
8. Bouwmeester HJM, Burggraaf AJ. Dense ceramic membranes for oxygen separation. In: Burggraff AJ, Cot L. editors. Fundamentals of inorganic membrane science and technology. New York: Elsevier Science B.V.; 1996. pp. 435–528.
9. Hamakawa S, Lin L, Li A, Iglesia E. Synthesis and hydrogen permeation properties of membranes based on $SrCeYbO_3$ thin films. Solid State Ionics. 2002;48:71.
10. Qi X, Lin YS. Electrical conduction and hydrogen permeation through mixed proton-electron conducting strontium cerate membranes. Solid State Ionics. 2000;130:149.
11. Hamakawa S, Hibino T, Iwahara H. Electrochemical hydrogen permeation in a proton-hole mixed conductor and its application to a membrane reactor. J Electrochem Soc. 1994;141(7):1720.
12. Wachsman ED, Jiang N. US Patent 6,296,687, 2 Oct 2001.
13. Iwahara H, Uchida H, Ono K, Ogaki K. Proton conduction in sintered oxides based on $BaCeO_3$. J. Electrochem. Soc.: Solid-State Sci Technol. 1988;135(2):529.
14. Liu JF, Nowick AS. The incorporation and migration of protons in nd-doped $BaCeO_3$. Solid State Ionics. 1992;50:131.
15. Song SJ, Wachsman ED, Rhodes J, Dorris SE, Balachandran U. Numerical modeling of hydrogen permeation in chemical potential gradients. Solid State Ionics. 2003;164:107–16.
16. Schober T, Schilling W, Wenzl H. Defect model of proton insertion into oxides. Solid State Ionics. 1996;86–88:653–8.
17. Schober T, Friedrich J, Condon JB. Effective hydrogen diffusivity in $SrCe_{0.95}Yb_{0.05}O_{3-a}$ and $SrZr_{0.95}Yb_{0.05Y}b0.05O_{3-a}$. Solid State Ionics. 1995;77:175–9.
18. Tan X, Liu S, Li K, Hughs R. Theoretical analysis of ion permeation through mixed conducting membranes and its application on dehydrogenation reactions. Solid State Ionics. 2000;138:149–59.
19. Krug F, Schober T, Springer T. In situ measurements of the water uptake in yb doped SrCeO3. Solid State Ionics. 1995;81:111–8.
20. Guan J, Dorris SE, Balachandran U, Liu M. Transport properties of SrCeYO3 and its applications for hydrogen separation. Solid State Ionics. 1998;110:303–10.

21. Yajima T, Iwahara H. Studies on behavior and mobility of protons in doped perovskite-type oxides (I) in-situ measurement of hydrogen concentration in $SrCe_{0.95}Yb_{0.05}O_{3-a}$ at high temperatures. Solid State Ionics. 1992;50:281–6.
22. Li L, Iglesia E. Modeling and analysis of hydrogen permeation in mixed proton-electronic conductors. Chem Eng Sci. 2003;58:1977–88.

Part II
Metal and Alloy Membranes

Chapter 8
Hydrogen Separation Using Dense Composite Membranes: Part 1 Fundamentals

Michael V. Mundschau

Introduction

This chapter reviews some fundamental science critical for the understanding, development and operation of many classes of dense composite inorganic membrane used for transport of hydrogen. A companion paper follows in this volume discussing some of the engineering issues of membrane scale-up.

By the term dense, it is implied that there are no intentional interconnected pores in the membranes other than atomic interstices, atomic vacancies and dislocations. Such void spaces are too small to accommodate even molecular hydrogen, and dense membranes, of the type reviewed, transport hydrogen only in a dissociated form. Dense membranes block transport even of helium, and the absence of larger pores gives dense membranes hydrogen selectivity approaching 100%. Transport of hydrogen in a dissociated form implies that dense membranes must possess adequate catalytic activity for the adsorption and dissociation of H_2 on the feed-side surface (retentate) as well as for the subsequent recombination and desorption from the permeate-side surface.

The term composite membrane, as defined in membrane technology, refers to membranes with two or more distinct layers. A layer within a composite membrane could itself be a composite material possessing two or more distinct components, as in membranes employing a layer of palladium cermet (ceramic-metal) supported by a layer of porous ceramic. The layers need not be composite, as in membranes using films of palladium on both sides of foils of niobium, tantalum, vanadium or zirconium.

A classic example of a composite membrane is that patented in 1916 by Snelling who used porous ceramics to support dense layers of palladium, 25 μm thick [1]. Variations of his theme remain at the forefront of research [2, 3]. Examples include use of porous alumina, silica, perovskites, and stainless steel to support thin layers of Pd and its alloys [2, 3]. Snelling addressed issues of perforations and pinholes

M.V. Mundschau (✉)
Eltron Research & Development Inc., 4600 Nautilus Court South, Boulder, CO 80301-3241, USA
e-mail: mundschau@eltronresearch.com

A.C. Bose (ed.), *Inorganic Membranes for Energy and Environmental Applications*,
DOI 10.1007/978-0-387-34526-0_8, © Springer Science+Business Media, LLC 2009

by electrodepositing thin layers of Pd onto relatively thick and robust foils of non-porous copper. The thick Pd/Cu laminates, which were much less susceptible to damage from handling, were then attached to the porous supports. After assembly, the foil of copper was acid etched and dissolved to expose the underlying Pd. Snelling patented membranes to extract hydrogen from coal-derived producer gas, recommending operation above at least 800°C [1].

Another class of composite membrane, patented by Makrides, Wright and Jewett, in 1967 employs exceptionally thin layers (100–1000 nm) of palladium and its alloys (Pd-Ag, Pd-Au, Pd-B) deposited onto both sides of dense foils of the highly hydrogen permeable metals, niobium, tantalum and vanadium [4]. Later variations used substrates of zirconium [5, 6] or titanium and its alloys [7]. The coatings of Pd function as hydrogen dissociation catalysts and also protect the reactive metal substrates from oxidation and other chemical reactions. Dense composite membranes of this type have been used in the nuclear industry to separate hydrogen and its isotopes from helium and other gases [5, 8]. For example, Buxbaum and Kinney discuss use of Pd-coated Zr to separate deuterium from helium [5]. They also suggested the use of Pd-coated Nb and Ta heat exchanger tubes for use in large-scale hydrogen separation systems [5]. These authors state that such membranes have very high tolerance to pinhole defects in the Pd because the non-porous substrates, usually greater than 100 μm thick, are very unlikely to possess pinholes and will thus block transport of other gases [5]. Such composite membranes, now using improved Pd alloys as catalysts and novel alloys of Group IVB (Zr, Ti, Hf) and Group VB (V, Nb, Ta) elements as substrates, are also at the forefront of membrane research [8, 9].

Composite membranes also employ dense cermets fabricated by sintering together mixed powders of metal and ceramic [10–12]. Examples include powders of Pd and its alloys sintered with powders of perovskites [11, 12], niobium sintered together with Al_2O_3 [12], and nickel sintered with proton-conducting perovskites. Layers of dense cermets, 25–100 μm thick, are supported by porous ceramic tubes. Cermets employing chemically reactive metals, Nb, Ta, Ti, V, Zr, and their alloys, are typically coated with Pd and alloys thereof [11, 12].

Dense composite membranes also include a class of material formed by sintering together powders of two ceramics. For many proton-conducting perovskites [13, 14], proton transport is limited by electron transport. In Ni-perovskite cermets, for example, electron transport is augmented by the metal phase. Alternatively, electron transport can be provided by an electron-conducting perovskite phase. In ceramic-ceramic composites, as well as in cermets, powders are mixed to contain approximately 40–60 volume percent of each phase. According to percolation theory, this gives a high probability that both phases will form continuous matrices in the composite material, ensuring conducting pathways for both protons and electrons.

In yet another variation, composite membranes are fabricated by sintering together powders of highly hydrogen permeable metals, Pd, Nb, Ta, Ti, V, Zr and their alloys, with powders of a second metal or alloy that is non-permeable to hydrogen [12]. The function of the non-permeable metal is to provide mechanical support for the hydrogen transport materials, especially if the latter are to be operated under conditions that might lead to hydrogen embrittlement. Although one may deem these all-metal systems as alloys, they are classified here as composite materials because of their method of fabrication and composite microstructure. For the reactive metals, protective catalytic layers are required on both sides of the membranes.

Guide to Selection of Materials for Fabrication of Composite Membranes

In the design of composite membranes containing metallic elements, it is fruitful to consider the equations for hydrogen permeability compiled by S.A. Steward at Lawrence Livermore Laboratory [15]. Selected equations of Steward are plotted in Fig. 8.1. Hydrogen permeability, P, is derived from $P = D \cdot S$, where D is the diffusivity and S is the hydrogen solubility.

Figure 8.1 shows that hydrogen permeability of Nb, V and Ta far exceeds that of Pd, which has hitherto been one of the most successful membrane materials [2, 3]. In accord with values calculated by Steward, many workers [5, 8, 9, 11,16–24] have reported permeabilities exceeding 1×10^{-7} mol $\cdot$ m^{-1} $\cdot$ s^{-1} $\cdot$ Pa$^{-0.5}$. These are partially summarized in Table 8.1. Buxbaum and Marker [16] and Peachey, Snow and Dye [17] also plot permeability of Zr and show that it can exceed that of Nb above about 550°C. Hill plots the permeability of β-Ti (the body centered cubic allotrope of Ti) in an alloy of Ti-13V-11Cr-3Al (mass%), which exceeds that of unalloyed Pd in the range of 300–425°C [7]. These membranes have been used in the nuclear industry [5, 7] and were operated continuously for more than 10,000 hours (416 days) according to Buxbaum and Kinney [5]. Buxbaum et al. also discuss use of membranes of V-4Cr-4Ti and V-7.5Cr-15Ti (mass%) for isotope extraction and for extraction of hydrogen from synthesis gas at 430°C [18]. Addition of Ti to V increases hydrogen solubility in the alloy relative to unalloyed vanadium [18].

According to the solution-diffusion theory [9], hydrogen permeability, $P = D \cdot S$, depends both on the diffusivity, which gives an indication of how rapidly hydrogen moves through the crystal lattice and the solubility, which is related to the number of hydrogen atoms that can be accommodated by the lattice. It is instructive to consider each of these factors in order to understand the underlying features contributing to high hydrogen permeability of materials and as a guide to production of new materials.

Diffusivity and diffusion coefficient are synonymous [25]. In dense membranes, diffusivity is associated with the rate of movement of dissociated hydrogen from site to site within a crystal lattice. In general, diffusivity of hydrogen is greater in metals with the body centered cubic structure (bcc) relative to metals with the face centered cubic structure (fcc) [26]. According to Wipf, dissociated hydrogen occupies tetrahedral interstitial sites in bcc metals and hops between such sites. Tetrahedral interstitial sites are only 1.01–1.17 Å apart in common bcc metals [26]. This relatively short distance permits quantum mechanical tunneling, which accord-

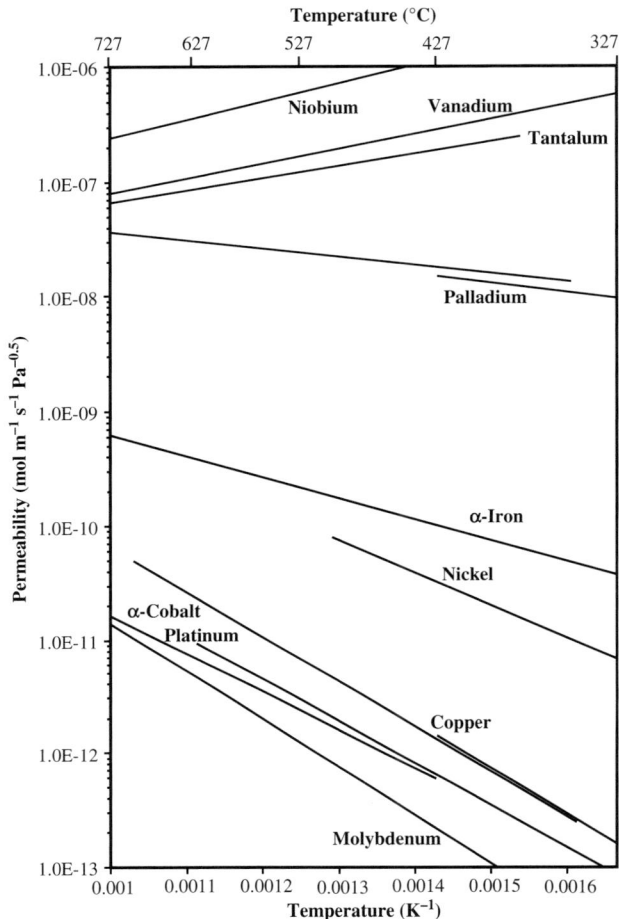

Fig. 8.1 Plots of hydrogen permeability of selected elements based upon the equations of Steward [15] (Copyright Wiley-VCH Verlag, GmbH & Co. KGaA, 2006. Adapted with permission from [8], Nonporous Inorganic Membranes.)

ing to Wipf is largely responsible for the high diffusivity of hydrogen in bcc metals [26]. For reference, the theoretical Bohr radius of a ground-state hydrogen atom is 0.5292 Å yielding an atomic diameter of 1.0584 Å. In common metals with the fcc structure, hydrogen occupies octahedral interstitial sites with nearest-neighbor distances ranging from 2.5 to 2.9 Å, making quantum mechanical tunneling less probable and diffusion between such sites more difficult [26].

According to Wipf, vanadium (together with iron) has the highest hydrogen diffusivity of all metals [26]. Quantum mechanical tunneling has been verified for hydrogen in V using deuterium and tritium isotopes [26]. It is noteworthy that V is practically transparent to thermal neutrons, which have nearly the same rest mass and De Broglie wavelengths as that of atomic hydrogen.

Table 8.1 Measured values of hydrogen permeability for membranes based on group IVB-VB elements

Membrane Material	Temperature (°C)	Permeability (mol m^{-1} s^{-1} Pa$^{-0.5}$)	Reference
Zr	350	1×10^{-6}	Buxbaum, Marker [16]
V	250	5.0×10^{-7}	Ozaki, Zhang, Komaki, Nishimura [19]
Nb Tube	>420	3.6×10^{-7}	Buxbaum, Kinney [5]
Nb	420	$3.26–4.26 \times 10^{-7}$	Balachandran [22]
Group IV-VB Composite	440	3.2×10^{-7}	Mundschau, Xie, Sammells [11]
Nb	425	3.2×10^{-7}	Buxbaum, Marker [16]
V-10Al	250	2.0×10^{-7}	Nishimura, Ozaki, Komaki, Zhang [20]
V	300	1.48×10^{-7}	Moss, Peachey, Snow, Dye [23]
Ta Tube	420	1.45×10^{-7}	Buxbaum, Kinney [5]
Ta	350–420	$1.07–1.45 \times 10^{-7}$	Rothenberger, Howard, Killmeyer, Cugini, Buxbaum, et al. [24]
Ta Disk	420	1.0×10^{-7}	Buxbaum, Marker [16]
V-10.5Ni-4.5Al	350	6.29×10^{-8}	Nishimura, Ozaki, Komaki, Zhang [21]

In the case of iron, high diffusivity allows hydrogen to rapidly diffuse even at room temperature. This can lead to well-known hydrogen embrittlement of bcc iron and its alloys. However, as shown in Fig. 8.1, the hydrogen permeability of iron is relatively poor despite its exceptionally high diffusivity. This is attributed to the relatively low hydrogen solubility of iron [27].

The diffusivity of hydrogen and its isotopes in Nb is also extremely high, relative to most elements, but is slightly lower relative to V [26]. The higher permeability of niobium at higher temperatures (Fig. 8.1), therefore, is attributed to higher hydrogen solubility of niobium at higher temperatures.

Peachey, Snow and Dye also attribute the high permeability of Ta, relative to that of Pd, as due to the much higher solubility of hydrogen in Ta [28]. Tantalum also has diffusivity greater than that of Pd, just below that of Nb [26], so that both factors contribute to the higher permeability of Ta over Pd.

The solubility of hydrogen in zirconium and α-titanium (hexagonal structure) far exceeds that of Nb, Ta and V, according to Smithells, who summarized much of the pioneering work of Sieverts [27]. At 300°C, Smithells lists the solubility of Ti as 40,000 cm^3 of hydrogen per 100 g of metal and that of Zr as 27,000. For comparison, V is listed as 6,000 and Pd as only 330 cm^3 of hydrogen per 100 g of metal at 300°C [27]. The exceptionally high hydrogen solubility of Ti and Zr is largely responsible for their very high permeabilities [5–7].

For a wide range of temperatures, the metals, Ti, Zr, Nb, V and Ta, have a much higher hydrogen solubility relative to Pd [27], which in turn far exceeds that of most common metals [2, 3]. It is noteworthy, from the early work of Smithells that the rare earth elements, for example lanthanum, neodymium, cerium, and radioactive elements such as thorium, were shown to have very high solubility for hydrogen exceeding that of Nb, Ta and V [27]. Hydrogen solubility in the range of 300–400°C (of interest for membranes placed downstream from water-gas shift reactors) follows the order: Ti > Zr > La > Ce > Th > Nb > V > Ta > Pd [27].

Metals with extremely high hydrogen solubility form solid solutions in which the hydrogen-to-metal atomic ratios can approach stoichiometric compositions [27]. This is far in excess of ideal solid solutions of non-hydride forming metals that might ordinarily contain only about one atom of hydrogen for every 1,000 atoms of metal, according to Smithells [27]. In hydrides of the solid-solution type, mobile hydrogen atoms occupy interstitial sites as in ideal solutions, and the lattice constants can be little changed from that of the host lattice. The hydride phases forming solid solutions are non-stoichiometric [29, 30]. However, phase transformations between the solid-solution type of hydride, which retains the crystal structure of the host metal lattice, to ordered hydrides (often of different crystal structure) can cause large changes in the volume of a membrane material and, in general, must be avoided. Ordered hydrides are often brittle and salt-like and can lead to membrane failure. Understanding of hydride formation and their phase changes are critical for composite membrane technology.

Considering Pd, its lower hydrogen permeability, relative to the metals discussed above (see Fig. 8.1), is due both to its lower hydrogen solubility as well as to its lower diffusivity. Lower diffusivity is attributed to longer hopping distance between fcc octahedral interstitial sites, greater activation barrier between sites, and low probability for quantum mechanical tunneling. The high density of fcc palladium limits hydrogen solubility relative to the more open bcc metals.

The permeability of Pd can be improved by alloying [2, 31–34]. For example, according to McKinley, alloying with silver to produce a composition, nominally Pd-27Ag (mass%), improves permeability 1.40 times relative to unalloyed Pd [32, 34]. This composition contains three atoms of Pd for each atom of silver (Pd_3Ag) and is equivalent to Pd-25Ag (atom%). The alloy containing nominally Pd-10Ag (mass%) improves permeability 1.76 times, according to McKinley [32, 34]. Alloys of Pd-5Au (mass%) improve permeability by 1.07 times [32, 34].

Both silver and gold form ideal solid solutions with palladium. However, stoichiometric compositions with unique properties, such as in Hunter's preferred membrane composition of Pd_3Ag, [31], might suggest the possibility of intermetallic compounds or ordered structures differing from that of the ideal solutions [35]. Palladium and copper also form ideal solid solutions, but in this system phase diagrams clearly show additional phases with crystal structures differing from the parent fcc phase of the solid solutions.

The cube edge of elemental Pd is 3.89 Å (see Table 8.2); that of Ag and Au are larger, 4.09 Å and 4.08 Å respectively [36]. From x-ray diffraction, it is shown that the addition of Ag or Au to Pd continuously expands the palladium lattice from its unalloyed cube edge of 3.89 Å to a maximum of 4.08–4.09 Å for pure Au and Ag, respectively [36, 37]. If Vegard's Law [37] is obeyed, then adding 25 atom% Ag to palladium, as in Hunter's preferred membrane composition [31], would increase the Pd lattice parameter by $(0.25)(4.09–3.89 \text{ Å}) = 0.05$ Å, yielding a cube edge of 3.94 Å, close to the experimentally quoted value of 3.93 Å.

Table 8.2 Some useful lattice parameters

Material	Structure	Cube edge (Å)
Pd	fcc	3.89
Pd-Au	fcc	3.88–4.08
Pd-Ag	fcc	3.88–4.09
Pd-H	–	3.88–4.05
Cu	fcc	3.61
Pt	fcc	3.92
Au	fcc	4.08
Ag	fcc	4.09
Fe	bcc	2.87
Cr	bcc	2.88
Pd-Cu	bcc	2.99
V	bcc	3.04
Mo	bcc	3.15
W	bcc	3.16
Nb	bcc	3.30
Ta	bcc	3.30
ß-Ti	bcc	3.33
U	bcc	3.44

Increase in the lattice parameters of Pd increases the octahedral interstitial distances and thus the hopping distance for hydrogen. Züchner has shown that the hydrogen diffusion coefficients of Pd-Ag alloys decrease with increasing silver content [38]. Thus, the improved permeability of Pd-Ag is not due to increased diffusivity but rather to increased solubility. Increased solubility has been shown for Pd-Ag [2]. In general, permeability of Pd is improved by addition of elements that increase hydrogen solubility, both by expanding the lattice and by forming more stable hydrides.

In addition to increasing solubility, alloying of Pd with elements expanding the lattice can reduce damage caused by the α-β hydride phase transformation. The crystal structure of the palladium α-hydride phase is simply that of a solid solution of dissociated hydrogen occupying octahedral interstitial sites of the fcc phase of Pd [30]. The lattice parameters of the α-hydride phase differ little from the pure Pd parent phase [30]. According to Darling, expansion and formation of the brittle β-hydride phase is the mechanism for failure of Pd membranes lowered below the α-β hydride phase transformation temperature in the presence of hydrogen [39]. According to Darling, the palladium α-hydride phase has a cube edge of 3.894 Å throughout its composition range, but swells to a structure with cube edge of 4.018 Å as the brittle β-hydride phase forms [39]. Using these parameters, calculations show that the cube edge increases by about 3% and the volume by about 9% during the phase change. For reference, this volume change exceeds even that of water, which expands about 8% when it freezes.

If unalloyed palladium is to be used in membranes, Darling recommends use at temperatures no lower than 310°C and preferentially no lower than 350°C, which are well above the critical temperature for the α-β phase transformation [39]. As indicated by Pd-H phase diagrams, the transformation occurs just under 300°C in unalloyed Pd [2, 27, 29, 38, 40]. Above 300°C, only the α-phase is found for hydrogen partial pressures up to 1,000 atm [30]. For Pd membranes, Darling recommends a very thorough purging of systems from hydrogen before membranes can be safely lowered below the transformation temperature. This advice remains valid for composite membranes employing palladium in forms that can undergo large chemical expansion. Similar rules apply for Nb, Ta, Ti, V and Zr.

Another advantage of alloying is that compositions such as Pd-23Ag (atom%) can reduce the critical temperature of the hydride phase change to near room temperature, according to Paglieri and Way [2]. A similar lowering of the critical temperature from just below 300°C in unalloyed Pd to near room temperature is observed for Pd-19Pt (atom%) [2]. This, in principle, allows membranes of such alloys to be operated well below 300°C without chemical expansion and hydrogen embrittlement.

It is noteworthy that Pd combined with Sc, Y and the lanthanide elements form ordered intermetallic compounds with the Cu_3Au structure that have cube edges in the range of 3.981–4.114 Å (see Table 8.3) [35]. In principle, these lattices are capable of better accommodating the palladium β-hydride. The lanthanides, along with scandium and yttrium, form hydrides that increase hydrogen solubility and thus hydrogen permeability relative to unalloyed Pd [27, 30]. Hydrogen permeability through some Pd-lanthanide alloys is more than double that of the best Pd-Ag alloys. On the negative side, lanthanide hydrides react with water and possible oxidation of the chemically reactive lanthanide elements must be considered if used in membranes.

Alloying of Pd with Cu is complex, due to the various phases that form. The cube edge of elemental fcc Cu is 3.61 Å, which is smaller than that of elemental Pd. Addition of copper, for compositions and temperatures that retain the fcc structure and that persists as ideal solutions, contracts the Pd lattice and lowers hydrogen solubility and permeability [32, 34]. However, Pd-Cu phase diagrams indicate

Table 8.3 Lattice parameters of Pd_3R compounds

System	Cube edge (Å) Cu_3Au structure
Pd_3Sc	3.981
Pd_3Y	4.068
Pd_3Ce	4.114
Pd_3Sm	4.110
Pd_3Eu	4.087
Pd_3Gd	4.081
Pd_3Dy	4.076
Pd_3Ho	4.064
Pd_3Er	4.056
Pd_3Yb	4.030
Pd_3Lu	4.028

Adapted from Savitsky, Polyakova, Gorina and Roshan [35].

that a body centered cubic structure with a cube edge of 2.994 Å [36] forms for temperatures below 598°C and for compositions centered around approximately 40 atomic percent Pd and 60 atomic percent Cu (40Pd–60Cu (atom%)). It is noteworthy that the bcc form of Pd–Cu has a cube edge intermediate to that of V (3.04 Å) and Fe (2.87 Å), both of which allow quantum mechanical tunneling.

The composition, 40Pd-60Cu (atom%) is equivalent to 53Pd-47Cu (mass%), and compositions within ± a few percent of this central value will have only the bcc crystal structure, if the system is brought to equilibrium. The much studied 60Pd-40Cu (mass%) composition, which is equivalent to 47.2Pd-52.8Cu (atom%), is just beyond the phase boundary of the single-phase bcc region and at equilibrium will contain a mixture of crystallites, some with the fcc and some with the bcc structure at temperatures from 300°C to above 500°C. For a fixed composition, the ratio of bcc/fcc crystallites in the material will depend upon the temperature; more bcc phase favored at lower temperature. Higher concentration of Pd beyond 60 mass% will favor the fcc structure. The 60Pd-40Cu (mass%) composition has permeability 1.08 times greater than that of unalloyed Pd [32, 34].

Readers should carefully note the distinction between 40Pd-60Cu (atom%) and 60Pd-40Cu (mass%). The former is in the center of the bcc phase region whereas the latter is just beyond the phase boundary and in the region having mixed crystallites, some with bcc and some with fcc crystal structure. Lowering the Pd concentration to slightly less than 60 mass% will place the material in the region having only the bcc crystal structure. Some authors use mass% and some atom%. Confusion can arise if this is not carefully noted.

For the stoichiometric composition of Pd–50Cu (atom%), the ordered Pd_1Cu_1, bcc structure is classified as having the more-specific CsCl crystal structure [30]. However, the disordered stoichiometric phase and compositions deviating from exact stoichiometry are better classified as having the body centered cubic structure. The Pd-50Cu (atom%) composition corresponds to Pd-37.39Cu (mass%). This places the 1 : 1 stoichiometric composition well into the region having mixed crystallites of bcc and fcc structure.

As in the case of Pd, other highly permeable metals are also alloyed to alter diffusivity, solubility, chemical expansion and mechanical properties [9]. According to Mackay, the highly permeable elements, Nb, Ta and V all form α-phase hydrides, which are solid solutions with hydrogen occupying tetrahedral interstitial sites [30]. A variety of complex hydrides can form at lower temperatures, which greatly expand the lattices [30, 41]. According to Mackay, the β-phase hydride of V disappears above about 200°C and the α + β-phase region of Nb disappears above 140°C [30].

As with Pd, these metals can be alloyed to lower the phase transformation temperatures and to inhibit nucleation of brittle hydrides. For example, according to Nishimura et al., the addition of nickel to vanadium to form V-15Ni (atom%), reduces the linear expansion from 3.3% in unalloyed V to 1.3% in the alloys under the same operating conditions [42]. This is attributed to reduced hydrogen solubility in the vanadium-nickel alloys [42].

Nishimura, Komaki and Amano find that the addition of molybdenum expands the vanadium lattice linearly, in accordance with Vegard's Law [43]. The hydrogen solubility and permeability are found to drop [43]. Zhang et al. find that aluminum, which has the fcc structure, also expands the vanadium lattice [44] and also lowers hydrogen solubility and permeability [20]. Expansion of the lattice and increased distance between interstitial sites is expected also to decrease the hydrogen diffusivity so that lower permeability is attributed to both lower solubility and lower diffusivity [44]. Cobalt and titanium expand the lattice of vanadium as well [45].

According to Nishimura et al., addition of Ni, Co, Fe, and Cr to vanadium lowers hydrogen solubility [42]. In part, lower hydrogen solubility might be attributed to diluting the vanadium lattice with weak or non-hydride forming elements. Contraction of the lattice can lower hydrogen solubility as well. In contrast, addition of Ti, which both forms stable hydrides and increases the lattice parameters, increases the hydrogen solubility in V-Ti alloys [18, 42]. On the negative side, greater hydrogen solubility results in greater chemical expansion relative to that of unalloyed V—as high as 4.5% for some V-Ti alloys [18].

If Vegard's Law is obeyed, the lattice parameters as summarized in Table 8.2 imply that W, Nb, Ta and U also expand the lattice of vanadium. The non-hydride forming element W may reduce solubility and permeability as in the case of Mo [43]. The elements Nb, Ta and U that form strong hydrides, and which increase lattice parameters as well, should increase solubility.

If it is assumed that the lattice parameters of unalloyed vanadium (and iron) are near the optimum for quantum mechanical tunneling and, therefore, hydrogen diffusivity, then expanding the lattice of vanadium will decrease the diffusivity and thus the permeability unless solubility is increased by adding elements that form stronger hydrides. Lower permeability may be acceptable if the alloying elements reduce subsequent hydrogen expansion, nucleation of brittle hydrides, and the hydride transformation temperature, or increase mechanical strength and so on.

In the case of iron, which has very high diffusivity but very low solubility, alloying with elements that form strong hydrides will increase solubility and thus permeability. Alloying candidates include V, Ta, Nb, Mg, the lanthanides, or even the alkali elements—if the material could be protected from adverse chemical reactions.

For Nb and Ta, which have larger lattice parameters relative to V, alloying with elements such as Fe, V, Mo and W will decrease the lattice parameters and increase diffusivity. However, solubility will decrease.

Ternary and quaternary alloys are employed to further adjust lattice parameters, diffusivity, solubility, chemical expansion, and other properties [9, 12]. Some reported examples include Ti–V–Cr–Al [7], V–4Cr–4Ti [18], V-7.5Cr–15Ti [18], and V–Ni–Al [21]). From Table 8.2 it is predicted that the addition of V and Cr will contract the cube edge of β-Ti, increase diffusivity and decrease solubility. Contraction of the lattice by V and Cr, however, can aggravate subsequent hydrogen expansion in these alloys [18]. Addition of aluminum to Nb, Ta and V, can be used as a getter for oxygen, reducing susceptibility of materials to oxygen embrittlement or reaction between hydrogen and oxygen at voids to form internal steam [8]. Addition of Mo or W can inhibit undesired excess grain growth during high-temperature processing and degassing of oxygen. Segregation of non-hydride forming elements to grain boundaries may protect against intergranular embrittlement. Alloying with various metals, including Pd, may be beneficial for lowering the driving force for interdiffusion between surface catalysts or supports [8].

Some of the concepts used in the design of alloys for hydrogen storage can be used as guides for the development of new hydrogen membrane materials. Hydrogen storage materials typically contain elements for adsorbing and dissociating hydrogen, such as Ni, Co, Fe, Mo, Mn, Pt, and Rh, along with elements forming strong hydrides, Zr, Ti, V, Nb, Ta, Ca, Mg, and La. Examples include Zr–Ni, Mg–Ni, La–Ni, Ti–Fe, Zr–Rh, La–Ni–Co, Mg–Fe, Mg–Co, Ca-Ni and allied materials. If palladium could be eliminated and replaced with less expensive hydrogen dissociation catalysts, such as nickel, lower cost membrane materials could be produced. As in the case of membranes, hydrogen storage materials typically face the same issues of chemical expansion, hydrogen embrittlement, and susceptibility to oxidation.

Surface and Interface Effects

In the design of composite membranes utilizing metals and ceramics, it is critical that the metal and ceramic wet. The terms wetting and nonwetting are used as defined in textbooks on surface chemistry and physics [46, 47]. By wetting it is meant that the contact angle between a liquid and solid is zero or close to zero so that the liquid spreads over the solid surface. By nonwetting it is meant that the contact angle is greater than 90° so that the liquid tends to ball up and run off the surface easily [46]. In both cases, overall interfacial energy is minimized [46, 47]. The concepts of wetting and nonwetting have been extended to solid–solid interfaces, especially in areas of thin film growth and in the joining of metals to ceramics [48].

For ill-designed composite membranes, for example, formed by depositing palladium onto substrates which it does not wet, surface tension will force the thin film to contract and ball up if the palladium atoms acquire sufficient surface mobility. Pinholes may form as a prelude to complete de-wetting, or pinholes may remain from the initial fabrication if the palladium did not fully wet its substrate. Kinetics of de-wetting is accelerated at elevated temperature and in the presence of adsorbates such as CO, which increase surface mobility of Pd. If molten metals do not wet ceramics, they will be expelled from ceramic pores. During sintering of cermets, Pd and other metals will not adhere to the ceramic phase, if the metal and ceramic do not wet.

The phenomena of wetting and nonwetting are both driven by the tendency of the system to lower its overall surface free energy. For a thin solid film to remain spread on its substrate, it is desired that the surface free energy of the gas-film surface be less than the sum of the free energies of the uncovered substrate and of the solid-solid interface.

To minimize the surface free energy of the solid-solid interfaces, one strategy is to design systems with coherent interfaces [11, 12]. The metal and ceramic are chosen

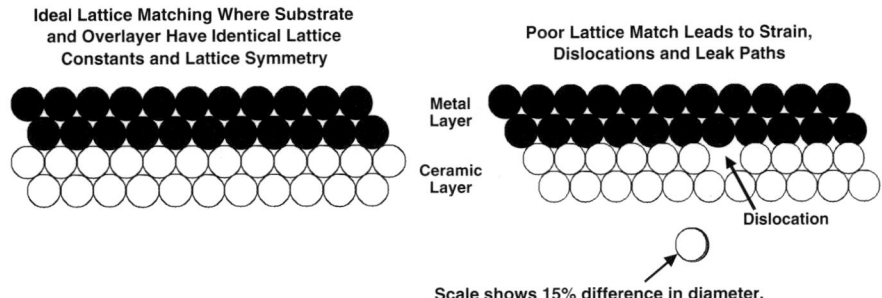

Fig. 8.2 Lattice matching, a concept borrowed from the semiconductor industry, matches lattice symmetry and lattice constants of metals and ceramics at the atomic level to minimize interfacial stress and minimize formation of dislocations. Lattice matching allows production of more stable cermets and allows better deposition of metal films atop porous ceramics

so that crystal lattices match at the atomic level. In lattice matching, both lattice symmetry and lattice constants are matched. This concept is widely used in the semiconductor industry to produce stable thin films with a minimum of dislocations at the over layer-substrate interface [49]. In the joining of metals to ceramics, lattice matching is also used in selecting compatible materials which wet [48].

Figure 8.2 shows schematically an example of a perfect lattice matched system in which atomic diameters are identical, and an example of a mismatch in which atomic diameters are drawn differing by 15% and for which a dislocation has formed. In a poorly matched system, atoms across the interface are in poor registry. This leads to high interfacial stress and high interfacial energy. Interfacial stress is relieved by periodic formation of dislocations [49]. In general, a lattice match of better than approximately 15% is required for a coherent interface [49].

In designing porous ceramic substrates for supporting Pd and in designing palladium cermets, it is found that many oxide ceramics with the perovskite crystal structure (see Fig. 8.3) match the lattice of palladium and its alloys extremely well. Some examples are listed in Table 8.4. By doping the perovskites and by adjusting the composition of Pd alloys, near perfect lattice matches can be formed. This aids wetting. Figure 8.4 shows an example of a well-adhered palladium film, which was electrolessly deposited onto a porous perovskite substrate of $LaFe_{0.90}Cr_{0.10}O_{3-x}$ [11, 12]. Figures 8.5 and 8.6 show a Pd-perovskite cermet fabricated by sintering together powders of Pd and $LaFe_{0.90}Cr_{0.10}O_{3-x}$ [11, 12].

Palladium cermets have a number of advantages over thin films of Pd supported by porous ceramics. For systems which wet, sintering cermets at very high temperatures (well above membrane operating temperatures) produces dense, pinhole-free composites (see Fig. 8.5). Because Pd is closely confined within a matrix of ceramic and because small, individual, micron-size Pd crystallites already possess a small surface-to-volume ratio of low surface energy, the Pd has relatively low driving

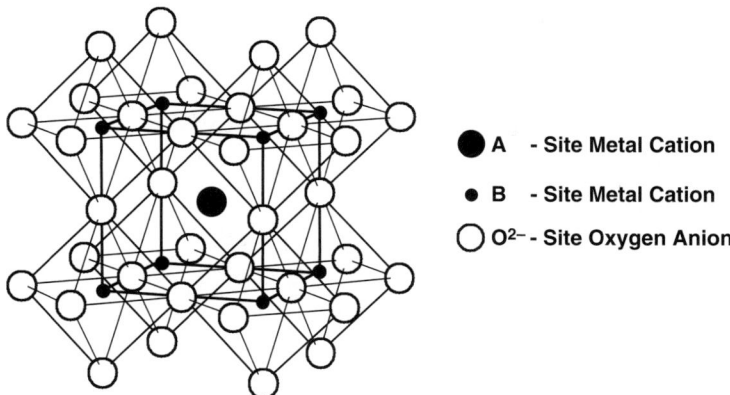

Fig. 8.3 Perovskite crystal structure of oxide ceramic materials used to fabricate composite membranes. A very large fraction of the metals in the periodic table can be substituted into the A and B lattice sites. A-sites contain larger cations such as alkaline earth and rare earths, including Ca, Sr and La, whereas the B-sites contain smaller transition metal cations such as Ti, Nb, V, Fe, Cr, Cu and Co. A near infinite variety of materials can be synthesized

Table 8.4 Lattice matching of Pd to perovskite ceramics (Pd: 3.89Å)

Perovskite Formula	Lattice Constant Å	% Mismatch
$LaCoO_{3-x}$	3.82	1.8
$La_{0.6}Ca_{0.4}MnO_{3-x}$	3.83	1.6
$CaTiO_{3-x}$	3.85	0.97
$SrFeO_{3-x}$	3.87	0.55
$La_{0.6}Sr_{0.4}MnO_{3-x}$	3.87	0.52
$LaCrO_{3-x}$	3.88	0.26
$LaMnO_{3-x}$	3.88	0.26
$LaFeO_{3-x}$	3.89	0
$SrTiO_{3-x}$	3.89	0
$La_{0.6}Ba_{0.4}MnO_{3-x}$	3.90	−0.25
$BaTiO_{3-x}$	3.98	−2.3

force to further contract or to form pinholes even at elevated temperatures, unlike the case for very thin films of Pd supported on porous materials.

For some applications, as in membrane-assisted steam reforming of methane, it is highly desired to operate Pd-based membranes at 600°C or above to assist decomposition of methane. Figure 8.6 shows a surface of Pd (on the right) after a cermet had been used at >600°C in a water-gas shift mixture containing H_2, H_2O, CO, and CO_2. Although the Pd surface was etched and underwent a change in morphology, pinholes were not observed, and interfaces remained coherent. Interdiffusion between the ceramic and Pd phases at sintering temperatures can be an issue. Interdiffusion at operating temperatures and under highly reducing conditions must be considered, especially if the ceramic phase could be partially reduced by hydrogen or carbon monoxide and if metal ions could migrate and poison the catalytic activity of Pd.

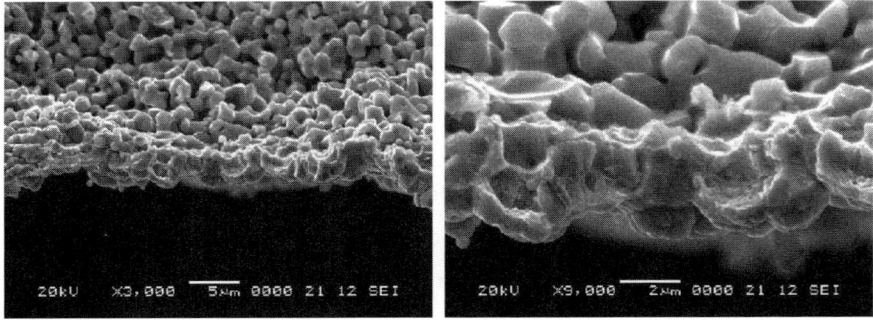

Fig. 8.4 Scanning electron microscope images at two magnifications of a three micron thick palladium layer deposited onto a fine porous layer of $LaFe_{0.90}Cr_{0.10}O_{3-x}$ perovskite ceramic. Metal and ceramic are lattice matched to minimize strain and dislocations and to aid nucleation and growth of the metal on the ceramic

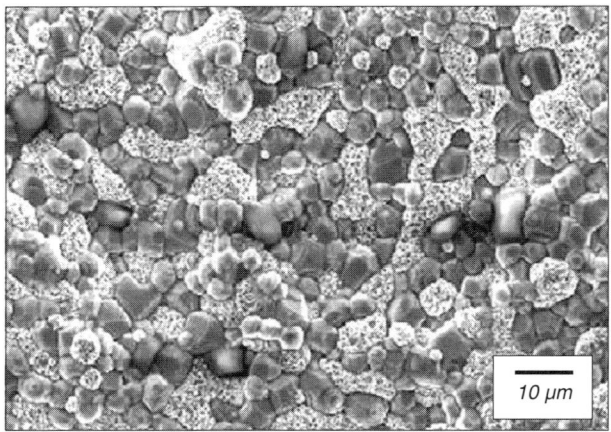

Fig. 8.5 Scanning Electron Microscope image of a perovskite-palladium cermet (ceramic-metal) made by sintering together $LaFe_{0.90}Cr_{0.10}O_{3-x}$ and Pd powder to form dense continuous matrices of both metal and ceramic. The palladium and ceramic were lattice matched to minimize strain and interfacial dislocations. (S. Rolfe, Eltron Research) (Copyright Elsevier, 2005. Adapted with permission from [11], Carbon Dioxide Capture and Storage in Deep Geological Formations.)

U. Balachandran of Argonne National Laboratories has operated Pd-alloy cermets successfully to temperatures as high as 900°C [10, 22]. At such extreme temperatures, most thin films of supported Pd and its alloys would be hard pressed not to ball up or form pinholes. Operation at elevated temperatures is desired, for example, in the extraction of hydrogen from gas mixtures containing compounds of sulfur. At such elevated temperatures, and in the presence of sufficient hydrogen, bulk sulfides of Pd are not thermodynamically stable, and the Pd-based cermets acquire some degree of tolerance to sulfur [22, 50]. In addition, permeability of Pd increases at elevated temperatures and becomes comparable to that of Ta and V (see Fig. 8.1).

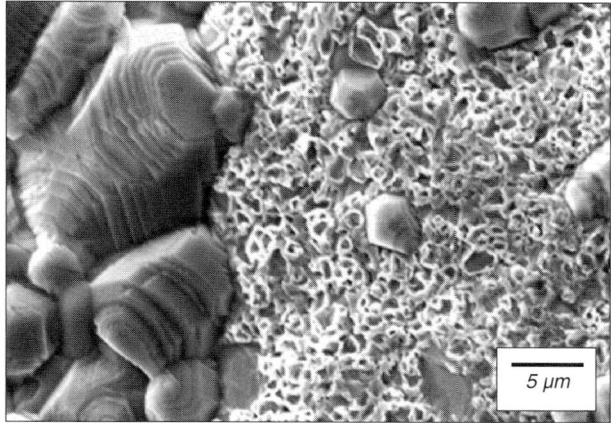

Fig. 8.6 Surface of a palladium/LaFe$_{0.90}$Cr$_{0.10}$O$_{3-x}$ cermet. Ceramic phase is on the left, metal on the right (Copyright Elsevier, 2005. Reproduced by permission. Adapted from [11], Carbon Dioxide Capture and Storage in Deep Geological Formations.)

For the highly permeable metals, excellent lattice matches as well as good matches of coefficient of thermal expansion are found, for example, between Nb and Al$_2$O$_3$, Ta and CaAl$_2$O$_4$, V and SrTiO$_3$ and Zr and Al$_2$O$_3$ [11, 12]. The system Nb/Al$_2$O$_3$ is a classic lattice-matched system and has long been used in joining of alumina to metals in metal-ceramic seals [48]. Figure 8.7 shows a cermet formed from a structural ceramic that was lattice matched to a Group VB metal. Note the excellent wetting between interfaces.

In the formation of cermets, it is also desirable that both components have compatible sintering temperatures. For particles of the order of 1 μm in diam-

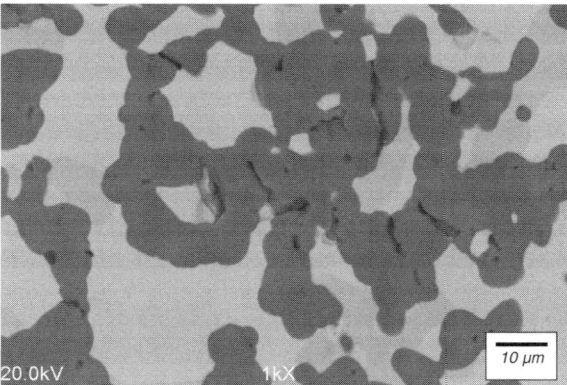

Fig. 8.7 Cermets formed by sintering together powders of a ceramic (darker grey) and a Group VB element. Note the coherent interfaces and excellent wetting. (Kleiner, Stephan, Anderson, CoorsTek)

eter, a reasonable sintering temperature is approximately three-fourths of the absolute melting point. This implies that it is desirable that the ceramic and the metal have similar melting points. Because melting temperatures and coefficients of thermal expansion both depend upon the strength of inter-atomic forces, which regulate lattice vibrations in the materials [25], melting temperatures also give a very rough guide to compatibility of coefficients of thermal expansion. Ceramic-metal cermet systems matched by absolute melting temperature include: $Nb(2741K)–Al_2O_3(2345K)$; $Zr(2125K)–Al_2O_3(2345K)$; $Ta(2969K)–ZrO_2(2973K)$; $V(2163K)–SrTiO_3(2353K)$; $V(2163K)–CaTiO_3(2248K)$; $Pd(1825K)–BaTiO_3(1883K)$. In general, melting points of metals are lowered by alloying so that alloys can be designed to lower sintering temperatures and can be adjusted to better match the sintering temperature of a compatible structural ceramic.

For reactive metals, sintering is performed under vacuum or under inert atmospheres containing sufficiently low partial pressures of gases containing oxygen, carbon and nitrogen in order to minimize formation of brittle bulk oxides, carbides and nitrides. In some cases, however, residual thin adherent oxides of the reactive metals may be beneficial and act as intermediary layers to aid wetting with the structural ceramic oxide in the cermet.

In the case of all-metal composite membranes formed by depositing palladium and its alloys onto metal substrates based on Nb, Ta, and V, strong coherent interfaces are highly desired not only for wetting and adherence, but also to ensure, at the atomic level, that the dissociated hydrogen is rapidly transferred between the catalyst and substrate layers. The rules of epitaxial growth and lattice matching are also used to optimize these solid-solid interfaces.

In the work of Peachey, Dye, Snow and Birdsell at Los Alamos National Laboratory, the authors report advantages of using oriented tantalum foils coated with oriented palladium [17, 51]. The authors, however, have not reported details outlining underlying reasons for this effect. Therefore, some of the fundamental science is discussed as follows. For metals with the body centered cubic structure, it has been known since at least the 1930s, that stress on the metal as it is cold-rolled to form foils causes the crystallites in the polycrystalline foil to become aligned [52, 53]. This is evidenced by x-ray diffraction, which shows large diffraction peak intensities for reflections from specific crystallographic planes, which are not as pronounced for well-annealed, equiaxed, polycrystalline foils. For foils of bcc iron-based alloys, the alignment of the crystallites in directions of easy magnetization is exploited in the fabrication of magnetic materials [52, 53]. This orientation effect produced by cold-rolling foils of bcc metals is most often discussed in textbooks on magnetic materials [52, 53]. According to Chen, foils of cold-rolled body centered cubic metals can assume the (110)[001] orientation [53]. In such oriented foils, the [001] crystallographic direction is aligned parallel to the rolling direction, and (110) planes are parallel to the rolled surfaces and are exposed on both surfaces of the foil [52, 53]. Crystallites of (100)[001] orientation may also be present.

The exposed bcc (110) surfaces have the highest atomic density of any bcc surface. They, therefore, have the lowest surface free energy of any bcc surface. This implies that foils with clean exposed (110) surfaces will be stable with respect to faceting. If annealed in vacuum, (110) planes grow at the expense of (100) planes. However, if annealed in the presence of impurities such as oxygen or sulfur, (100) planes may have lower surface energy and grow at the expense of (110) planes in the case of iron-silicon alloys. Because of its lower surface free energy, (110) surfaces have the smallest driving force for the segregation of impurities to the buried interface beneath the catalyst layers. This is important for minimizing formation of hydrogen diffusion barriers that could form from impurities diffusing to the buried interface.

Furthermore, it follows from the rules of epitaxial growth [49], that if a face-centered cubic metal such as palladium is deposited onto the (110) surfaces of body centered cubic Nb, Ta and V, the palladium will assume an orientation with the Pd(111)//bcc(110) and Pd[110]//bcc[111]. This metal-metal interface is expected to be coherent and of minimum interfacial energy. Because the Pd(111) surface is the closest packed, it will also have the lowest surface free energy at both the solid-solid interface and the gas-surface interface. The Pd(111) surface at the gas-solid interface will be the most resistant to faceting, to de-wetting, to poisoning by adsorption from the gas phase, and to poisoning by impurity segregation from the bulk.

A number of excellent epitaxial matches exist in the unalloyed systems. For example: Pd(111)//V(110) and Pd[110]//V[111]; Pd(111)//Nb(110) and Pd[110]// Nb[111]; Pd(111)//Ta(110) and Pd[110]//Ta[111]. Using the lattice parameters of Table 8.2, the cube-face diagonal for unalloyed Pd in the closest-packed fcc [110] direction is 5.50 Å. The cube diagonal for unalloyed V in the [111] direction is 5.26 Å, and that of unalloyed Nb and Ta is 5.72 Å. Calculating the epitaxial mismatches only along the closest-packed directions, using ([over layer − substrate]/[substrate]) × 100%, yields mismatches of +4.6% for Pd/V and −3.8% for both Pd/Ta and Pd/Nb. This is quite good if it is assumed that mismatch of up to 15% is considered adequate for reasonable epitaxial fit [49].

For the Pd–25Ag (atom%) system, assuming cube edge of 3.94 Å, the cube-face diagonal would expand to 5.57 Å, yielding a still better epitaxial fit of −2.6% on Ta and Nb, and a slightly poorer fit of +5.9% on V. In general, the epitaxial fit of many of the preferred Pd alloys including Pd–5Au (mass%), Pd-40Au (mass%) and Pd–19Pt (mass%) are excellent on the (110) surface of Nb, Ta and V. Expanding the vanadium lattice and contracting the Ta and Nb lattices by alloying makes the epitaxial fit even better with the above Pd alloys.

For the Pd–Cu single-phase composition, which has the bcc structure, the preferred epitaxial arrangements will be: Pd-Cu(110)//V(110) and Pd-Cu[111]//V[111]; Pd-Cu(110)//Nb(110) and Pd-Cu[111]//Nb[111]; Pd-Cu(110)// Ta(110) and Pd–Cu[111]//Ta[111]. Assuming a cube edge of 2.994 Å for bcc Pd–Cu [36] and a cube diagonal of 5.18 Å, the epitaxial mismatches along the closest-packed [111] direction will be −1.5% for Pd-Cu/V(110) and +9.4% for Pd–Cu/Nb(110) and Pd–Cu/Ta(110). For the case of Pd–Cu, contracting the vanadium lattice, for example with nickel, will improve the match. Other lattice constants might be assumed for bcc Pd–Cu having various ratios of Pd: Cu, but in general, the epitaxial fit will be more than adequate on the (110) surfaces of Nb, Ta and V.

A number of workers report use of various Pd alloys deposited onto substrates of the Group VB elements and their alloys. Zhang et al. discuss sputter deposition of Pd–Ag alloys onto membranes of V–15Ni (atom%) [54–56]. Yang et al. discuss use of Pd–Cu alloys deposited onto V–15Ni (atom%), including Pd–40Cu (mass%) [57, 58]. Paglieri et al. discuss use of Pd–Ag and Pd–Cu catalysts atop membranes of V–12Cu (atom%), V–5Ti (atom%), V–6Ni–5Co (atom%) and atop unalloyed Ta [59, 60].

Paglieri et al. at Los Alamos note that interdiffusion between Pd catalyst layers and substrates of Ta and V is severe if membranes are operated much above 400°C [59, 60]. This has been noted by a number of authors [8]. Assuming that intermetallic diffusion is an activated process with Arrhenius-type behavior, then the rate of interdiffusion will depend exponentially on temperature. At 300°C, Moss et al., also at Los Alamos, state that no interdiffusion of Pd with V or Ta was detected by Rutherford backscattering, and that membrane performance was stable for 775 h [23]. According to the group of Nishimura, alloying of V with Ni lowers the hydrogen embrittling temperature and allows membrane operation at 200°C [19, 21, 42]. This will further decrease interdiffusion exponentially. In principle, alloying of both catalyst and substrate lowers the driving force for interdiffusion as metals become saturated with alloying elements [8, 61, 62].

Surface Catalysis

The need for surface catalysis on dense membranes to adsorb and dissociate molecular hydrogen has been alluded to above. Figure 8.8 shows schematically some aspects of surface catalysis of dense membranes used to extract hydrogen from water-gas shift mixtures containing H_2, H_2O, CO, CO_2, H_2S and other impurities. According to Smithells [27], William Ramsey (winner of the Nobel Prize in chemistry in 1904) had realized already by 1894 that hydrogen must diffuse through palladium in a dissociated form [63]. This was based upon the fact that nascent hydrogen produced by electrolysis or acid hydrolysis in aqueous solution diffused into Pd more rapidly relative to molecular H_2 in solution at the same temperature [27]. By the 1930s, it was very clear from the work of Sieverts, Langmuir and others that there could be no diffusion of hydrogen through dense metals without the preliminary steps of adsorption and dissociation [27].

The requirement for the adsorption, dissociation and later recombination and desorption of molecular hydrogen (Fig. 8.8), implies that membrane catalysts must be present on both membrane surfaces and that catalytic activity must be maintained. As with any catalyst, membrane catalysts must be protected from catalyst poisons adsorbing from the gas phase or segregating to the surface by diffusion from the bulk.

In the case of palladium, common catalyst poisons originating from the gas phase include CO, alkenes, alkynes and unsaturated organic compounds, elemental carbon, S, Te, P, As, Sb, Bi, Hg, Cd, Cl, Br, I, and Si. Metal carbonyls and

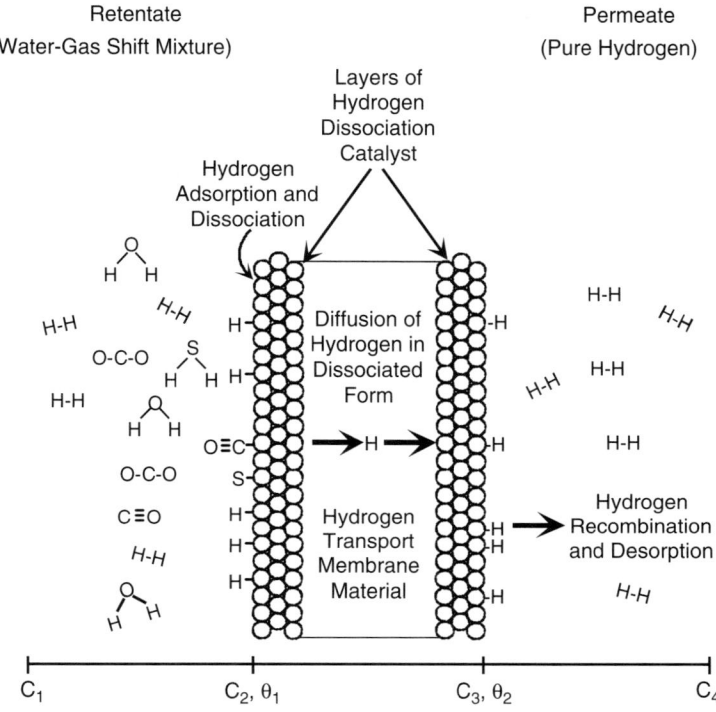

Fig. 8.8 Schematic of a composite membrane coated with hydrogen dissociation catalysts on both sides. Catalyst poisoning by sulfur and competitive adsorption by molecules such as CO must be considered

oxyhydroxides can also transport metals such as Fe, Ni and Cr. Particulates of many types, originating from the process gas or reactor walls, can encrust and deactivate membranes. Impurities originating from adhering particulates can diffuse and contaminate large surface areas of membrane. If the palladium is supported, elements from the support can diffuse into the palladium and segregate to the Pd catalytic surfaces. If the membranes are used in steam, stable refractory oxides may form from diffusing species, which are trapped at the surface and which can block adsorption of molecular hydrogen. Oxide forming elements of concern include, but are not limited to, Fe, Cr, Si, Ca, Mg, Nb, Ta, Ti, V, and Zr.

At lower temperatures ($< 250°C$), CO can readily form a monolayer on Pd and other Pt-group metals, effectively blocking adsorption of hydrogen and poisoning the surface [64, 65]. At elevated temperatures ($> 300°C$), CO rapidly desorbs allowing hydrogen to adsorb and dissociate [50]. In general, CO and hydrogen will be in dynamic equilibrium and will compete for surface sites. The fraction of the membrane surface covered by CO will depend upon the temperature and partial pressures of CO, H_2, and other components of the gas mixture.

Adsorbed CO can dissociate via the Boudouard reaction: $2CO = C + CO_2$. Elemental carbon can accumulate on the surface and block adsorption of hydrogen.

Adsorbed organic compounds can also compete with hydrogen for surface sites or can crack, depositing a residue of elemental carbon. Accumulation of elemental carbon is prevented by ensuring that the feed gas mixture contains adequate steam, thus removing carbon by the reaction: $C + H_2O = H_2 + CO$ [50].

Palladium and Pd–Ag are very susceptible to poisoning by sulfur [33, 50]. Under some operating conditions, a few parts per billion by volume H_2S is sufficient to form bulk Pd_4S [50]. To increase tolerance to sulfur, researchers have employed Pd–Cu [32–34, 57–59, 66–71], Pd–Au [4, 32–34, 66, 72, 73] Pd-Pt [2, 62], Pd–Rh, Pd–Ru, and various ternary Pd alloys such as Pd–25Au–5Pt (mass%) [69, 74]. According to McKinley, a Pd–40Au (mass%) alloy is superior to a Pd–40Cu (mass%) alloy towards poisoning by H_2S [33]. For measurements at 350°C, pressures up to 300 psig (20.7 barg) and using stainless steel porous supports, unalloyed palladium and Pd–27%Ag (mass%) were rapidly poisoned, but Pd–40Au (mass%) retained 80% of its permeability after 6 days with 4 ppmv H_2S and 40% of its permeability with 20 ppmv H_2S [33]. The Pd–40Cu (mass%) was not satisfactory according to McKinley [33]. Paglieri and Way discuss a Pd–19Pt (atom%) alloy for its benefits of lowering the α–β phase transformation [2], and Buxbaum discusses use of platinum black atop Pd layers [62]. Buxbaum and Kinney state that replacing the Pd surface with fresh palladium restores activity lost by deterioration [5]. In general, the sulfides of Pt, Ir, Rh and Ru are much less stable than those of Pd, and these elements show better tolerance to poisoning by sulfur.

To protect Pd membranes from the entire gamut of potential gas phase catalyst poisons, McBride et al. [75] and McBride and McKinley [76] successfully employed guard bed adsorbents. Guard beds were used to adequately remove solids, hydrocarbons, oils, compounds of sulfur and other impurities from the feed [75]. Parallel guard beds were used. One was kept in service while the second was regenerated or replaced [75]. Iron oxides were used to partially remove H_2S [76]. Zeolites and other adsorbents were also employed. Nine plants based upon Pd membranes were in operation by Union Carbide by mid–1965 [76]. The largest plant produced 9,000,000 ft^3 (250,000 m^3) of hydrogen per day. Total capacity of all 9 plants was 34,000,000 ft^3 (960,000 m^3) per day [76]. Recovery of hydrogen from coke-oven gas was envisioned.

Membranes were supported by porous stainless steel and were run at 350–400°C [76]). Operation at relatively low temperature likely limited interdiffusion with supports and seals. Copper gaskets and stainless steel frames were used to form compression seals [75]. Pressure vessels and the porous stainless steel supports were designed to handle feed pressures of 650 psig (44.8 barg) [75]. Purging with high-purity nitrogen was used to remove hydrogen and to bring systems up and down through temperatures otherwise favoring the brittle palladium β-hydride phase. Membranes were not permitted to be exposed to hydrogen below 330°C [75]. A pilot plant built in South Charlestown, West Virginia (U.S.A) was in continuous operation for more than three years [76].

Palladium surfaces and their catalytic activity must be protected from impurities diffusing through the bulk and originating from supports. For example, although porous stainless steel is an excellent support for Pd [72], and was used successfully in the large plants built by Union Carbide [75, 76], Fe and Cr can diffuse into Pd at elevated temperatures. For some applications this may not be an issue. However, if steam is used in the retentate, the Fe and Cr may be oxidized and trapped at the Pd–gas interface poisoning the hydrogen dissociation reaction. To prevent this, attempts have been made to employ diffusion barriers on the stainless steel surface [2]. Issues of wetting must be considered in the choice of diffusion barrier.

If SiO_2, glass or other silicon-bearing materials are used in substrates, there is the potential that a small fraction of silicon will be reduced in the highly reducing environment of the pure hydrogen permeate and then react to form palladium silicides. Silicon can diffuse into the Pd via dislocations or spread on the Pd by surface diffusion from pinholes. If steam is present in the feed, stable oxides of silicon may form in the more oxidizing environment and block the hydrogen dissociation reaction.

Other oxides used as supports, and normally considered refractory, may also be reduced to a minor extent depending upon temperature and pressure of hydrogen [77]. Metals originating from the ceramic supports can diffuse through the Pd layer and react with steam in the retentate, forming oxide layers that block the adsorption of hydrogen.

Reactive metals Nb, Ta, Ti, V and Zr can diffuse into Pd at temperatures above about 400°C, especially in the presence of hydrogen [59, 60]. These can diffuse through thin layers of Pd and react with steam in the retentate to form refractory oxides blocking the adsorption of hydrogen.

In general, oxides of Pd are not thermodynamically stable in the partially reducing atmospheres of water-gas shift mixtures, and pure Pd is not poisoned by high-pressure steam alone [50]. However, if impurities are present in Pd, which segregate to the gas–Pd interface and react with steam to form stable oxides, then the Pd can be poisoned. Likewise, hydrothermal transport of silicon and metals by high-pressure steam via volatile oxyhydroxides, originating from ill-chosen reactor wall materials, can also poison the catalytic activity of Pd.

Although hydrogen predominantly diffuses through Pd via interstitial sites, the much larger metal atoms must diffuse through the dense Pd by a vacancy mechanism or through edge dislocations [78–80] spanning the Pd crystallites or through the edge dislocations at grain boundaries [81]. Atoms of carbon and oxygen can also diffuse through edge dislocations of Pd and recombine on the permeate surface to form CO or CO_2 or react with hydrogen to form H_2O or CH_4. These are well-established mechanisms for contamination of hydrogen transported through dense Pd membranes.

Formation of vacancies in metals and alloys is an activated process [82]. Diffusion of metals into Pd is decreased exponentially by simply decreasing the absolute temperature at which the membranes are operated. Effects of interdiffusion can also be decreased linearly by increasing the thickness of the Pd. In general, doubling the thickness of a layer of Pd will double the time required for metals to diffuse through the layer and thus double the operation time before decay of the catalytic activity. This assumes that Pd is not poisoned by surface diffusion of impurities originating from pinholes or the membrane seals. It follows that poisoning of supported membranes by diffusion from the bulk will be most severe for the thinnest Pd layers operated at the highest temperatures.

To inhibit impurity diffusion through edge dislocations, both at grain boundaries and within crystallites, it is highly desirable that grain boundaries and dislocations be minimized or that immobile species are formed from reaction of two elements, meeting and precipitating in the dislocations to block diffusion of impurities. It is imperative that additives do not weaken cohesion of grain boundaries or foster intergranular fracture or embrittlement.

Membrane Flux Measurements

Referring to Fig 8.8, hydrogen flux, J, through the gas phase to the retentate-side surface of the membrane is given by Fick's First Law of Diffusion [83, 84]: $J = -D(dc/dx)$ where J is in mol m^{-2} s^{-1}; D is the diffusivity at a given temperature in m^2 s^{-1}; $dc = (c_1 - c_2)$, the change in concentration in mol m^{-3} from the gas phase to the membrane surface; and $dx = x_1 - x_2$, the geometric path length in meters, m. If the flux is limited by mass transport through the gas phase, which can easily happen for very thin Pd membranes or for super-permeable membranes using Nb, Ta or V [8, 85], the hydrogen flux will be directly proportional to the first power (or approach the first power) of the hydrogen partial pressure in the feed. Indications that gas phase mass transport is limiting flux include increase of flux with increase in feed flow (without change in hydrogen partial pressure), increase in flux with increase in gas feed turbulence [85], and negligible or anomalously low activation energies (i.e. little change in flux with change in temperature). If flux is limited by porous support materials or porous guard bed materials, flux will be directly proportional to the hydrogen partial pressure if pore size is such that Darcy's Law holds [86].

Fick's First Law of Diffusion is fundamental and will also describe flux through a dense membrane if the concentration of hydrogen on each surface is known. The most common approximation for the concentration of hydrogen on surfaces is given by Langmuir's Isotherm [87, 88]. Assuming dissociative adsorption of hydrogen on the surface, two surface sites are required to accommodate two hydrogen atoms. Adsorption of hydrogen can be represented by the equation.

$$H_2 + 2 \text{ (adsorption sites)} \rightleftarrows 2 \text{ H (adsorbed)}.$$

The rate of hydrogen adsorption is proportional to the partial pressure of hydrogen in the gas phase and the square of the number of empty adsorption sites on the membrane surface, and can be written thus:

$$\text{rate(adsorption)} = k_1 p_{H2}(1 - \theta)^2$$

where k_1 is the rate constant for adsorption, p_{H2} is the partial pressure of H_2 in the gas phase above the membrane and θ is the fraction of the surface covered by hydrogen. The term $(1-\theta)$ is the fraction of the surface uncovered by hydrogen. The rate of desorption from the surface is given by:

$$\text{rate(desorption)} = k_{-1}\theta^2,$$

where k_{-1} is the rate constant for desorption of molecular hydrogen.

At equilibrium, the two rates are equal and thus:

$$k_1 p_{H2}(1 - \theta)^2 = k_{-1}\theta^2.$$

Solving for θ, the surface coverage, gives:

$$\theta = \frac{(k_1/k_{-1})^{1/2}p_{H2}^{1/2}}{1 + (k_1/k_{-1})^{1/2}p_{H2}^{1/2}}$$

If the partial pressure of hydrogen is low then $1 \gg (k_1/k_{-1})^{1/2}p_{H2}^{1/2}$, and $\theta \propto p_{H2}^{1/2}$. It is thus seen that the coverage of hydrogen on the membrane surfaces will be proportional to the square root of the hydrogen partial pressure. Historically, it was the square root dependence that indicated dissociation of molecular hydrogen through dense membranes. From this, one can justify the origin of the square root dependence of hydrogen flux according to Sieverts' Law:

$$J = P \cdot d^{-1}(p_f^{0.5} - p_p^{0.5})$$

where P is the permeability in $mol \cdot m \cdot m^{-2} \cdot s^{-1} \cdot Pa^{-0.5}$, d is the membrane thickness in m, and p_f and p_p are the partial pressures of hydrogen in Pa on the feed and permeate side of the membrane.

However, if the hydrogen partial pressure in the feed is high, $1 \ll (k_1/k_{-1})^{1/2}p_{H2}^{1/2}$, and then $\theta_1 = 1$. The feed-side surface will become saturated with hydrogen at some pressure, and the flux will no longer be a function of the hydrogen partial pressure but will remain a constant. Thus at high hydrogen feed pressure the equation for hydrogen flux may become $J = $ constant (or $J \propto p_{H2}^0$), independent of the partial pressure of hydrogen in the feed. For a gradient to exist, obviously the surface concentration, θ_2, on the permeate side of the membrane cannot also equal 1.

For intermediate partial pressures of hydrogen, the full Langmuir expression would be needed to determine the surface concentration and the rate of diffusion as a function of hydrogen partial pressure. The exponent of the partial pressure of hydrogen would lie between 0 and 0.5.

The Langmuir Adsorption Isotherm assumes that all surface sites are identical, that the heat of adsorption does not change with coverage, and that the system reaches equilibrium [87, 88]. In practice, the surface of the membranes will be heterogeneous, and heats of adsorption will vary with hydrogen coverage and the system may not reach equilibrium. The pressure dependence may then follow a Freundlich or other isotherm and be a function of pressure to some power other

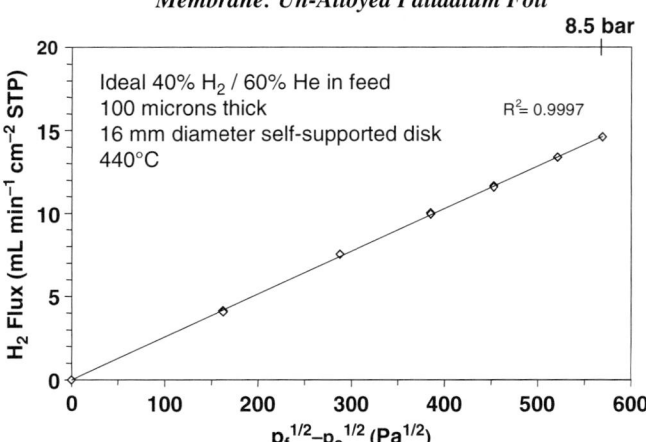

Fig. 8.9 Hydrogen flux data from a 100 μm thick unalloyed palladium foil. Sieverts' Law is followed very well, and the square root dependence implies that hydrogen is transported through the membrane in a dissociated form. The membrane, sealed by copper gaskets, was essentially 100% selective towards hydrogen with no leak to helium detected. Permeability at 440°C was 1.9×10^{-8} mol m^{-1} s^{-1} Pa$^{-0.5}$

than 0.5 (i.e., $p^{1/n}$) or the pressure dependence may be logarithmic [89]. Finally, the Langmuir approximation does not take into account exchange of adsorbed hydrogen with the bulk.

Figure 8.9 shows a plot of hydrogen flux through a 100 μm thick membrane of unalloyed palladium (non-composite and unsupported) measured at Eltron Research & Development Inc. Sieverts' Law is followed extremely well, implying that hydrogen is transported in a dissociated form and that hydrogen flux is limited by mass transport through the Pd bulk. Permeability at 440°C was 1.9×10^{-8} mol m^{-1} s^{-1} Pa$^{-0.5}$, in good agreement with the calculations of Steward (Fig. 8.1) [15]. For this membrane thickness, leaks towards helium were below detection limits and the membrane was essentially 100% selective towards hydrogen.

Figure 8.10 plots hydrogen flux through an all-metal composite membrane fabricated by using one of the Group IVB-VB elements. Permeability was 2.3×10^{-7} mol m^{-1} s^{-1} Pa$^{-0.5}$ also at 440°C. Higher mechanical strength relative to the soft unalloyed Pd allowed the unsupported membrane to withstand 33 bar differential pressure and to achieve hydrogen flux exceeding 400 mL min^{-1} cm^{-2} (STP) in an ideal H_2/He feed mixture with close to 100% H_2 in the feed to minimize gas phase mass transport limitations. Such composite membranes show great promise if they can be scaled up to industrial size and used under industrial conditions. Efforts at scale up are discussed in a companion paper in this volume.

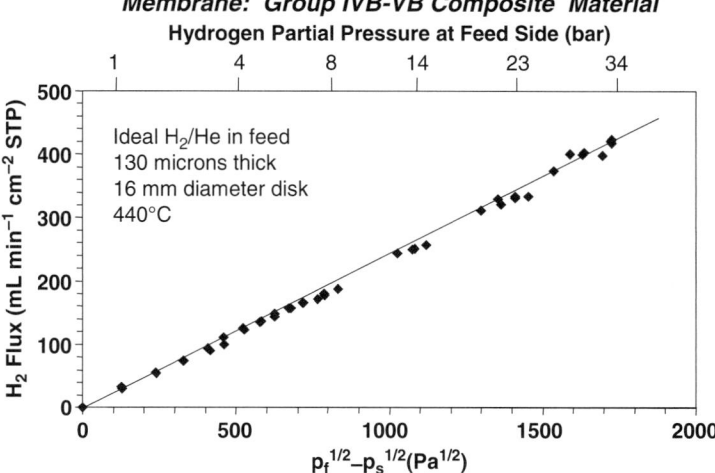

Fig. 8.10 Hydrogen flux data of a composite membrane incorporating a Group IVB-VB material. Sieverts' Law is followed very well and a permeability at 440°C of $2.3 \cdot 10^{-7}$ mol m^{-1} s^{-1} Pa$^{-0.5}$ was achieved. The membrane, sealed with copper gaskets, was essentially 100% selective towards hydrogen showing no detectable leak to helium. The disk withstood 33 bar differential pressure (Copyright Wiley-VCH Verlag, GmbH & Co. KGaA, 2006. Adapted with permission from [8], Nonporous Inorganic Membranes.)

Summary

In order to improve composite membranes for hydrogen separation and purification, fundamental scientific principles are applied. For dense composite membranes approaching essentially 100% selectivity for hydrogen, hydrogen must be transported in a dissociated form. This implies the need for adequate catalytic activity on the feed side of the membrane for adsorption and dissociation of molecular hydrogen as well as for catalytic activity on the permeate side surface for recombination and desorption. Membrane catalytic activity must be adequately protected from poisoning and inhibition from impurities originating from the gas phase as well as segregating from the bulk. Permeability, P, of membranes is dependent upon diffusivity D, and solubility, S. Diffusivity can be altered by varying lattice parameters and thus the hydrogen hopping distance and activation energies. Solubility is increased by adding elements forming stable hydrides and decreased by diluting lattices with elements not forming stable hydrides. For fabrication of composite membranes, components must be compatible to allow wetting while minimizing interdiffusion. Components of composite materials must possess compatible coefficients of thermal and chemical expansion. Many elements including especially Nb, Ta, V and Zr can exceed the permeability of palladium and its alloys.

References

1. Snelling, Walter O. Apparatus for Separating Gases. US Patent 1,174,631, 7 March 1916.
2. Paglieri, S. N, Way, J. D. Innovations in palladium membrane research. Separation and Purification Methods 2002;31(1):1–169, and references therein.
3. Paglieri, S. N. Palladium membranes. In: Sammells, A. F, Mundschau, M. V, editors. Nonporous inorganic membranes. Weinheim, Germany: Wiley-VCH; 2006. pp. 77–105, and references therein.
4. Makrides, Alkis C, Wright, Maurice A, Jewett, David N. Separation of Hydrogen by Permeation. US Patent 3,350,846, 7 Nov 1967.
5. Buxbaum, Robert E, Kinney, Andrew B. Hydrogen transport through tubular membranes of palladium-coated tantalum and niobium. Ind Eng Chem Res. 1996;35:530–7.
6. Buxbaum, Robert E, Hsu, Peter C. Method for Plating Palladium. US Patent 5,149,420, 22 Sep 1992.
7. Hill, Eugene F. Hydrogen Separation Using Coated Titanium Alloys. US Patent 4,468,235, 28 Aug 28, 1984.
8. Mundschau, Michael V, Xie, Xiaobing, Evenson IV Carl R. Superpermeable hydrogen transport membranes. In: Sammells, A. F, Mundschau, M. V, editors. Nonporous inorganic membranes. Weinheim, Germany: Wiley-VCH; 2006. pp. 107–38, and references therein.
9. Phair, John W, Donelson, Richard. Developments and design of novel (non-palladium-based) metal membranes for hydrogen separation. Ind Eng Chem Res. 2006;45:5657–74, and references therein.
10. Dorris, Stephen E, Lee, Tae H, Balachandran, Uthamalingam. Metal/Ceramic Composites with High Hydrogen Permeability. US Patent 6,569,226 B1, 27 May 2003.
11. Mundschau, Michael V, Xie, Xiaobing, Sammells, Anthony F. Hydrogen transport membrane technology for simultaneous carbon dioxide capture and hydrogen separation in a membrane shift reactor. In: Thomas, D. C, Benson, S. M, editors. Carbon dioxide capture for storage in deep geologic formations. Vol. 1. Amsterdam: Elsevier; 2005. pp. 291–306.
12. Mundschau, Michael V. Hydrogen Transport Membranes. US Patent 6,899,744 B2, 31 May 2005.
13. Norby, T, Haugsrud, R. Dense ceramic membranes for hydrogen separation. In: Sammells, A. F, Mundschau, M. V, editors. Nonporous inorganic membranes. Weinheim Germany: Wiley-VCH; 2006. pp. 1–48.
14. Gupta, V. K, Lin, Y. S. Ceramic proton conductors. In: Sammells, A. F, Mundschau, M. V, editors, Nonporous inorganic membranes. Weinheim Germany: Wiley-VCH;. 2006. pp. 49–76.
15. Steward, S. A. Review of hydrogen isotope permeability through materials. Lawrence Livermore National Laboratory Report UCRL-53441; DE84 007362, Available from: National Technical Information Service, US Department of Commerce, Springfield, VA, USA, 1984.
16. Buxbaum, Robert E, Marker, Terry L. Hydrogen transport through non-porous membranes of palladium-coated niobium, tantalum and vanadium. J Membr Sci. 1993;85:29–38.
17. Peachey, N. M, Snow, R. C, Dye, R. C. Composite Pd/Ta metal membranes for hydrogen separation. J Membr Sci. 1996;111:123–33.
18. Buxbaum, R. E, Subramanian, R, Park, J. H, Smith, D. L. Hydrogen transport and embrittlement for palladium coated vanadium-chromium-titanium alloys. J Nucl Mater. 1996; 233–237:510–2.
19. Ozaki, T, Zhang, Y, Komaki, M, Nishimura, C. Preparation of palladium-coated V and V-15Ni membranes for hydrogen purification by electroless plating technique. Int J Hydrogen Energy. 2002;28:297–302.
20. Nishimura, C, Ozaki, T, Komaki, M, Zhang, Y. Hydrogen permeation and transmission electron microscope observations of V-Al alloys. J Alloys Compd. 2003;356–357:295–9.
21. Ozaki, Tetsuya, Zhang, Yi, Komaki, Masao, Nishimura, Chikashi. Hydrogen Permeation Characteristics of V-Ni-Al alloys. Int J Hydrogen Energy. 2003;28:1229–35.

22. Balachandran, Uthamalingam. Argonne National Laboratory, private communication; seminar presented at Eltron Research, Feb 2007.
23. Moss, T. S, Peachey, N. M, Snow, R. C, Dye, R. C. Multilayer metal membranes for hydrogen separation. Int J Hydrogen Energy. 1998;23:99–106.
24. Rothenberger, K. S, Howard, B. H, Killmeyer, R. P, Cugini, A. V, Enick, R. M, Bustamante F, et al. Evaluation of tantalum-based materials for hydrogen separation at elevated temperatures and pressures. J Membr Sci. 2003;218:19–37.
25. Askeland, Donald R. The science and engineering of materials. 3rd ed. Boston: PWS Publishing; 1994.
26. Wipf, H. Diffusion of hydrogen in metals. In: Wipf, H, editor. Hydrogen in metals III, Topics in applied physics. Vol. 73. Berlin: Springer; 1997. pp. 51–91.
27. Smithells, Colin J. Gases and metals. New York: Wiley; 1937.
28. Peachey, N. M, Snow, R. C, Dye, R. C. Composite Pd/Ta metal membranes for hydrogen separation. J Membr Sci. 1996;111:123–33.
29. Katz, O. M, Gulbransen, E. A. Occluded gases in transition metals. In: Mandelcorn, L, editor. Non-stoichiometric compounds. New York: Academic Press; 1964. p. 244.
30. Mackay, K. M. Hydrogen compounds of the metallic elements. London: E & F N. Spon; 1966.
31. Hunter, James B. Silver-Palladium Film for Separation and Purification of Hydrogen. US Patent 2,773,561, 11 Dec 1956.
32. McKinley, David L. Method for Hydrogen Separation and Purification. US Patent 3,247,648, 6 Dec 1966.
33. McKinley, David L. Metal Alloy for Hydrogen Separation and Purification. US Patent 3,350,845, 7 Nov 1967.
34. McKinley, David L. Method for Hydrogen Separation and Purification. US Patent 3,439,474, 22 April 1969.
35. Savitsky, E, Polyakova, V, Gorina, N, Roshan, N. Physical metallurgy of platinum metals. Moscow: Mir; 1978.
36. Donnay, J. D. H, Donnay, Gabrielle, Cox, E. G, Kennard, Olga, King, Murray Vernon, editors. Crystal data determinative tables. 2nd ed. Washington DC: American Crystallographic Association; 1963.
37. Clark, George L. Applied X-rays, 4th ed. New York: McGraw Hill; 1955. p. 558.
38. Züchner Harald. Ewald Wicke and his work on metal-hydrogen systems. J Alloys Compd. 2002;330–332:2–7, and references therein.
39. Darling, Alan Sydney. Treatment of Hydrogen or Gaseous Mixtures Containing Hydrogen. US Patent 2,962,123, 29 Nov 1960.
40. Hansen, M. Constitution of binary alloys. New York: McGraw-Hill; 1958. p. 353.
41. Smith, J. F. Peterson, D. T (Hydrogen-Vanadium). In: Smith, J. F, editor. Phase diagrams of binary vanadium alloys. Metals Park, OH: ASM International; 1989.
42. Nishimura, Chikashi, Komaki, Masao, Amano, Muneyuki. Hydrogen permeation characteristics of vanadium-nickel alloys. Mater Trans JIM. 1991;32(5):501–7.
43. Nishimura, C, Komaki, M, Amano, M. Hydrogen permeation characteristics of vanadium-molybdenum alloys. Trans Mat Res Soc Jpn. 1994;18B:1273–6.
44. Zhang, Y, Ozaki, T, Komaki, M, Nishimura, C. Hydrogen permeation characteristics of vanadium-aluminum alloys. Scr Mater. 2002;47:601–6.
45. Rostoker, W. The metallurgy of vanadium. New York: Wiley; 1958.
46. Adamson, Arthur W. Physical chemistry of surfaces. 2nd ed. New York: Interscience; 1967.
47. Adam, Neil K. The physics and chemistry of surfaces. 3rd ed. Oxford: London; 1941.
48. Suganuma, K, Miyamoto, Y, Koizumi, M. Joining of ceramics and metals. In: Huggins, R. A, editor. Annual review of materials science. Palo Alto, CA: Annual Reviews, Inc.; 1988. pp. 47–73.
49. van der Merwe, J. H. Recent developments in the theory of epitaxy. In: Vanselow, R, Howe, R, editors. Chemistry and physics of solid surfaces V. Berlin: Springer; 1984. pp. 365–401.

50. Mundschau, M. V, Xie, X, Evenson IV, C. R, Sammells, A. F. Dense inorganic membranes for production of hydrogen from methane and coal with carbon dioxide sequestration. Catal Today. 2006;118:12–23.

51. Peachey, Nathaniel M, Dye, Robert C, Snow, Ronny C, Birdsell, Stephan A. Composite Metal Membrane. US Patent 5,738,708, 14 April 1998.

52. Brailsford, F. Ferromagnetic theory. In: Say, M. G, editor. Magnetic alloys and ferrites. London: George Newnes Ltd; 1954. pp. 1–36.

53. Chen, Chih-Wen. Magnetism and metallurgy of soft magnetic material. New York: Dover; 1986. p. 302.

54. Zhang, Y, Ozaki, T, Komaki, M, Nishimura, C. Hydrogen permeation characteristics of V-15Ni membrane with Pd/Ag overlayer by sputtering. J Alloys Compd. 2003;356–357:553–6.

55. Zhang, Y, Ozaki, T, Komaki, M, Nishimura, C. Hydrogen permeation of Pd-Ag coated V-15Ni composite membrane: effects of overlayer composition. J Membr Sci. 2003;224:81–91.

56. Zhang, Y, Komaki, M, Nishimura, C. Morphological study of supported thin Pd and Pd-25Ag membranes upon hydrogen permeation. J Membr Sci. 2005;246:173–80.

57. Yang, J. Y, Nishimura, C, Komaki, M. Effect of overlayer composition on hydrogen permeation of Pd-Cu alloy coated V-15Ni composite membrane. J Membr Sci. 2006;282:337–41.

58. Yang, J. Y, Komaki, M, Nishimura, C. Effect of overlayer thickness on hydrogen permeation of $Pd_{60}Cu_{40}$/V-15Ni composite membranes. Int J Hydrogen Energy. 2007. doi. 10.1016/j.ijhydene.2006.12.015.

59. Paglieri, S. N, Pesiri, D. R, Dye, R. C, Tewell, C. R, Snow, R. C, Smith, F. M, et al. Influence of surface coating on the performance of V-Cu, V-Ti, and Ta membranes for hydrogen separation. In: Akin, F. T, Lin, Y. S, editors. Inorganic membranes: proceedings of the eighth international conference on inorganic membranes. Chicago: Adams Press; 2004. pp. 363–6.

60. Paglieri, Steven N, Anderson, Iver E, Terpstra, Robert L, Venhaus, Thomas J, Wang, Yongqiang, Buxbaum, Robert E, et al. Metal membranes for hydrogen separation. 20th Annual Conf. Fossil Energy Mater, Knoxville, Tennessee, USA,US DOE, 12–14 June 2006, p. 31.

61. Behr, Friedrich, Schulten, Rudolf, Weirich, Walter. Diffusion Membrane and Process for Separating Hydrogen from Gas Mixture. US Patent 4,496,373, 29 Jan 1985.

62. Buxbaum, Robert E. Apparatus and Methods for Gas Extraction. US Patent 6,183,543 B1, 6 Feb 2001.

63. Ramsey, William. Phil Mag. 1894;38:206.

64. Amandusson, H, Ekedahl, L-G, Dannetun, H. The effect of CO and O_2 on hydrogen permeation through a palladium membrane. Appl Surf Sci. 2000;153:259–67.

65. Mundschau, M, Kordesch, M. E, Rausenberger, B, Engel, W, Bradshaw, A. M, Zeitler, E. Real-time observation of the nucleation and propagation of reaction fronts on surfaces using photoemission electron microscopy. Surf Sci. 1990;227:246–60.

66. Hale, William J. Production of Catalytic Septa. US Patent 2,206,773, 2 July 1940.

67. Morreale, Bryan David. The influence of H_2S on palladium and palladium-copper alloy membranes. Dissertation, University of Pittsburgh, Pittsburgh, P. A, August 2006, and references therein.

68. Morreale, B. D, Ciocco, M. V, Howard, B. H, Killmeyer, R. P, Cugini, A. V, Enick, R. M. Effect of hydrogen-sulfide on the hydrogen permeance of palladium-copper alloys at elevated temperatures. J Membr Sci. 2004;241:219–24.

69. Alfonso, Dominic R, Cugini, Anthony V, Sholl, David S. Density functional theory studies of sulfur binding on Pd, Cu and Ag and their alloys. Surf Sci. 2003;546:12–26.

70. Kulprathipanja, A, Alptekin, G. O, Falconer, J. L, Way, J. D. Effects of water-gas shift gases on Pd-Cu alloy membrane surface morphology and separation properties. Ind Eng Chem Res. 2004;43:4188–98.

71. Kulprathipanja, A, Alptekin, G. O, Falconer, J. L, Way, J. D. Pd and Pd-Cu membranes: inhibition of H_2 permeation by H_2S. J Membr Sci. 2005;254:49–62.

72. de Rosset, Armand J. Purification of Hydrogen Utilizing Hydrogen-Permeable Membranes. US Patents 2,824,620, 25 Feb 1958 and 2,958,391, 1 Nov 1960.

73. Gade, Sabina K, Keeling, Matthew K, Steele, Daniel K, Thoen, Paul M, Way, J. Douglas, DeVoss, Sarah, et al. Sulfur resistant Pd-Au composite membranes for H_2 separations. Proc. 9th Int Conf. Inorganic Membranes, Lillehammer, Norway, 25–29 June 2006.
74. Yamamoto, Yuzo, Goto, Ryosuke. Palladium Alloy Permeable Wall for the Separation and Purification of Hydrogen. US Patent 3,155,467, 3 Nov 1964.
75. McBride, Robert B, Nelson Robert T, McKinley, David L, Hovey, Roger S. Hydrogen Continuous Production Method and Apparatus. US Patent 3,336,730, 22 Aug 1967.
76. McBride, R. B, McKinley D. L. A new hydrogen recovery route. Chem Eng Progr. 1965;61(3):81–5.
77. Beamish, F. E, McBryde, W. A. E, Barefoot, R. R. The platinum metals. In: Hampel, Clifford A, editor. Rare metals handbook. New York: Reinhold, NY, 1954. pp. 291–328.
78. Read, W. T. Jr. Dislocations in crystals. New York: McGraw-Hill; 1953.
79. Cottrell, A. H. Dislocations and plastic flow in crystals. Oxford: Oxford University Press; 1953.
80. Gjostein, N. A. Short circuit diffusion. In: Diffusion. Metals Park, OH: American Society for Metals. 1973. pp. 241–74.
81. McLean, D. Grain boundaries in metals. Oxford: Oxford University Press; 1957.
82. Broom, T, Ham, R. K. The effects of lattice defects on some physical properties of metals. In: Vacancies and other point defects in metals and alloys. London: The Institute of Metals; 1958. pp. 41–78.
83. Jost, W. Diffusion in solids, liquids and gases. New York: Academic Press; 1952.
84. Crank, J. The mathematics of diffusion. Oxford: Oxford University Press; 1967.
85. Buxbaum, Robert E. Hydrogen Generator. US Patent 6,461,408 B2, 8 Oct 2002.
86. Collins, R. E. Flow of fluids through porous materials. New York: Reinhold; 1961.
87. Brunauer, S. The adsorption of gases and vapors. Princeton: Princeton University Press; 1945.
88. Brunauer, S, Copeland, L. E, Kantro, D. L. The langmuir and BET theories. In: Flood, E. Alison, editor. The solid-gas interface; flood. New York: Marcel Dekker; 1966. pp. 77–103.
89. Laidler, K. L. Chemisorption. In: Emmett, P. H, editor. Catalysis, fundamental principles. Vol. 1. New York: Reinhold ; 1954. pp. 75–118.

Chapter 9
Hydrogen Separation Using Dense Composite Membranes

Part 2: Process Integration and Scale-Up for H_2 Production and CO_2 Sequestration

David H. Anderson, Carl R. Evenson IV, Todd H. Harkins, Douglas S. Jack, Richard Mackay, and Michael V. Mundschau

Introduction

This chapter describes the scale-up of dense composite hydrogen transport membranes that can separate hydrogen from carbon dioxide and other components of water-gas shift (WGS) mixtures derived from coal gasification. The primary application considered is a hydrogen separation system for production of hydrogen from synthesis gas produced from a coal-fired Integrated Gasification Combined Cycle (IGCC) power plant. A companion chapter reviews the fundamentals of dense composite inorganic membranes used for the transport of hydrogen [1]. Another benefit of the hydrogen membranes is the ability to retain carbon dioxide (CO_2) at high pressure from the balance of the membrane feed. This feature reduces the compression cost for the CO_2 captured.

Hydrogen derived from coal could be used as a non-polluting fuel in hydrogen-powered turbine engines or in fuel cells. Hydrogen could also be used to produce very large quantities of synthetic fuels including very low-sulfur diesel fuel, substitute natural gas, and methanol. In many parts of the world, sequestration of CO_2 is now being considered. The petroleum industry also has the need for large quantities of high-pressure CO_2 for injection into oil wells and for enhanced oil recovery from marginal wells.

Advantages of membranes over conventional systems include the ability to retain both hydrogen and carbon dioxide at high pressures, thus saving on capital and operating costs for compression. Potential hydrogen separation technologies include dense membranes, porous or size exclusion membranes, pressure swing adsorption (PSA), and solvent recovery methods. Hydrogen is separated from

D.H. Anderson (✉)
Eltron Research & Development Inc., 4600 Nautilus Court South, Boulder, CO 80301-3241, USA
e-mail: danderson@eltronresearch.com

A.C. Bose (ed.), *Inorganic Membranes for Energy and Environmental Applications*,
DOI 10.1007/978-0-387-34526-0_9, © Springer Science+Business Media, LLC 2009

syngas after coal gasification and pre-combustion, whereas CO_2 may be separated pre- or post-combustion. Amine absorption is the current method of choice for post-combustion CO_2 capture from flue gas [2]. Use of methanol (Rectisol) or glycol (Selexol) for physical absorption/regeneration are established solvent-based techniques that may be used to capture CO_2 from synthesis gas pre-combustion (Chiesa et al. [3], for example). Solvent-based recovery may require refrigeration and/or additional heating that can reduce thermal efficiency. In addition, another disadvantage of solvent absorption/regeneration is that CO_2 is recovered at reduced pressure. For transport through pipelines and for geological sequestration, the CO_2 must be compressed.

The technology under development for dense hydrogen membranes has reached the point where important consideration must be given to the integration of membranes into actual applications such as DOE's FutureGen initiative that is based on a coal fired IGCC power plant [4]. Eltron is taking the approach of developing a complete hydrogen separation system rather than individual hydrogen separation membranes. The difference between a hydrogen separation membrane unit and a complete hydrogen separation system is illustrated in Fig. 9.1. The distinction is critical in that hydrogen separation membranes must be integrated with upstream WGS and gas cleaning technologies and downstream technologies such as fuel cells and/or hydrogen turbines for power production. Technologies such as warm gas cleaning and hydrogen turbines are also currently under development in other programs. Therefore, it is more practical to design an entire hydrogen separation unit as a whole to allow for integration of these technologies. For example, by considering an entire hydrogen separation process, Eltron can optimize hydrogen separation membranes for integration with WGS catalysis, warm gas cleaning, and CO_2 sequestration on the feed side of the membrane, and simultaneously match operating condition for hydrogen turbines or high pressure hydrogen recovery on the permeate side of the membrane. Development of a complete hydrogen separation system includes demonstration of lifetime and scale-up issues such as membrane configuration and sealing.

Eltron is working to ensure that membranes will integrate seamlessly with the gas clean-up and turbine technologies being developed by others. Membrane

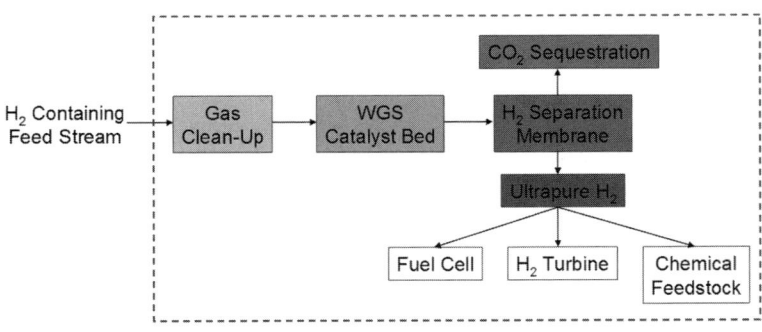

Hydrogen Separation System

Fig. 9.1 Application of hydrogen separation membrane for hydrogen production and CO_2 sequestration

development includes new membrane materials, optimized catalysts and deposition techniques for the dissociation and desorption of hydrogen on respective sides of the membrane, and the testing of membranes under expected operating conditions including the effect of potential poisons such as mercury and hydrogen sulfide.

Hydrogen Membrane Applications

Industrial sources of hydrogen include

- steam reforming of hydrocarbons, mainly methane in natural gas;
- partial oxidation of coal or hydrocarbons;
- dissociation of ammonia;
- electrolysis of water.

Hydrogen is used in a large number of chemical processes, and may be used as a fuel itself or as a reactant in the production of synthetic fuels such as in the Fischer–Tropsch hydrocarbon synthesis process, for example. In applications where hydrogen purification is required, membranes can be used for hydrogen separation. Other hydrogen purification methods include pressure swing adsorption and cryogenic separation.

Process Description for Hydrogen from Coal

One possible process scenario incorporating membranes for sequestration of CO_2 is shown schematically in Fig. 9.2. Synthesis gas is produced from coal gasification with oxygen and steam. The synthesis gas (or syngas) product streams from these hydrogen generation processes largely consist of water, carbon monoxide, carbon dioxide, and hydrogen. These are components of the WGS reaction, which further converts carbon monoxide with steam to carbon dioxide and hydrogen. Natural gas or other hydrocarbon may also be used as fuel for syngas production. The syngas product may also include unreacted fuel or hydrocarbons, particulate matter, and impurities such as hydrogen sulfide (H_2S).

The objectives of process integration and scale-up are to:

1. Provide hydrogen source from steam reforming or partial oxidation of coal.
2. Convert CO in the feed stream to CO_2 and hydrogen using the WGS reaction.
3. Separate hydrogen and CO_2 rich streams.
4. Supply hydrogen for use as:

   (i) clean burning fuel for gas turbine;
   (ii) feed to fuel cell;
   (iii) chemical feedstock;
   (iv) fuel for mobile users.

5. Separate CO_2 rich stream at high pressure for sequestration.

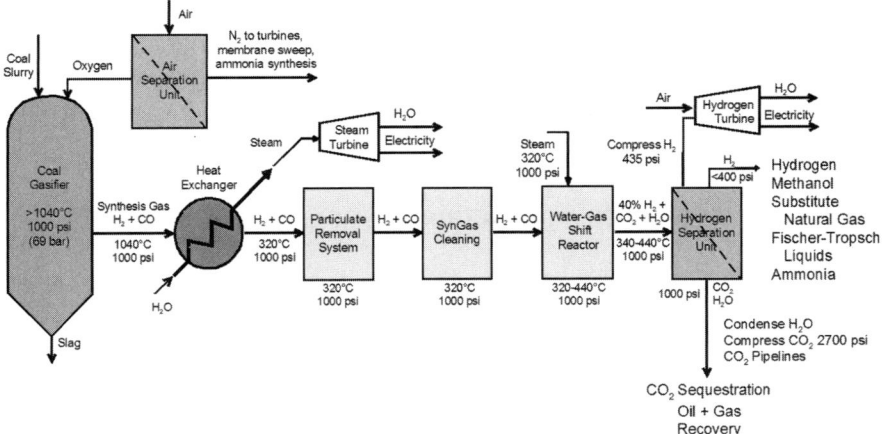

Fig. 9.2 Scenario for integrating hydrogen separation membranes with coal gasifiers and WGS reactors for producing hydrogen at high purity while retaining CO_2 at high pressure and concentration for economic sequestration

Hydrogen for Power Production

In one configuration envisioned for the plant, the hydrogen would be fed to hydrogen-powered turbine engines, which would produce electricity and steam as exhaust. An estimated hydrogen inlet pressure of 435 psia (30 bar) is assumed. Another possibility is that solid oxide fuel cells may be used in the plants. Conventional solid oxide fuel cells would require hydrogen at ambient pressures or at a few atmospheres pressure at most. Use of solid oxide fuel cells would avoid the need to compress hydrogen to high pressures, saving on such compression costs. However, solid oxide fuel cells will likely operate at 850–1,000°C, requiring re-heating of hydrogen as well as heating oxygen to the fuel cell operating temperature. Large quantities of oxygen (or air) would need to be circulated past the cathode side of the fuel cells.

Hydrogen for Production of Synthetic Fuels

In the future, coal-fired electric power plants, in their off-peak hours, may also need to serve as chemical production facilities for various synthetic fuels. Coal-fired power plants have the capital-intensive infrastructure for mining, transport and handling the massive quantities of coal that will be required. At least 1 billion metric tons per year of coal would be required to replace present U.S. consumption of petroleum distillates by synthetic fuels [5]. For perspective, approximately 1 billion metric tons of coal are now produced annually by the United States, 92% of which is used by electric power plants. All synthetic fuels will require large quantities of hydrogen at high pressures.

Once coal is gasified to synthesis gas, the H_2 and CO can be utilized in a number of synthetic routes. Fischer-Tropsch liquids can be synthesized at approximately 300 psia (20.7 bar) according to the reaction: $(2n+1)H_2 + nCO = C_nH_{2n+2} + nH_2O$. This route can produce low-sulfur diesel fuel. Methanol, CH_3OH, can be produced at 600–1,160 psi (40–80 bar) by the reaction: $2H_2 + CO = CH_3OH$. Substitute natural gas can be produced at 600 psia (41.4 bar) by the reaction: $3H_2 + CO = CH_4 + H_2O$. Finally, in the Bergius Process, coal can be reacted directly with hydrogen (produced from synthesis gas) at pressures of 3,000–10,000 psia (207–690 bar) to produce liquid hydrocarbons.

IGCC Plants

The process scenario incorporating membranes for hydrogen separation and sequestration of CO_2 in an IGCC plant is shown schematically in Fig. 9.2. Instead of burning coal directly in air as in older conventional power plants, coal is reacted with steam and pure oxygen in gasifiers typically operating above approximately $1,040°C$ to form synthesis gas. In order to avoid diluting downstream CO_2 with tons of atmospheric nitrogen (that will add to the costs of separating and sequestering CO_2), oxygen for the gasifier needs to be separated from the other components of air using conventional cryogenic technology or possibly using advanced oxygen separation membranes or chemical looping. The oxygen will be produced with approximately 95% purity by a conventional Vacuum Swing Absorption (VSA) Air Separation Unit (ASU) and the bulk of the inert will be argon. Coal gasifiers equipped with conventional oxygen separation technology are already used in practice in advanced IGCC electric power plants now in place in some plants operating around the world.

From the perspective of the membrane system, the major impact of selecting oxygen as opposed to an air blown gasifier is that it improves membrane performance by increasing the hydrogen partial pressure in the membrane separator. IGCC plants also have the advantage of gas clean-up at high pressure prior to combustion, to reduce or eliminate potential pollutants such as sulfur dioxide (SO_2).

Coal Gasifiers

Coal gasifiers are designed to operate at pressures of up to 1,000 psia (69 bar) [6] and will likely operate above about $1,040°C$. Retaining CO_2 near 1,000 psia (69 bar) would save the bulk of compression costs that would otherwise be required to compress CO_2 collected at lower pressure. In addition to delivering CO_2 at high pressures, higher gasifier pressure also increases hydrogen partial pressure in the syngas. Production and retention of hydrogen at high pressure would save on the compression costs for utilization of hydrogen in hydrogen-powered turbine engines, projected to operate with H_2 at about 435 psia (30 bar) inlet pressure. Although coal gasifiers with operating pressures of 1,000 psia (69 bar) would allow considerable downstream savings in capital costs for compressors for both CO_2 and H_2 and for the compressor operating costs, operation at 1,000 psia (69 bar) would add to the capital costs of the coal gasifiers and all system components that would be required to operate at 1,000 psia (69 bar). In addition, pure oxygen would need to be compressed to 1,000 psia (69 bar). In the final selection of the operating pressure of the coal gasifiers, costs must be weighed between the capital and operating costs of the coal gasifiers and system components operating at 1,000 psia versus capital and operating costs of compressing both CO_2 and H_2.

From a process analysis perspective, higher gasifier pressures favor hydrogen transport membranes (HTMs) because permeate H_2 can be obtained at increasingly higher pressures for a given recovery. The General Electric (GE) gasifier has demonstrated operation at 1,000 psia (69.0 bar) whereas other entrained-flow gasifiers operate below 500–600 psia (34.5–41.4 bar) [6]. The GE gasifier is also commonly outfitted with quench syngas cooling, thereby avoiding a more costly syngas cooler and steam addition upstream of the sour WGS reactor. Methane (CH_4) conversion in the GE gasifier is also typically greater than in staged-fed gasifiers.

WGS Reactors

In addition to efficient production of electricity and sequestration of CO_2, another objective is the production of low cost, high purity hydrogen. To maximize the production of H_2 from synthesis gas, the CO in the syngas will be reacted with steam in a commercial WGS reactor: $CO + H_2O = CO_2 + H_2$. To avoid sintering and deactivation of commercial WGS catalysts such as Fe_3O_4/Cr_2O_3 or Co-Mo, temperature of the syngas would need to be adjusted from gasifier temperatures above 1,040°C to approximately 320°C at the inlet to the WGS reactor in order to maintain temperatures well below 440°C for Fe_3O_4/Cr_2O_3 or 550°C for Co-Mo as heat is released by the exothermic WGS reaction. Heat extracted from cooling the syngas from 1,040°C to 320°C would be used to raise clean steam for powering non-polluting steam turbines. Membranes can also operate in the 320–440°C temperature range of the WGS reactors.

Lowering the syngas temperature to 320°C would allow insertion of particulate removal systems to remove impurities that might clog the entrance to beds of WGS catalyst. High temperature (i.e. 320–440°C) WGS catalysts, such as Fe_3O_4/Cr_2O_3, however, are fairly robust and have been used with coal-derived synthesis gas since the beginning of the twentieth century. Dual guard beds of sacrificial Fe_3O_4/Cr_2O_3 may also be employed to protect Fe_3O_4/Cr_2O_3 placed downstream. Sulfur-tolerant (sour) shift reactors may be desirable depending on the impurity removal method used (see later section).

Role of Membranes in CO_2 Sequestration and Hydrogen Production

From the FutureGen vision, the plant will be designed to sequester on the order of 1,000,000 metric tons of CO_2 per year. It is estimated that a plant will produce approximately 265 metric tons of hydrogen per day. This is equivalent to

5.9 million standard liters of hydrogen per day $(4100 \, L \, H_2 \, min^{-1})$. Membranes, therefore, must be capable of separating this capacity of hydrogen. However, membranes do not presently exist at the scale required for the projected FutureGen plant. Eltron Research is working to develop appropriate membranes that can withstand coal-derived WGS conditions at pressures of up to 1,000 psia (69 bar) and can economically extract hydrogen.

Membranes would extract hydrogen from WGS mixtures while retaining the CO_2-rich stream near the pressure of the coal gasifier. After condensation of steam in the retentate, the CO_2, at very high concentration and remaining at high pressure, would then be further pressurized to CO_2 pipeline pressures of 2,200–2,700 psia (151–186 bar). Some additional processing may be required to reduce impurities in the final CO_2 stream. After transport through the pipelines, CO_2 pressure would need to be further increased to force it into final storage sites. Carbon dioxide will be in the liquid phase above its critical pressure of 1,071 psia (73.9 bar) if cooled below its critical temperature of 304.2 K (31.1°C). It should be appreciated that pressurization of liquid CO_2 at 1,071 psia (73.9 bar) to pressures of 2,700 psia (186 bar) and above will consume much less energy than that required to compress the gas from ambient pressure to 1,071 psia (73.9 bar) [7]. Carbon dioxide capture and compression cost is estimated as adding 25% to the cost of electricity when using Selexol process for CO_2 separation [8].

In principle, hydrogen pressure in any membrane permeate cannot exceed the partial pressure of hydrogen in the membrane feed. At best, membranes can only bring partial pressures of hydrogen on either side of the membrane to equilibrium. Assuming a partial pressure of hydrogen of 400 psia (27.6 bar) in the membrane feed, obtained using a gasifier operating at 1,000 psia (69 bar), the maximum pressure of pure hydrogen in the permeate, therefore, cannot exceed 400 psia (27.6 bar). In addition, a hydrogen pressure of 400 psia (27.6 bar) in the permeate can only be achieved if the membrane brings the partial pressures of hydrogen to equilibrium at equal partial pressures on both sides of the membrane. If gasifiers are to operate at a total pressure of 450 psia (31 bar), the partial pressure of hydrogen downstream from a WGS reactor would be approximately 40% of this or 180 psia (12.4 bar) and the pressure of hydrogen in a membrane permeate could not exceed this value. As in the case with CO_2, production and retention of hydrogen in the membrane permeate at elevated pressures will save on compression capital and compression operating costs relative to the case of hydrogen collected at lower pressure.

Impact of Impurities and Gas Cleaning Methods

Significant impurities in the syngas emerging from the gasifier include particulate matter, hydrogen sulfide (H_2S), carbonyl sulfide (COS), nitrogen compounds, and mercury. Many of these impurities may be poisons for the WGS catalyst and/or hydrogen transport membrane. Gas cleaning methods are used to remove the impurities. Cold and warm cleaning systems are available or under development. Cold gas cleaning implies cooling the gas to a point where the bulk of diluent water is condensed. This often means conducting gas cleaning at approximately 105°F

(41°C), although some systems such as Rectisol (that uses chilled methanol) operate as low as −40 to −80°F (−40 to −62°C).

Several cold gas cleaning technologies are commercially available, such as amine absorbers. Cold gas cleaning technologies are often complex and lose thermal efficiency due to the additional heating and cooling duties that are required. This thermal efficiency penalty is greater when using air blown gasifiers due to large amounts of nitrogen handling. Cold gas cleaning will, however, benefit hydrogen membrane performance by raising hydrogen partial pressures due to the removal of water vapor and CO_2. Also, cold gas cleaning is capable of achieving low total sulfur concentrations if required.

For sulfur treatment, various cold gas systems have been developed to remove the bulk of sulfurous contaminants. The technologies and the levels of gas cleaning that they can achieve were summarized by Korens et al. [9]:

- Chemical solvents such as methyldiethylamine (MDEA) that are suitable for sulfur removal (H_2S) to approximately 10–20 ppmv.
- Selexol, a physical solvent that can remove H_2S to between 10 and 20 ppmv and lower levels are possible by chilling.
- Rectisol, a chilled methanol solution that, by operating between −40 and −80°F (−40 and −62°C), can accomplish deep gas cleaning down to as low as 0.1 ppmv. Although high sulfur removals are possible, this technology is expensive because of the refrigeration requirements.

Most of these systems, except Rectisol, will typically be preceded by a catalytic hydrolysis stage to convert COS into H_2S [9]. The hydrolysis occurs over an activated alumina catalyst at approximately 400°F (204°C) and is typically required because COS has poorer absorption capacity in most of the solvents relative to H_2S.

Overall, the implication is that a system incorporating such a membrane and cold gas cleaning will require the following steps:

1. Gas cooling;
2. Carbonyl sulphide hydrolysis to H_2S;
3. Primary acid gas removal using, for example, an amine;
4. Secondary clean-up using packed bed systems to remove metal carbonyls, mercury and trace acid gas;
5. Gas reheats to an appropriate temperature for membrane operation;
6. Membrane hydrogen separation.

The Selexol method of acid gas removal may be used for combined removal of H_2S and CO_2 [3, 9], in which case it is desirable to employ sulfur-tolerant shift reactors before the Selexol system. On the other hand, if H_2S alone is removed with a cold gas cleaning system, it is preferred to do so before shift reactor to avoid loss of CO_2.

It is also desirable for process efficiency to choose gas cleaning techniques with operating temperatures that avoid large amounts of heating and/or cooling. R&D projects are underway on warm gas (300–700°F) cleaning technology [10]. It simplifies the reheat cycle and has higher thermal efficiency.

Membrane Performance and Testing

Eltron Research & Development Inc. has the ability to test membranes under a variety of experimental conditions including high pressures and simulated WGS feed streams. As a step to scale-up, Eltron designed a reactor to be flexible for testing a variety of membrane configurations including multiple planar or tubular membranes up to a hydrogen separation rate of several pounds of H_2 per day. Separation membranes are housed in a containment vessel that is 5.7 feet (1.7 m) long and 7.75 inches (19.7 cm) in diameter. The containment vessel is surrounded by a tube furnace with a 3.5 foot (1.07 m) hot zone. Gas flows are supplied to the hydrogen transport membrane using mass flow controllers. Feed gases including hydrogen, carbon monoxide, carbon dioxide, helium, methane, nitrogen, and steam are controlled individually and manifolded together into a single feed stream prior to entering the reactor. Any ratio of feed gases is possible up to a total feed rate of 30 SLPM. Membranes can be operated with no sweep gas or with sweep gases of nitrogen or steam. Under dry hydrogen/inert gas mixtures, the unit is capable of feed pressures up to 1,000 psig (70 bar). Under gas mixtures that include steam, the unit can be operated up to 500 psig (35.5 bar). Gas chromatographs and mass flow meters monitor both permeate and retentate streams. Appropriate safety protocols including over-pressure, over-temperature, and carbon monoxide alarms are integrated into the system and monitored with a programmable logic controller (PLC).

The hydrogen membrane reactor was initially evaluated by measuring the performance of a pure palladium membrane. The permeability of pure palladium under various conditions has been well established and therefore can be used as a baseline test to verify if a new hydrogen membrane reactor is calibrated properly. At 440°C and under a simulated WGS feed stream, the pure palladium membrane exhibited a permeability of 2.0×10^{-8} mol $\cdot$ m^{-1} $\cdot$ s^{-1} $\cdot$ Pa$^{-0.5}$ that is consistent with literature reports.

To establish the effect of higher pressure of full WGS feed streams compared to a H_2/N_2 feed stream, two equivalent membranes were tested. Both membranes were 150 μm thick. The first membrane was tested at 380°C with a total feed flow rate of 10 L $\cdot$ min^{-1}. The composition of the feed stream was 40 mol% hydrogen, 54 mol% nitrogen, and 6 mol% helium for leak checking. 2.5 L $\cdot$ min^{-1} of argon sweep gas was used on the permeate side of the membrane. The second membrane was tested at 390°C using the same total feed flow rate of a simulated WGS mixture. The composition was 40 mol% hydrogen, 3.2 mol% carbon monoxide, 17.6 mol% carbon dioxide, 37.4 mol% steam, and 1.8 mol% helium for leak checking. The same 2.5 L $\cdot$ min^{-1} of argon sweep gas was used on the permeate side of the membrane. For each membrane the feed pressure was increased to greater than 200 psig (14.8 bar) and the hydrogen flux across the membrane was recorded. The results of these two experiments are shown in Fig. 9.3.

The solid diamonds in Fig. 9.3 represent data collected for the first membrane tested with a feed stream composed of hydrogen, nitrogen, and helium. Hydrogen flux increased as the pressure on the feed side of the membrane was increased. At 250 psig (18.3 bar) feed pressure, the hydrogen flux rate was 85 mL $\cdot$ min^{-1} $\cdot$cm^{-2}.

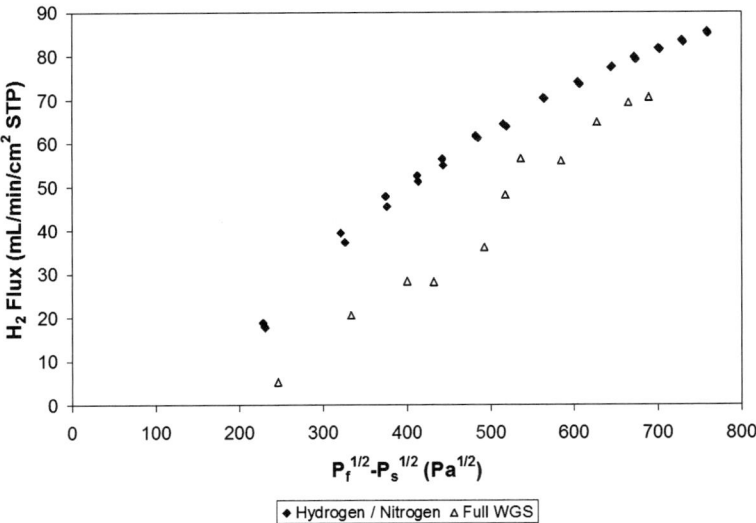

Fig. 9.3 H_2 flux vs. the difference in the square roots of hydrogen partial pressure across the membrane for equivalent membranes tested under H_2/N_2 and full WGS feed streams

The triangles in Fig. 9.3 represent data collected on the second membrane that was tested using a full WGS feed stream. As shown, the initial permeation rate was lower than when a hydrogen/nitrogen feed stream was used. The drop in flux was attributed to competitive adsorption of CO on the surface of the membrane. The feed pressure was increased to 200 psig (14.8 bar) and the average hydrogen flux rate reached 70 mL $\cdot$ min^{-1} $\cdot$cm^{-2}. For comparison, at the same feed pressure the membrane tested with the hydrogen/nitrogen feed had a hydrogen flux of 80 mL $\cdot$ min^{-1} $\cdot$cm^{-2}.

The reactor described above also has the ability to operate without a sweep gas. In this mode of operation 100% pure hydrogen can be separated at high pressures to avoid significant compression costs. One of Eltron's membranes was placed in the reactor and exposed to the simulated WGS stream at 440°C. Measured membrane thickness was 514 μm in this case. The feed stream was composed of 41.8 mol% H_2, 3.3 mol% CO, 17.3 mol% CO_2, 35.4 mol% steam, and the balance He for leak checking. Results of the permeation experiment are shown in Fig. 9.4. The feed pressure was steadily increased to 325 psig (23.4 bar) while maintaining a constant pressure of 9 psig (1.6 bar) on the sweep side of the membrane. At this point the sweep gas was removed. The feed pressure was increased to 450 psig (32.0 bar) and the permeate pressure increased to 28 psig (2.9 bar). Under these conditions the membrane achieved a hydrogen flux rate of 47 mL $\cdot$ min^{-1} $\cdot$cm^{-2}. With the appropriate feed side driving force Eltron has been able to operate membranes with 100% pure hydrogen at 450 psig (32.0 bar) on the permeate side of the membrane.

A scale-up hydrogen separation unit was designed with a membrane surface area of 63 cm^2. This unit, shown in Fig. 9.5, allows investigation into planar membrane scale-up variables such as supports, thermal expansion, and sealing. Outer diameter of the vessel is 15 cm. The unit was tested at 380°C, and, under full

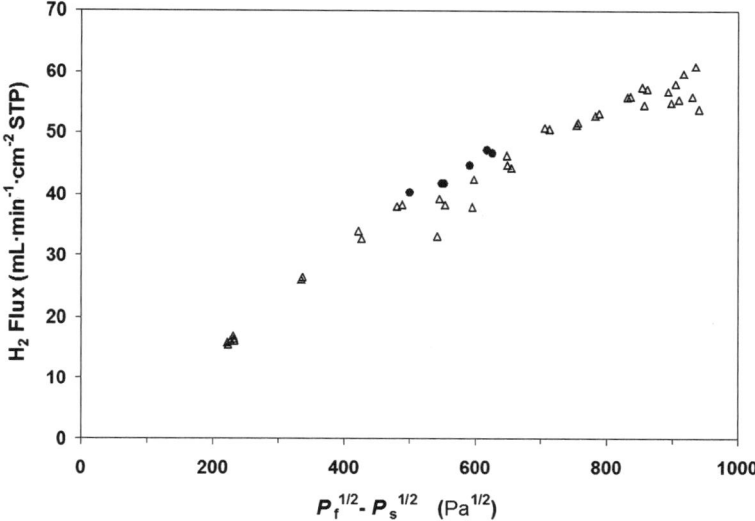

Fig. 9.4 H$_2$ flux vs. the difference in the square roots of hydrogen partial pressure across the membrane for a membrane tested under a WGS feed stream with and without a permeate sweep gas. Open triangles represent data collected with N$_2$ sweep gas. Filled circles represent data collected with a 100% H$_2$ permeate stream

Fig. 9.5 Hydrogen membrane module to separate 1.5 pounds (0.7 kg) of H$_2$ per day from a simulated WGS feed stream

WGS composition used in Fig. 9.4 (previous paragraph), a permeation rate of 82 mL $\cdot$ min^{-1} $\cdot$cm^{-2} was recorded. At this rate this unit was able to separate hydrogen at 5.2 L $\cdot$ min^{-1} or 1.5 lbs H$_2$/day (0.66 kg/day) as shown in Fig. 9.6.

The membrane permeation rate is based on an activated, power law driving force as shown below:

$$Q = \frac{AQ_o}{t} \exp\left(\frac{-E}{RT}\right) (P_{H2\ ret}^n - P_{H2\ perm}^n)$$

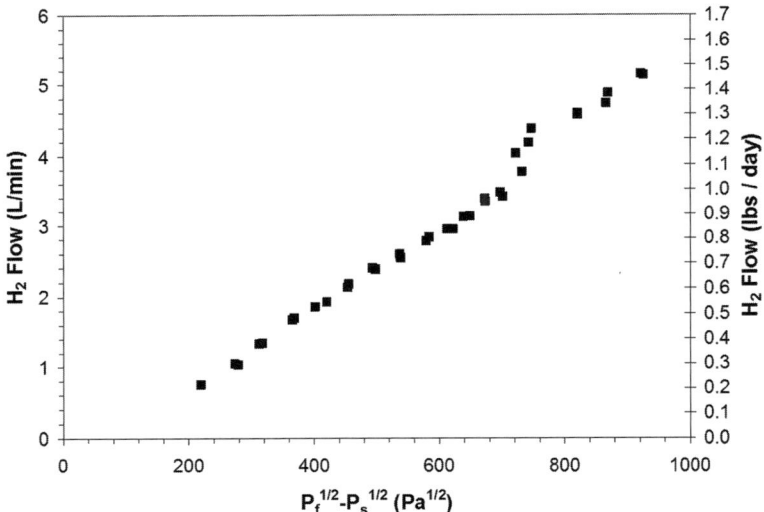

Fig. 9.6 H_2 flow vs. the difference in the square roots of hydrogen partial pressure across the membrane for a $63\,cm^2$ membrane tested under a simulated WGS feed stream

where Q is the permeation rate, Q_o is the permeability pre-exponential factor, E is the permeability activation energy, A is the membrane surface area, t is the membrane thickness, R is the gas constant, $P_{H2\,ret}$ and $P_{H2\,perm}$ are the retentate and permeate H_2 partial pressures and n is the power law exponent. Sieverts' Law follows n = 0.5.

Concepts for Membrane Scale-Up

Scale-up to meet production quantities require larger and/or more membranes in the process. General membrane scale-up was discussed by Edlund [11]. Here we discuss the types of configurations that may be employed for the current system.

Composite membranes may be scaled up in various geometries. For the all-metal membranes, both planar (Fig. 9.7) and closed-one-ended tubular configurations (Figs. 9.8 and 9.9) are candidates for larger scale systems. Closed-one-ended tubes (Fig. 9.10) are the preferred configuration for the cermet-type membranes. Closed-one-ended tubes have the advantage of allowing free thermal and chemical expansion and of minimizing membrane seal areas. The planar configuration has possible advantages of simplicity, ease of fabrication, and precedence in chemical industry.

The design shown in Fig. 9.7 is from a patent filed by Union Carbide in 1964 (U.S. Patent 3,336,730) [12]. According to a paper published by the inventors [13], Union Carbide had a membrane capacity for separating 10 million cubic feet (25 tons) of hydrogen per day in operation by 1965. At that time, Union Carbide used sheets of palladium-silver alloys approximately 1 ft × 1 ft square (30.5 × 30.5 cm)

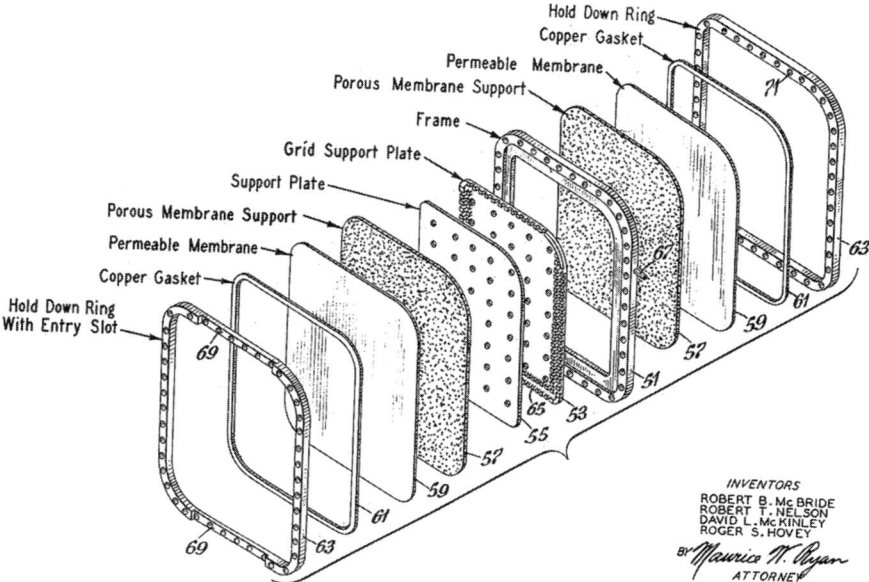

Fig. 9.7 Scale-up design for a planar, all metal membrane system from Union Carbide Patent 3,336,730, 22 August 1967 [12]. Precedence for a 10 million cubic feet (25 tons) per day hydrogen plant using metal membranes

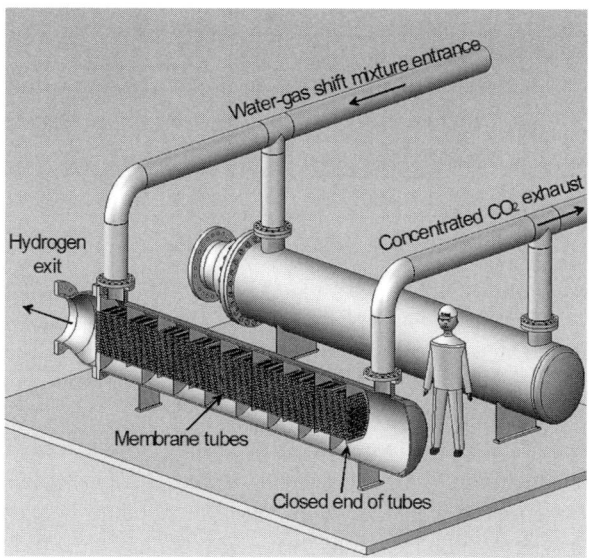

Fig. 9.8 NORAM conceptual design of a commercial metal membrane unit capable of separating 25 tons per day (∼10 MMSCFD) of hydrogen based upon closed-one-ended all-metal membrane tubes. Sizing is based upon syngas at 1,000 psig (69 barg), 450°C, 50 vol.% H_2 in feed (Courtesy C. Brereton, J. Lockhart, and W. Wolfs, NORAM)

Fig. 9.9 A tantalum metal heat exchanger shown as a concept for scale-up of closed-one-ended all-metal membrane tubes (NORAM) (Courtesy C. Brereton, J. Lockhart, and W. Wolfs, NORAM)

Fig. 9.10 (a) Concept for closed-one-ended ceramic tubes bundled for cermet membrane applications. (b) Bundle of ceramic membrane tubes produced by CoorsTek for an earlier industrial separation application (Courtesy R. Kleiner, J. Stephan, and F. Anderson, CoorsTek)

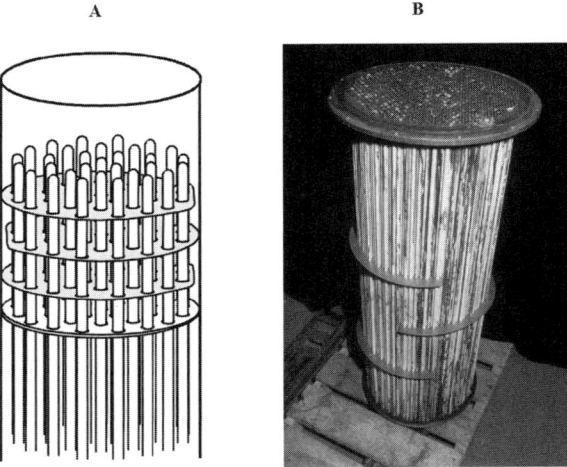

A   B

sealed by copper compression gaskets. In one patent design, planar membranes were stacked to yield 1,000 ft^2 (93 m^2) of membrane, which operated at 500 psia (34.5 bar) feed pressure. The cost of thick palladium-silver membranes for industrial scale use has since become prohibitive. However, this or similar configurations could be adapted for new all-metal membrane materials.

Figure 9.8 shows a closed-one-ended tubular configuration designed by NORAM for the all-metal membranes. The envisioned tubes are approximately 20 feet in length. This configuration is similar to that used in tantalum heat exchangers such as that shown in Fig. 9.9 that have long been used in the nuclear industry.

Figure 9.10a shows a concept for scale-up of cermet (ceramic-metal) type membranes. Ceramic closed-one-ended tubes coated with thin, dense hydrogen permeable materials are envisioned for the cermet-type membranes (Fig. 9.10a) As with the all-metal systems, the closed-one-ended tube arrangement allows free chemical and thermal expansion and minimizes the seal area. CoorsTek has had previous experience in scale-up of ceramic membranes for various industrial separations. Figure 9.10b shows an example of a bundle of open-ended ceramic tubes of porous aluminum oxide shipped by CoorsTek to an industrial customer in a previous project. Tubes shown are approximately three feet long and are sealed to headers. CoorsTek also has extensive experience in the mass production of ceramic tubes. Ceramic-supported cermet membranes may have a low cost of scale-up.

Plant Design Issues

Process simulation and economic evaluation studies (techno-economic analysis) are used to compare use of hydrogen transport membranes (HTM) with conventional separation technologies. Both power-focused and hydrogen-focused process schemes can be evaluated. Ultimately, HTM-based processes are most likely to prove advantageous by enabling the elimination of conventional CO_2 absorption (e.g. MEA, Selexol, Rectisol) while maintaining reasonable permeate total pressures and minimizing H_2 losses in the CO_2-rich retentate. In their 2005 book [2], the Carbon Capture Project team stated that "The team believes that membrane reactors for hydrogen production have the potential for significant cost reduction and gave this technology its top priority."

DOE has provided guidelines for techno-economic analyses of CO_2 capture systems [14]. The guidelines include a representative basis for plant size and operating requirements, such as fuel specifications, fraction of CO_2 captured, plant emissions, power generation, and the like. Process economics depend on assumptions for equipment costs, feed and product prices, and so on. Previously published techno-economic studies on coal-derived power and H_2 include those by Mitretek [15], Parsons [16–18], Nexant [19, 20], and others [21–23]. Reviews of existing coal-based chemicals plants can also provide useful insight. Such plants demonstrate relevant technologies such as sour WGS reactors, absorption-based H_2S and CO_2 removal and PSA-based H_2 production in a gasification environment. Furthermore, they demonstrate that downstream processes involving sensitive catalysts can be successfully operated given appropriate impurity mitigation steps [24, 25]. Analogs could be drawn for HTM-based systems. Specific examples of such plants include production of coal-derived substitute natural gas at Great Plains Synfuels in North Dakota, methanol at Eastman's Kingsport, Tennessee complex and Fischer-Tropsch liquids at Sasol facilities in South Africa. Also, coal-derived NH_3 and urea is produced in the US (Coffeyville, Kansas), Japan (Ube) and China (e.g. Lunan, Weihe, Huainan, and Haolinghe).

Summary

Hydrogen transport membranes provide advantages in separating streams of hydrogen and CO_2 from synthesis gas produced in coal gasifiers. Key advantages of membranes over conventional systems include the ability to retain both hydrogen and carbon dioxide at high pressures, thus saving on capital and operating costs for compression, and membrane separation can occur at temperatures of the WGS reaction. The membranes will work with synthesis gas generated from any source including coal, petroleum coke, natural gas, or biomass after gas cleaning. Scale-up of hydrogen transport membranes is under development based on planar and tubular configurations of similar systems.

Acknowledgments This chapter is based on work conducted for the U.S. Department of Energy, Contract Number DE-FC26-05NT42469 (Arun Bose, Program Manager). The chapter includes contributions from our partners at NORAM (Clive Brereton, James Lockhart, and Warren Wolfs), Praxair (Aqil Jamal, Gregory J. Panuccio, and Troy M. Raybold), and CoorsTek (Frank Anderson, Richard N. Kleiner, and James E. Stephan). Jarrod A. Benjamin, Matthew B. Post, Adam E. Calihman and David A. Gribble, Jr. of Eltron Research & Development Inc. prepared the hydrogen membranes for testing and conducted the high pressure membrane flux experiments.

References

1. Mundschau MV. Hydrogen separation using dense composite membranes. Part 1. Fundamentals. In: Bose A. editors. This Volume. Springer; 2007.
2. Thomas DC, Kerr HR. Introduction. In: Thomas DC, Benson SM. editors. Carbon dioxide capture for storage in deep geologic formations. Vol. 1. Amsterdam: Elsevier; 2005. pp. 37–46.
3. Chiesa P, Consonni S, Kreutz T, Williams R. Co-production of hydrogen, electricity and CO2 from coal with commercially ready technology. part a: performance and emissions. Int J Hydrogen Energy. 2005;30:747–67.
4. U.S. DOE, Office of Fossil Energy, FutureGen: Integrated Hydrogen, Electric Power Production and Carbon Sequestration Research Initiative (Energy Independence Through Carbon Sequestration and Hydrogen from Coal). 2004. http://www.fossil.energy.gov/programs/powersystems/futuregen/futuregen_report_march_04. pdf. Accessed May 2007.
5. Landrum M, Warzel J. Technical Session: Crude Oil Supply High Quality Competitive Distillate Fuels from Coal-to-Liquids Processing. NPRA Annual Meeting Presentation. March 2007. http://www. bakerobrien.com/images/uploads/NPRA%20CTL%20Presentation.pdf. Accessed May 2007.
6. Holt N. Gasification Process Selection – Trade-Offs and Ironies, Presented at the 2004 Gasification Technologies Conference. 3–6 Oct 2004. http://www.gasification.org/ Docs/2004_Papers/ 30HOLT.pdf.Accessed May 2007.
7. McCollum DL, Ogden JM. Techno-Economic Models for Carbon Dioxide Compression, Transport, and Storage, UC Davis. 2006. http://pubs.its.ucdavis.edu/download_pdf. php?id=1047. Accessed May 2007.
8. National Energy Technology Laboratory (NETL). Carbon Sequestration: CO2 Capture. http://www.netl.doe.gov/technologies/carbon_seq/core_rd/co2capture.html. Accessed May 2007.
9. Korens N, Simbeck DR, Wilhelm DJ. Process Screening Analysis of Alternative Gas Treating and Sulfur Removal for Gasification. Dec 2002. http://www. netl.doe.gov/technologies/coalpower/gasification/pubs/pdf/SFA%20Pacific_Process%20 Screening%20Analysis_Dec%202002.pdf. Accessed May 2007.
10. National Energy Technology Laboratory (NETL). Gasification – Gas Cleaning & Conditioning. http://www.netl.doe.gov/technologies/coalpower/gasification/gas-clean/index.html. Accessed May 2007.

11. Edlund DJ. Engineering scale-up for hydrogen transport membranes. In: Sammells AF, Mundschau MV, editors. Weinheim, Germany: Wiley-VCH; 2006. pp. 139–64.
12. McBride RB, Nelson RT, McKinley DL, Hovey RS. Hydrogen continuous production method and apparatus. U.S. Patent 3,336,730, 22 Aug 1967.
13. McBride RB, McKinley DL. A new hydrogen recovery route. Chem Eng Prog. 1965;61(3):81–5.
14. National Energy Technology Laboratory (NETL). Carbon Capture and Sequestration System Analysis Guidelines. April 2005. http://www.netl.doe.gov/ technologies/carbon_seq/Resources/Analysis/pubs/CO2CaptureGuidelines.pdf. Accessed May 2007.
15. Gray D, Tomlinson G. Hydrogen from Coal, MTR 2002-31, Contract DE-AM26-99FT40465. July 2002.
16. Rutkowski MD, Klett MG, White J, Schoff RL, Buchanan TL. Hydrogen production facilities plant performance and cost comparisons. Final Report (Compilation of Letter Reports from June 1999 to July 2001), Contract DE-AM26-99FT40465, Subcontract 990700362, March 2002.
17. Rutkowski MD, Buchanan TL, Klett MG, Schoff RL. Capital and operating cost of hydrogen production from coal gasification. Final Report, Contract DE-AM26-99FT40465, Subcontract 990700362, Task 50611, April 2003.
18. Rutkowski MD, Schoff RL, Holt NA, Booras G. Pre-Investment of IGCC for CO_2 Capture with the Potential for Hydrogen Co-Production, presented at the 2003 Gasification Technologies Conference, 12–15 Oct 2003.
19. Bechtel Corp., Global Energy, Inc., Nexant, Inc. Gasification Plant Cost and Performance Optimization, Contract DE-AC26-99FT40342, Sep 2003.
20. Schwartz M. Novel Composite Membranes for Hydrogen Separation in Gasification Processes in Vision 21 Energy Plants: Final Technical Progress Report 5/26/01-9/30/04, Contract DE-FC26-01NT40973, Dec 2004.
21. Collot A-G. Prospects for hydrogen from coal. IEA Clean Coal Centre, London ISBN 92-9029-393-4, 2003.
22. Doctor RD, Molburg JC, Brockmeier NF, Stiegel GJ. Designing for hydrogen, electricity, and CO_2 Recovery from a shell gasification-based system, presented at the 18th Annual International Pittsburgh Coal Conference, Newcastle, New South Wales, Australia, 14 Oct 2001.
23. Parsons EL, Shelton WW, Lyons JL. Advanced Fossil Power Systems Comparison Study, Final Report, Dec 2002.
24. Wang A. Task 3.5: Poison Resistant Catalyst Development and Testing Final Topical Report, Design and Construction of the Alternative Fuels Field Test Unit and Liquid Phase Methanol Feedstock and Catalyst Life Testing at Eastman Chemical Company, Kingsport, TN, Contract DE-FC22-95PC93052, March 1997.
25. Eastman Chemical Co. and Air Products and Chemicals, Inc. Removal of Trace Contaminants from Coal-Derived Synthesis Gas Topical Report, Contract DE-FC22-92PC90543, March 2003.

Chapter 10
Gasification and Associated Degradation Mechanisms Applicable to Dense Metal Hydrogen Membranes

Bryan Morreale, Jared Ciferno, Bret Howard, Michael Ciocco John Marano, Osemwengie Iyoha, and Robert Enick

Scope

Dense metal membranes have been identified as a promising technology for post-gasifier forward water-gas shift (WGS) membrane reactors or post-shift membrane separation processes. It is known that both major and minor gasification effluent constituents can have adverse effects on the mechanical and chemical stability of potential metal membranes. With this in mind, the scope of this chapter is to provide introductions to the gasification process and dense metal membranes, and the possible degradation mechanisms that dense metal hydrogen membranes may encounter in gasification environments. Degradation mechanisms of interest include catalytic poisoning, oxidation, sulfidation and hydrogen embrittlement. Additionally, the influence of in-situ and post-WGS gas compositions and CO conversions on the aforementioned degradation mechanisms will be addressed.

Gasification and Integrated Membrane Technologies

The US Department of Energy's National Energy Technology Laboratory (NETL) is aggressively pursuing research and development directed toward reducing apprehensions over fuel independence, energy availability and reliability, and environmental issues especially as related to global warming. The production of "synthesis gas" (syngas) via the gasification of indigenous carbonaceous feedstocks has the potential to address some of these energy and environmental concerns. The abundance of coal in the United States in conjunction with the flexibility of syngas, which can be converted into electricity, hydrogen and/or liquid fuels, is considered

B. Morreale (✉)
US DOE National Energy Technology Laboratory, 626 Cochrans Mill Road, PO Box 10940, Pittsburgh, PA 15236, USA
e-mail: bryan.morreale@netl.doe.gov

A.C. Bose (ed.), *Inorganic Membranes for Energy and Environmental Applications*,   173
DOI 10.1007/978-0-387-34526-0_10, © Springer Science+Business Media, LLC 2009

a promising near- to mid-term component in the transition to a renewable energy society.

Gasification of carbonaceous solids, such as biomass or coal, is generally accomplished in a fixed bed, fluidized bed or entrained flow reactor at operating temperatures and pressures ranging from 698 to 1,873 K and 0.1 to 7 MPa respectively. Gasification involves complex reaction mechanisms for partial and complete oxidation of the carbon in the feedstock as described in detail by Higman [1]. The resulting syngas is primarily composed of H_2, CO, CO_2, H_2O and CH_4. Additionally, the impurities often encountered in the feedstock can result in minor syngas constituents including H_2S, NH_3 and other low molecular weight S- and N-containing compounds such as COS and HCN. In general, the syngas must be cleaned to remove impurities and particulates before use as fuel or chemical. After gas cleaning, the syngas may be mixed with air and combusted in a gas turbine to produce electric power (topping-cycle). Waste heat is recovered from the combustion turbine exhaust by generating steam in a heat recovery steam generator (HRSG). The steam is then sent to a steam turbine for additional power production (bottoming-cycle). A simplified schematic of an integrated gasification combined cycle (IGCC) plant is illustrated in Fig. 10.1.

When compared to existing coal combustion systems, gasification systems have higher process efficiencies, feedstock and product flexibility, and the potential for achieving "near-zero" emission levels of SOx, NOx, particulate matter, and toxic compounds of Hg and As. Moreover, gasification has the potential to effectively generate electricity while minimizing greenhouse gas emissions through more facile CO_2 capture and sequestration. As shown in Fig. 10.2, the production of a sequestration ready CO_2 stream is accomplished by reacting to the CO-rich syngas with water to produce H_2 and CO_2, via the WGS reaction. Current gasification schemes use an

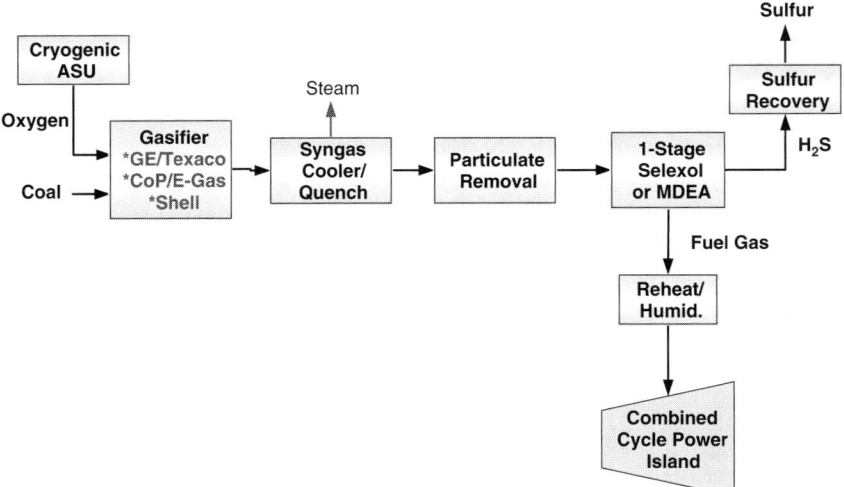

Fig. 10.1 IGCC process without CO_2 capture

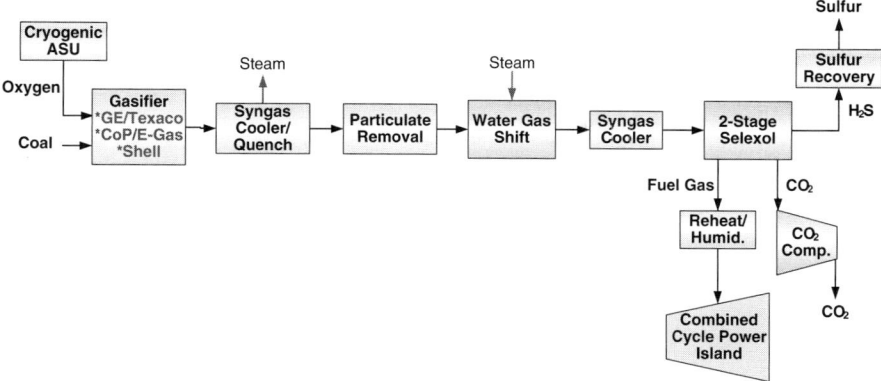

Fig. 10.2 IGCC process with CO_2 capture using Selexol^{TM}

acid gas removal (AGR) system, such as UOP's Selexol^{TM} absorption process, to selectively separate the CO_2 and trace amounts of H_2S from the H_2-rich syngas. The highly concentrated CO_2 is compressed for transport to a suitable sequestration site. Currently, 90% CO_2 capture using the state-of-the-art Selexol^{TM} scrubbing increases the cost of electricity by 30% relative to an IGCC plant with no-capture and reduces the net power plant efficiency by 7–10%.

In order to further reduce the cost of carbon capture with IGCC, a number of novel technologies are being investigated for separating H_2 and CO_2. In particular, hydrogen separation membranes have garnered considerable attention in recent years. Figure 10.3 identifies possible integrations of hydrogen separation membranes into an advanced IGCC power plant [2]. Integration of hydrogen separation membranes with IGCC for carbon capture has a number of potential advantages. Foremost is the production of CO_2 for sequestration at relatively high system pressures, thus minimizing the amount of further compression required. Generally, with hydrogen separation membranes, hydrogen is produced at low pressure, but with the proper selection of operating pressure and the use of a permeate sweep gas, a "turbine-ready" hydrogen stream can be produced with minimum re-compression requirements.

Process conditions, gas composition, pressure, and temperature are different at the various membrane locations identified in Fig. 10.3. The operating envelope for any given membrane technology should match the conditions at the position of the membrane in the system. Each location has unique advantages and disadvantages with regards to hydrogen separation and recovery. In addition, other technologies under development, such as warm-gas clean-up systems, will influence selection of a H_2-CO_2 separation technology and may or may not complement membrane separation. Proper placement of a membrane unit in the process is critical, and it is unlikely that one type of membrane material can perform adequately in all feasible locations in the process. The challenge to developers is to take advantage of the unique characteristics of individual membrane technologies, while minimizing any

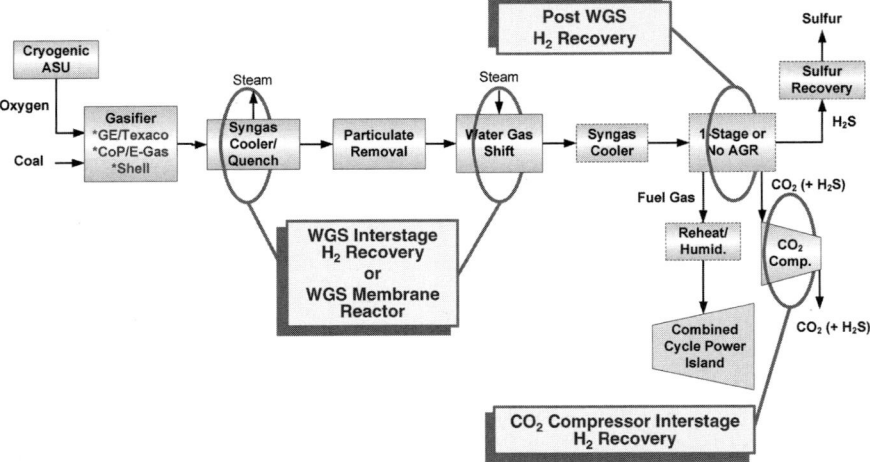

Fig. 10.3 Possible gas separation membrane integration with IGCC

Table 10.1 Approximate compositions and conditions of the gas streams of the gasification process where membrane technologies have potential impact

Constituent	Gasifier exit	WGS inlet	WGS outlet
Temperature (K)	1,300	500	500
Pressure (MPa)	4.2	3.9	3.6
H_2O (mol%)	18	45	26
CO (mol%)	34	20	0.5
CO_2 (mol%)	14	11	30
H_2 (mol%)	28	19	39
N_2 (mol%)	1	1	1
CH_4 (mol%)	4	2	2
COS (ppm)	400	200	5
H_2S (ppm)	7,000	4,000	4,000
NH_3 (ppm)	3,000	2,000	2,000

shortcomings of the conditions at each potential membrane location, as detailed in Table 10.1.

In principle, membranes can be employed anywhere downstream of the gasifier. Discussion here will be limited to placement of membranes upstream of acid-gas removal, the so-called "hot end" of the IGCC process because the focus of this article is on dense-metallic membranes. Upstream application of membranes could be solely for the separation of H_2 from CO_2, or could additionally incorporate the WGS reaction as a "WGS membrane reactor" (WGSMR). Operating temperatures in the hot end of the IGCC process range from as high as 973 to 1,173 K prior and during syngas cooling, to as low as 513–593 K exiting the last reactor stage used for the WGS. Nominal coal gasification pressures range from 2 to 6 MPa. However, higher pressures may be feasible. Operating at higher pressures will improve the driving force for gas separation.

The WGS reaction (Eq. 10.1) is typically carried out at elevated temperatures in order to improve reaction kinetics. However, the reaction is exothermic and equilibrium is favored at lower temperatures. Therefore, the high-temperature shift (573–823 K) is often followed by a low temperature shift (473–573 K) to maximize conversion of CO. In addition, excess steam ($H_2O/CO > 2.5$) is also used to enhance the equilibrium for the production of H_2 and to prevent the production of coke on the WGS catalysts.

$$CO + H_2O \leftrightarrow CO_2 + H_2, \quad \Delta H^o_{298} = -41 \text{ kJ/mol} \tag{10.1}$$

Clearly, the application of any type of membrane in the hot-end of the IGCC process can improve the conversion of CO if the membrane can selectively remove H_2 or CO_2 from the syngas. A highly selective membrane for H_2 relative to the other syngas components is required to enhance equilibrium as well as to prevent the loss of valuable reactants and to minimize the need for further purification of the H_2-rich permeate.

Combining the H_2/CO_2 separation with the WGS reaction can be achieved either by integrating membrane separators between WGS reactor stages or by combining the membrane module and WGS reactor in a single vessel, referred to as a WGSMR. The latter could be accomplished by taking advantage of the catalytic properties of the membrane, loading the WGSMR with WGS catalysts, or possibly by applying an additional catalytic coating to the membrane surface.

In addition to improved conversion as a result of Le Chatelier's principle, integration of the WGSMR can potentially provide other process improvements. These include lowering of the steam-to-carbon ratio, reduction of the number of reactor stages, improvement in reaction kinetics by operation at elevated temperatures and possible elimination of the need for additional external catalyst beds.

The integrated WGSMR concept presents a number of challenges due to operations at conditions typically beyond those normally encountered for hydrogen membrane separations. The feed gas will contain particulates and other impurities. High temperature corrosion, large temperature gradients along the length of the reactor, and potential side reactions (Boudouard, CO-hydrogenation and methanation) may be problematic.

As noted above, high hydrogen selectivity is a prerequisite for successful application of gas separation membranes in this service. The severity of the conditions encountered by a membrane will likely result in the use of advanced or exotic materials with associated high membrane production costs. The total membrane area required will need to be minimized to mitigate these costs. In order to reduce membrane area, the membrane will need to have a very high H_2 permeance. A sweep gas may also be required on the permeate side of the membrane to improve the driving force for separation. Although these membrane characteristics may be achievable, they are counter to gas separation membrane commercial experience, where mostly low-cost, relatively low permeance and moderately selective membranes have demonstrated a competitive advantage over other separation technologies. Finally, it may be difficult to achieve extremely high rates of H_2 recovery in this system since the WGSMR will still be subject to a hydrogen partial pressure pinch, limiting hydrogen transport at the outlet end of the unit, and some additional gas separation may be required downstream of the membrane reactor.

Dense Metal Membrane Materials, Configurations, Mechanisms of Transport, and Permeability

Palladium

Initial hydrogen diffusion experiments with metal membranes began as early as the middle of the nineteenth century with palladium [3–5]. Since these experiments, Pd has been the most widely investigated membrane material with the most acknowledged studies being conducted in the mid-twentieth century [6–11]. Additionally, several comprehensive reviews have been published that summarize the developments of palladium membrane technologies over the past 150 years [12, 13].

Interest in palladium as a membrane material stems from its relatively high permeability as compared to other pure metals (Fig. 10.4), its infinite selectivity for hydrogen, reasonable mechanical characteristics, and highly catalytic surface [14, 15]. Although palladium has been the most highly investigated membrane material, properties including cost and chemical stability inhibit its widespread application in large-scale processes such as gasification.

The cost of palladium, varying between about $180 and $1,100 per troy-ounce over the last decade, makes it economically impractical as a membrane material for large-scale processes in which a thickness of greater than ~25 microns is

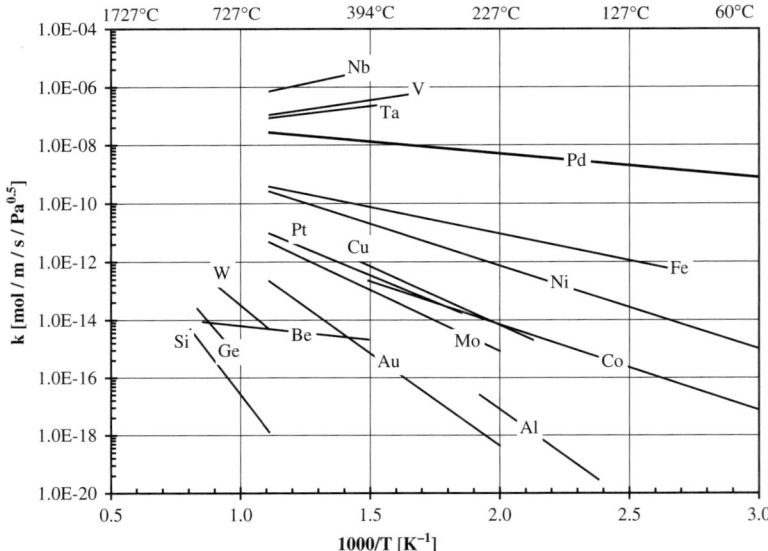

Fig. 10.4 Hydrogen permeability of selected pure metals as a function of temperature

employed. Current membrane research is focusing on the development of new and more efficient technologies to circumvent the high capital costs of palladium. These technologies include alloying palladium with lower cost metals, applying very thin films of palladium and palladium alloys to highly permeable substrates, and identifying efficient means of recycling "used" membranes [16–64].

Palladium Alloys

Palladium alloy research has been conducted in an effort to minimize the amount of palladium needed in membrane fabrication, increase the performance, promote chemical robustness and increase mechanical integrity by improving mechanical strength and reducing the effect of the hydride phase change [12, 65, 66]. Palladium has been alloyed with an array of elements including boron [67, 68], cerium [67–70], copper [26, 33, 57, 67, 68,71–76], gold [67, 68], iron [77], nickel [27, 43, 44, 67, 68, 78], silver [16, 20, 30, 41, 51, 57, 67, 68, 71,78–82], yttrium [70, 83, 84], and others. Although all the metals alloyed with palladium have some effect on performance and deformation resistance, cerium, copper, gold, silver and yttrium are unique in that these binary alloys can exhibit higher permeability values than each of the pure metal components exhibit individually [12, 66, 67], as shown in Fig. 10.5. The observed change in permeability of a metal alloy is attributed to increasing either the solubility and/or the diffusivity of hydrogen within the metal and thus increasing permeability [12,85–89].

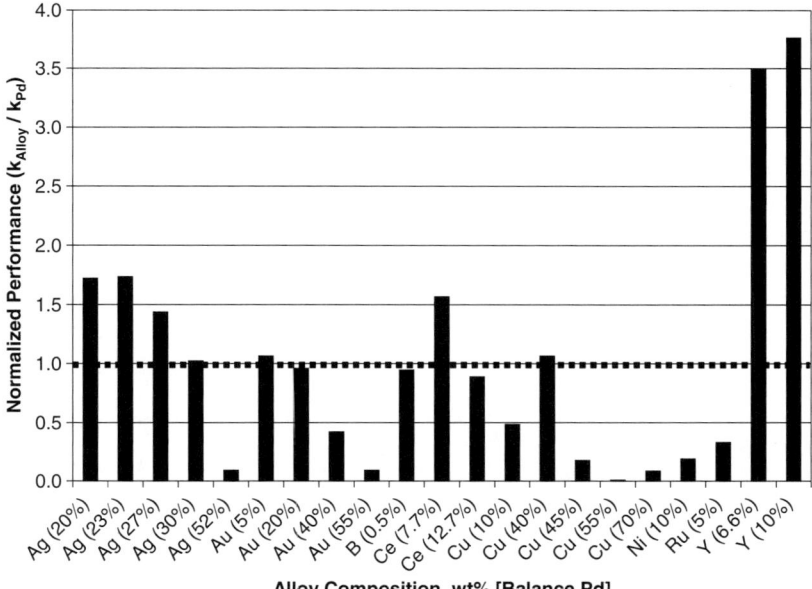

Fig. 10.5 Hydrogen permeability of various Pd-based alloys as compared to pure palladium

The palladium-silver alloy membrane system was successfully commercialized in the early 1960s [12], but the reduction of palladium content by the addition of silver would still not be a cost-effective alternative for large-scale processes [42] unless micron-scale films could be prepared, a goal currently being addressed by many researchers. In recent years, the Pd–Cu system has been the most heavily investigated alloy for hydrogen membrane applications due to the high permeability of select alloys [67, 90, 91], enhanced mechanical properties [92] and reported chemical resistance. The elevated permeability identified for select Pd–Cu alloys is attributed to an increase in both the solubility and diffusivity of the B2 crystalline phase [86–88] as compared to the face-centered-cubic (fcc) phase that exhibits permeability values proportional to the Pd-content [89, 91, 93].

Super-Permeable Metals

In the past 20 years, select refractory metals (niobium, tantalum, vanadium, and zirconium) have been of increasing interest due to their very high permeability (relative to palladium as shown in Fig. 10.4) and have been the focus of a comprehensive review by Mundschau and co-workers [94]. Although these "super-permeable" materials show very high performance potentials, the application of these materials to large scale processes has been hindered due to an insufficient catalytic surface (attributed to oxide formation associated with even brief exposure to dilute concentrations of oxygen during handling, transport, installation or operation), and embrittlement due to the formation of metal-hydrides [12,95–98]. Efforts by numerous researchers have focused on incorporating these super-permeable metals and super-permeable metal alloys (ex. $V_{85}Ni, V_{85}Cu$) into an effective composite membrane [15, 18, 21, 42,47–49,54, 95,98–111]. These composite membranes generally involve the application of a highly catalytic surface coating (such as palladium) on an oxide free super-permeable metal or metal alloy substrate. The resultant composite membranes can demonstrate a very high performance (governed by the super-permeability metal), reduced costs (reduced amounts of palladium), and increased mechanical strength (high strength of the refractory metals). Additionally, since the "super-permeable" substrate is utilized as a dense foil, a highly pure hydrogen product will be produced even in the presence of defects in the coatings. Conversely, a defect in a coating layer on a porous substrate can allow essentially unrestricted flow of all the gas components.

Transport Mechanism of Hydrogen Through Dense Metal Membranes and Governing Permeation Equations

Hydrogen transport through a dense metal membrane is usually envisioned as following an atomic transport mechanism. The atomic transport process, assuming transport from the high hydrogen pressure surface to the low pressure surface,

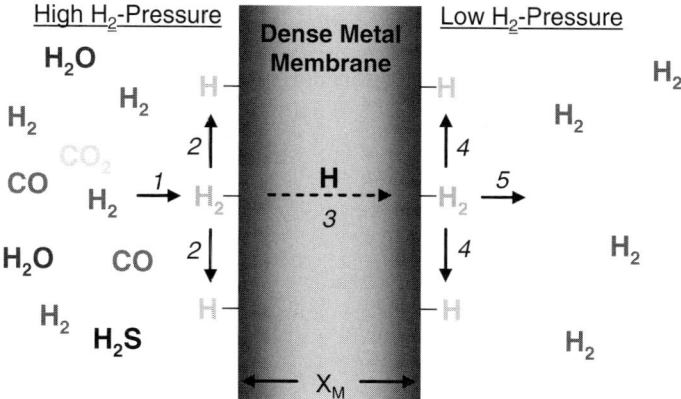

Fig. 10.6 Illustration of the accepted hydrogen transport mechanism through dense metal membranes

is illustrated in Fig. 10.6 and follows several sequential steps: (1) the adsorption of hydrogen molecules, (2) the dissociation of the adsorbed hydrogen molecules into hydrogen atoms, (3) the transport of atomic hydrogen through the bulk of the metal via hopping through defects and/or interstitial sites within the metal lattice, (4) recombination of atomic hydrogen to molecular hydrogen and (5) followed lastly by the desorbtion of molecular hydrogen from the membrane surface [79, 112, 113].

Defect free dense metal membranes are infinitely selective to hydrogen due to the atomic transport mechanism that can be described through derivation of Fick's Law (defining the transient mass transfer of hydrogen in the direction of decreasing concentration).

$$\frac{dN_H}{dt} = -D\frac{dC_H}{dx} \tag{10.2}$$

In relation to the hydrogen membrane processes, the direction of mass transfer would be defined from the retentate (hydrogen rich feed stream) to the permeate (hydrogen poor stream). Assuming a steady state permeation process and integrating Eq. 10.2 yields the governing, steady state form of Fick's law for membrane processes, Eq. 10.3,

$$N_H = 2N_{H_2} = -D_M\frac{\Delta C_H}{X_M} \tag{10.3}$$

where N_H is the atomic flux of hydrogen, N_{H2} is the molecular flux of hydrogen, D_M is the diffusivity of the membrane material, and ΔC_H is the atomic concentration gradient of hydrogen across the membrane thickness X_M.

For "thick" membranes, the rate-limiting step in the transport mechanism is assumed to be the diffusion of the atomic hydrogen through the membrane because the adsorption and dissociation of hydrogen on the catalytic surface is expected to be very rapid. As a result of the slow diffusion process, it can be assumed that the atomic concentrations at the membrane surfaces are at equilibrium with the retentate and permeate hydrogen gases. Application of Sieverts thermodynamic relation, Eq. 10.4, gives the atomic concentration of hydrogen in terms of the partial pressure of the hydrogen gas at the membrane surface.

$$C_H = K_S P_{H_2}^{0.5} \tag{10.4}$$

Combining Eqs. 10.3 and 10.4 gives the governing membrane permeation equation in terms of the hydrogen partial pressures of the retentate and permeate, the Richardson Eq. 10.5 [102].

$$N_{H_2} = -\frac{D_M K_S}{2} \frac{(P_{H_2,Ret}^{0.5} - P_{H_2,Per}^{0.5})}{X_M} \tag{10.5}$$

Defining the isothermal permeability constant of a membrane material by half of the product of the membrane diffusivity and the Sieverts constant, and substituting into the Richardson Eq. 10.5 yields Eq. 10.6, which is the governing equation for the diffusion limited, atomic transport membrane process.

$$N_{H_2} = -k_M \frac{(P_{H_2,Ret}^{0.5} - P_{H_2,Per}^{0.5})}{X_M} = k_M'(P_{H_2,Ret}^{0.5} - P_{H_2,Per}^{0.5}) \tag{10.6}$$

The rate of the molecular dissociation on the membrane surface and atomic-diffusion through the membrane has a large influence on the exponent of the partial pressure drop across the membrane. If the dissociation of hydrogen on the surface of the membrane is the rate-limiting step, then the exponent of the partial pressure is 1.0, while membranes governed by a rate-limiting step of hydrogen diffusion have an exponent of 0.5 in the partial pressure expression. However it has been demonstrated that for some atomic-diffusion processes the partial pressure exponent can range from 0.5 to 1.0 due to the competing rates of dissociation and diffusion [114]. Thus Eq. 10.7 represents a more global form of the membrane governing equation, which compensates for the competing "rate-determining steps" associated with the applicable mechanism.

$$N_{H_2} = -\frac{k_M}{X_M}(P_{H_2,Ret}^n - P_{H_2,Per}^n) \tag{10.7}$$

Additionally, the temperature dependence of the permeability of a dense metal membrane is observed to follow an Arrhenius-type expression, Eq. 10.8. E_p and k_o represent the activation energy of permeation and the pre-exponential constant respectively.

$$k_M = k_o \exp[-E_p/RT] \tag{10.8}$$

The permeability through a composite membrane consisting of several dense metal layers can be described by a "resistances in series" methodology, similar to that used in heat conduction through a composite wall [100]. The total permeability for a composite dense metal membrane can be estimated from Eq. 10.9 where $X_{M,Tot}$ is the total thickness of the membrane.

$$\frac{k_{M,Tot}}{X_{M,Tot}} = \frac{1}{\sum \frac{X_{M,i}}{k_{M,i}}} \tag{10.9}$$

Metal Membrane Fabrication Methods

Dense metal membranes have been of interest to researchers for many years and have been identified as a promising technology for producing highly pure hydrogen from mixed gas streams such as those resulting from the gasification of fossil fuels and biomass. Unfortunately, inadequacies including relatively high cost, susceptibility to surface poisoning and corrosion in the presence of gases such as H_2S, NH_3, CO, CO_2 and Hg, and hydride formation have prevented the widespread application of these dense metal membranes. In an effort to mitigate the costs associated with membranes containing precious metals and to increase the performance of dense metal membranes, researchers are focusing on the development of fabrication methods that would yield a commercially viable membrane configuration. Some of these approaches will be described in the following sections.

Unsupported Thin Foils

In an effort to reduce the concerns over the use of high-cost metals in membrane fabrication as well as to increase membrane performance, researchers are investigating the feasibility of "thin" foil development [25, 26, 115]. Limitations of "traditional" rolling techniques for the production of thin foils include the high potential for foil defects and thickness inconsistency over the foil width due to difficulties encountered in maintaining precise roller tolerances and avoiding roller deformation at increased sheet widths. Methods including vacuum deposition on removable substrates and acid etching are being investigated for the fabrication of foils as thin as 3 microns. However, applying these thin foil membranes to processes where elevated transmembrane pressure differentials are encountered would require the use of a mechanical support.

Thin Films on Porous Substrates

Several researchers are investigating the feasibility of Pd-based membranes deposited onto highly permeable, porous metal and/or ceramic substrates, Fig. 10.7a [76,116–119]. Currently, the most common method used for the deposition of the "active" membrane layer onto the porous substrate is electroless plating, although some researchers have utilized physical and chemical vapor deposition techniques.

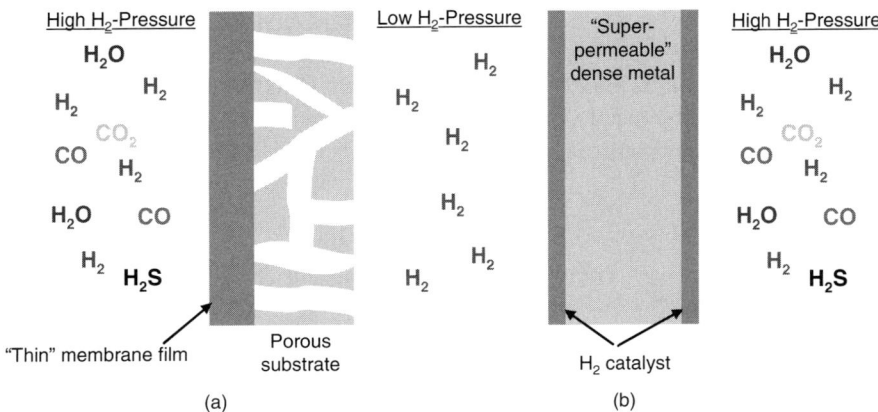

Fig. 10.7 Illustration of the configuration of a membrane deposited onto (a) a porous substrate, (b) a dense substrate

Membranes fabricated in this manner rely heavily on the characteristics of the substrate, such as pore size and pore size distribution. Support substrates having large pores often require much thicker Pd-layers leading to decreased performance and increased materials and fabrication costs. Several other factors, including spalling (loss of adhesion) of the coating material, intermetallic diffusion between the substrate and the coating, and coating defects can lead to less than desirable performance characteristics including membrane failure, decreased permeability, and the presence of unwanted constituents in the permeate stream. Membranes currently being fabricated by the abovementioned methods include Pd-based membrane layers ranging from 1 to 30 microns in thickness, with separation factors generally greater than 100.

Layered Membranes

Analogous to application of a Pd-based alloy on a highly permeable porous substrate, some researchers have focused efforts on applying hydrogen dissociation catalysts onto highly permeable dense metal substrates, Fig. 10.7b. These highly permeable dense metals, often referred to as "super-permeable" metals, generally utilize Nb, Ta, V, and/or Zr as major components. Unfortunately, with the formation of non-catalytic, highly stable surface oxides prevents these super-permeable metals from being used as stand-alone membranes. Applying a Pd-based metal layer by physical or chemical vaporization techniques to the surface of a highly clean super-permeable metal substrate provides a coating that will catalyze the dissociation of molecular hydrogen to atomic hydrogen as well as a barrier to prevent surface oxidation. Mundschau and co-workers recently published a comprehensive review on composite membranes utilizing Pd-based catalyst layers on super-permeable dense metal substrates [94].

Dense Metal Membrane Degradation Mechanisms

The successful implementation of a dense metal membrane in the gasification process would require the membrane to retain both its mechanical and chemical integrity in the presence of both major and minor gasification constituents such as CO, CO_2, H_2O and H_2S. As stated in a previous section, the hydrogen transport through dense metals involves catalytic and diffusion processes. Reductions in any of these steps can cause reductions in apparent performance. Therefore, the following sections will describe possible degradation mechanisms of dense metal membranes in the presence of relevant gasification constituents.

Catalytic Poisoning

Researchers have reported the influence of various chemical species on the performance of palladium-based membranes. Observed reductions in hydrogen flux through dense metal membranes are typically attributed to the blocking of the adsorption of molecular hydrogen, which is a necessary step in the atomic transport mechanism [13, 80, 112, 114,120–123].

Many palladium membrane studies initiate testing with an oxygen exposure prior to characterizing hydrogen permeability. The oxygen pretreatment has been shown to positively influence permeability values [80, 114, 124, 125]. The improvement in performance is thought to arise from an "activation" of the membrane by the removal of surface contaminants that can inhibit the surface reaction required in the hydrogen transport mechanism. Although the operation of a palladium membrane in an oxidizing atmosphere can remove surface contaminants, it has been revealed that such exposures can highly influence surface morphology even though palladium oxides are not observed [121].

Various studies have focused on the influence of major gasifier effluent species (CO, CO_2, and H_2O) on the performance and viability of palladium-based membrane materials [13, 57, 80, 120, 123, 124,126–130]. Although it is apparent that all major gasification constituents decrease the hydrogen flux through Pd-based membranes [126], the consensus from these studies is that CO has the largest effect [20, 123, 126, 131, 132]. Decreases in hydrogen flux as high as 75% have been observed in the presence of CO; however, as temperature increases, the apparent influence decreases [131, 132]. Although the favored explanation for the observed reductions in hydrogen flux in the presence of CO (as well as CO_2 and H_2O) for Pd-based membranes is competitive adsorption, phenomena including the formation of graphitic carbon within the interstitials of the palladium crystal structure decreasing the bulk diffusion rate have been noted [126].

Minor gasifier effluent constituents, such as S-containing gases, have been reported to deleteriously affect hydrogen flux through dense metal membranes. The decreases observed in the presence of S-containing gases are generally attributed to catalytic deactivation by adsorbed S [67, 72, 79, 80, 124], although a subsequent section will address corrosive decay. Catalytic poisoning by S-containing gases has been the focus of many experimental and theoretical studies [67, 72, 79, 80, 124,133–141]. Experiments using surface science techniques such as XPS, LEED, TPD, STM and AES have been utilized on various catalyst surfaces dosed with S-containing gases and have generally shown that hydrogen adsorption decreases with increasing S-coverage. Sulfur appears not only to block hydrogen adsorption sites, but also exhibits steric hindrance where only 0.28 monolayer of S was required to completely inhibit hydrogen adsorption on a Pd(100) surface [133–135].

Pd-based catalyst studies have focused on the influence of S-containing gases on reactions including thiophene hydrodesulphurization and methane oxidation. It appears that not only does the presence of sulfur-containing gases decreases the effectiveness of the Pd-based catalysts studied, but the formation of Pd-sulfides does as well [136, 137]. Feuerriegel and coworkers demonstrated complete Pd-catalyst deactivation for the oxidation of methane with as low as 5.4 ppm of H_2S present and as little as 0.08 ML of S completely poisoning the surface [137], while Vazquez and co-workers witnessed a 60% decrease in thiophene conversion for a sulfided catalyst as compared to a sulfide-free Pd catalyst [136].

The interaction of S with various Pd surfaces has also been the focus of numerous atomistic computational studies [138–142]. It is apparent from these theoretical studies that S coverage ranging from 0.33 to 0.5 ML is sufficient to completely inhibit the adsorption of hydrogen, but the coverage appears to be dependent on the crystallographic surface studied. Furthermore, Alfonso and co-workers demonstrated that the binding energy of S on various metals shows the following trend: $E_{Pd(111)} > E_{Cu(111)} > E_{Ag(111)}$, while Pd-alloys with Cu and Ag show weaker binding energies than pure Pd [140]. Additional studies conclude that while H_2S will adsorb and dissociate relatively easily on a Pd(111) surface, S is the most stable adsorbed species at low surface coverage ($\theta_S \leq 0.5$). Although at a surface coverage of sulfur greater than 0.5 ML, adsorbed sulfur has a tendency to form S–S bonds on the Pd surface [141, 142].

Many studies have been conducted on the influence of S-containing gas mixtures on the hydrogen flux of various Pd-based membranes over a wide range of temperatures, membrane thicknesses and H_2S concentrations [67, 68, 72, 77, 79, 80, 91, 111, 112, 114, 124,143–145]. The explanations of the observed changes in hydrogen membrane performance in the presence of S-containing gases vary considerably between studies. Several researchers attribute the observed decreases in hydrogen flux in the presence of S-containing gases to the blocking of the hydrogen surface reaction by adsorbed S, but the affect appears to be dependent on S concentration, temperature and alloy composition [67, 72, 80, 91, 124]. Recent work conducted by Morreale and co-workers at NETL suggest that the mechanism of poisoning can be derived from the shape of the flux degradation curve. An immediate decrease in flux to a lower and constant value coupled with a membrane surface free of bulk corrosion after testing is attributed to catalytic poisoning. Figure 10.8 shows an example of this phenomenon and is consistent with similar reported studies [67].

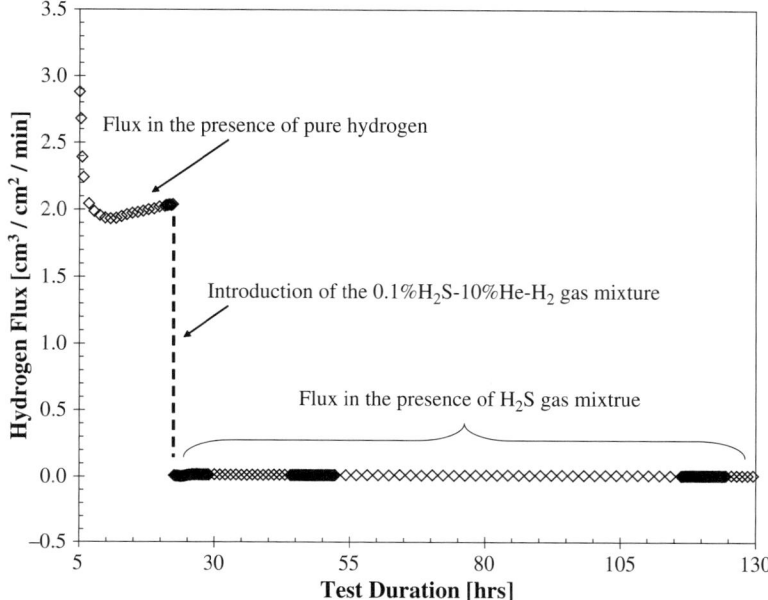

Fig. 10.8 Hydrogen flux degradation curve shape attributed to catalytic poisoning in the presence of H_2S for a Pd-based alloy

In summary, adsorbed sulfur on a metal surface can significantly impact the hydrogen adsorption and dissociation properties of the catalyst, with the influence increasing with increasing surface coverage. Additionally, adsorbed sulfur has been shown to decrease the reaction rate as well as conversion of several reactions involving a hydrogen dissociation step. These decreases in catalytic activity due to the presence of sulfur may have significant impacts on the performance of hydrogen membranes due to the inability of hydrogen to adsorb and dissociate on the membrane surface. However, the impact of the catalytic activity of the membrane surface does not account for all of the observed decreases in hydrogen flux. The effects of corrosion will be the focus of the following section.

Corrosive Decay

High temperature corrosion is a widespread problem affecting many industries and processes including power generation, aerospace, gas turbines, heat treating, mineral and metallurgical processing, chemical processing, refining and petrochemical, automotive, pulp and paper, as well as waste incineration. Types of corrosion that are important to these industries include oxidation, sulfidation and carburization. The thermodynamic stabilities of corrosion products can be estimated from the appropriate Gibbs free energy values (Figs 10.9 and 10.10), but this information does not give any insight on the growth rate of the corrosion product [146–148].

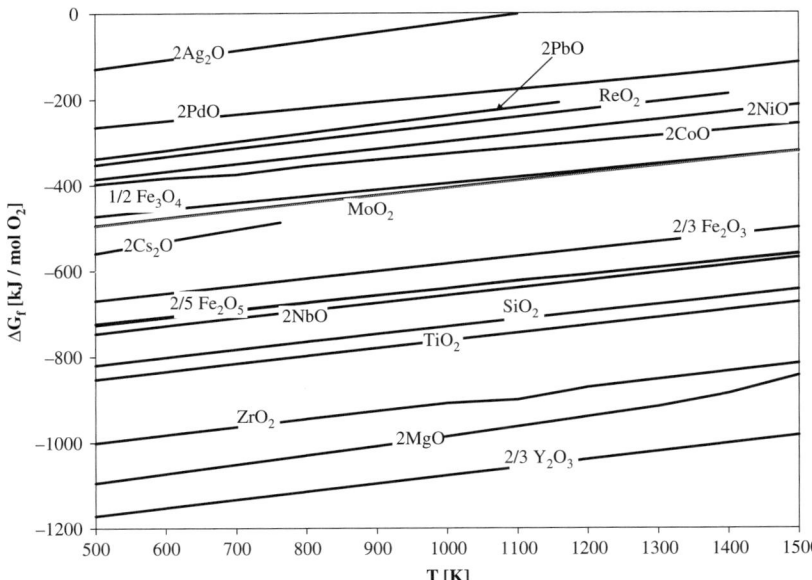

Fig. 10.9 Illustration of Gibbs free energy of formation for various metal oxides as a function of temperature

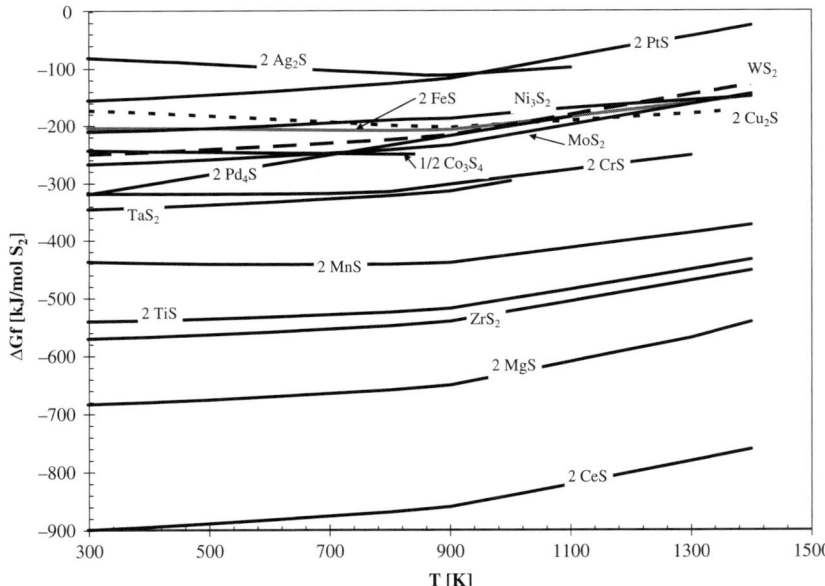

Fig. 10.10 Illustration of Gibbs free energy of formation for various metal sulfides as a function of temperature

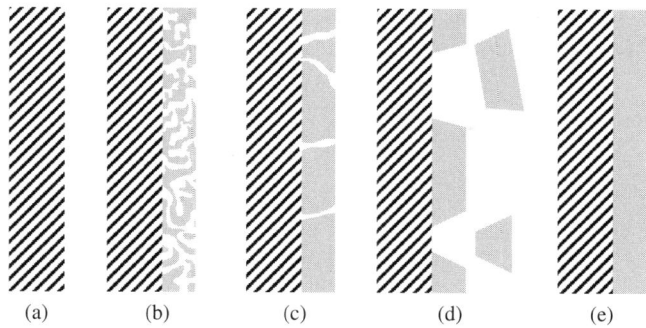

(a)   (b)   (c)   (d)   (e)

Fig. 10.11 Illustration of the possible scale phenomena as they pertain to membrane transport: (a) Cross section of a corrosion free membrane, (b) membrane with porous corrosion scale, (c) membrane with cracked corrosion scale, (d) membrane with spalling corrosion scale, (e) membrane with continuous, dense corrosion scale

The formation of a corrosion product layer on a membrane surface can have significant impacts on the overall membrane performance (and mechanical life), but the influence relies heavily on the structure/formation of the scale. For example, the surface scale may only partially cover the membrane surface, such as a porous scale or a scale exhibiting cracks resulting from relieving surface stresses (Fig. 10.11). The partially covered membrane surface may still exhibit enough catalytic sites to promote the adsorption and dissociation of hydrogen, but may yield a smaller amount of atomic hydrogen due to the reduced active metal surface area.

A second impact of a corrosion product may involve the formation of a continuous, dense scale on the membrane surface. This continuous scale would entirely cover the catalytic surface of the original membrane inhibiting the interaction of the gas phase molecular hydrogen with the original catalytic surface of the membrane. However, if the sulfide scale exhibits transport properties for molecular hydrogen (i.e. through grain boundaries) that permits hydrogen to contact the membrane at the metal/scale interface, hydrogen may still dissociate and diffuse through the membrane. If the sulfide scale exhibits catalytic properties for dissociating hydrogen [149–151] and diffusive properties for atomic hydrogen, the membrane may transport hydrogen to the scale-metal interface to allow atomic diffusion through the metal. The performance of the sulfide scale with respect to hydrogen transport would depend on the hydrogen solubility and diffusivity properties of the scale and the scale thickness (scale permeance).

Several membrane studies have reported the formation of corrosion scales on membrane surfaces or a significant change in surface appearance and morphology [67, 79, 91, 111, 114, 145]. A consistent observation in most of the studies reporting flux as a function of exposure time is that the apparent decrease in performance is gradual. Morreale and co-workers at NETL recently hypothesized and demonstrated, through corrosion growth rate data and observed decreases in flux in the presence of H_2S-containing gases, that the gradual flux reduction observed may be

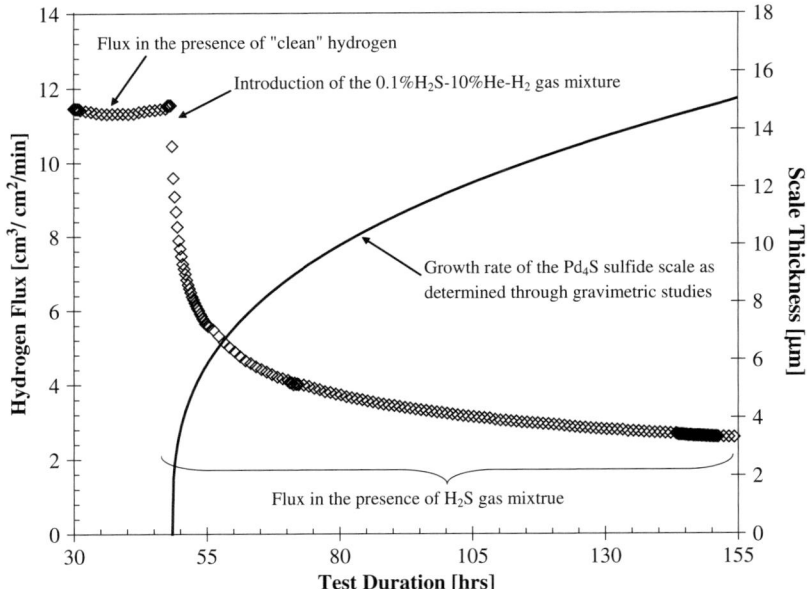

Fig. 10.12 Illustration of the apparent impact of scale growth as a function of S-exposure time

at least partially attributed to the low permeability of the growing surface scale, Fig. 10.12 [91].

Impact of WGSMR Operation on Corrosion

The Gibbs free energy of formation values, Figs. 10.9 and 10.10, can be used to estimate the dissociation pressure or the gas composition where surface scale formation will be thermodynamically feasible. Assuming the formation of the corrosion product follows a similar reaction as the decomposition of palladium-sulfide (Eq. 10.10), the correlation between the equilibrium constant and Gibbs free energy of formation is illustrated by Eq. 10.11.

$$Pd_4S(s) + H_2(g) \leftrightarrow H_2S(g) + 4Pd(s) \tag{10.10}$$

$$\Delta G_f = -RT \ln K_{EQ}^{Pd/Pd4S} = -RT \ln \left(\frac{a_{Pd}^4 a_{H2S}}{a_{Pd4S} a_{H2}} \right) \tag{10.11}$$

From Eq. 10.11, it is evident that the formation of corrosion products is dependent on the ratio of the partial pressures and activities of the products and reactants. This relation is analogous to the evaluation of the stability of oxides in CO_2-CO atmospheres, except that the driving force for oxide scale stability is proportional to the CO_2-to-CO partial pressure ratio. The activity for pure metals is often assumed to be unity, although the activity of metal species in alloys is often only identified through experimentation. However, by assuming that the alloy forms an ideal solution, the activity of an individual metal component can be estimated as the mole fraction ($Pd_{70}Cu_{30}$ alloy yields $a_{Pd} = 0.7$ and $a_{Cu} = 0.3$), an approximation often utilized when the actual activity is not known.

In the WGSMR concept, the syngas would be reacted with steam to enhance the conversion of CO to CO_2 and H_2, while simultaneously removing produced hydrogen and shifting the reaction towards complete conversion by Le Chatelier's principle. Therefore, in an effort to determine the corrosivity of the gaseous atmosphere within the membrane reactor, a COMSOL® multi-physics model was used to qualitatively illustrate the concentrations of the major gaseous constituents and H_2S along the length of a WGSMR (Fig. 10.13). The feed gas composition used in the model was based on values from Table 10.1 at the WGS inlet and consisted of approximately 45%H_2O, 20%CO, 19%H_2, 11%CO_2 and 0.4%H_2S. Furthermore, additional model assumptions included: isothermal operation, a membrane permeability comparable to pure palladium, membrane transport limited by atomic hydrogen diffusion, the membrane is infinitely selective for hydrogen, the use of a sweep gas on the permeate, countercurrent flow pattern between retentate and permeate, no additional catalyst, and kinetics verified through experimentation conducted at NETL.

As expected, the concentrations of H_2O, CO and H_2 decrease along the length of the WGSMR while the concentrations of H_2S and CO_2 increase. Although hydrogen is produced via the WGS reaction, the decreasing H_2 concentration profile within the WGSMR is attributed to the rate of hydrogen extraction being greater than the rate

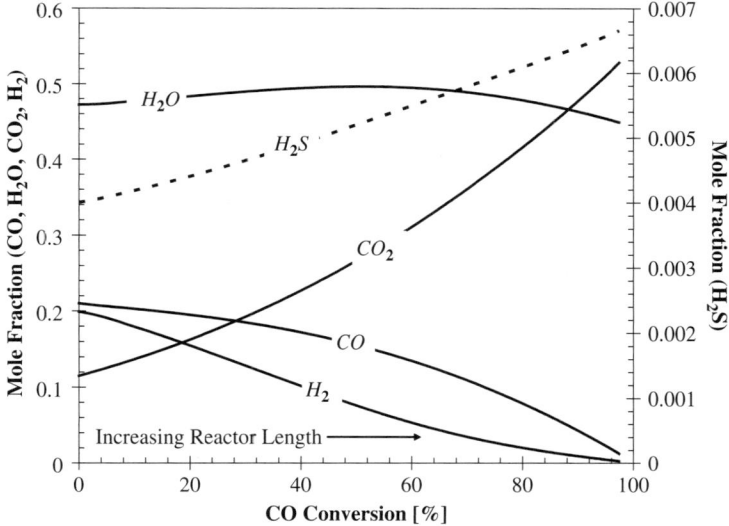

Fig. 10.13 COMSOL® simulated gas composition of a syngas mixture along the length of a WGSMR

of hydrogen production via the WGS reaction. Furthermore, the increase observed for the concentration of H_2S within the reactor is attributed to a decrease in total reactant mass within the reactor due to the selective extraction of hydrogen via the membrane. The changes in gaseous concentrations along the length of the reactor can significantly impact the corrosiveness of the environment. The increase in CO_2 concentration coupled with the decrease in CO can result in higher driving forces for oxidation, while similar trends with respect to H_2S and H_2 can result in higher driving forces for sulfidation. Figure 10.14 illustrates the change in the oxidation (CO_2-CO ratio) and sulfidation (H_2S-H_2 ratio) driving force along the length of the reactor as simulated by the COMSOL® model.

A qualitative depiction of the changing corrosiveness that could exist in a WGSMR is illustrated in Fig. 10.14. The driving force for oxidation and sulfidation increases substantially along the reactor length. Using the previously shown thermodynamic stability data (Figs. 10.9 and 10.10), the minimum gas ratio required for reaction or the dissociation pressure of various metal oxides and sulfides can be estimated by Eq. 10.11, assuming that the activities corresponding to the metal and oxide or sulfide product are unity. The resulting dissociation pressure of various metal oxides and sulfides are illustrated in Figs. 10.15 and 10.16 respectively as a function of temperature.

The equilibrium lines for the corrosion products can be used to qualitatively access the tendency for a pure metal to form a corrosion product in a given gaseous environment. If the driving force for oxidation (P_{CO2}–P_{CO}) is above the equilibrium line at a given temperature, then the specific metal would be expected to oxidize. The feasibility of sulfide formation can be evaluated in a similar fashion.

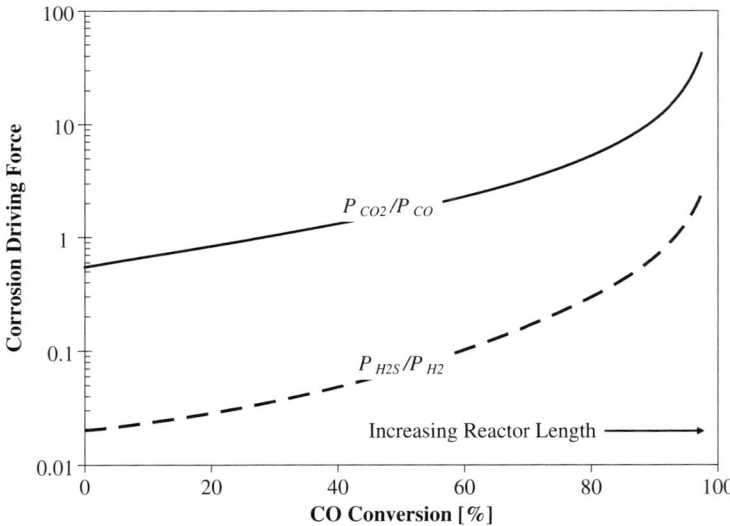

Fig. 10.14 Resulting oxidizing and sulfiding driving force along the length of the simulated WGSMR assuming a CO conversion and H_2 recovery of > 95%

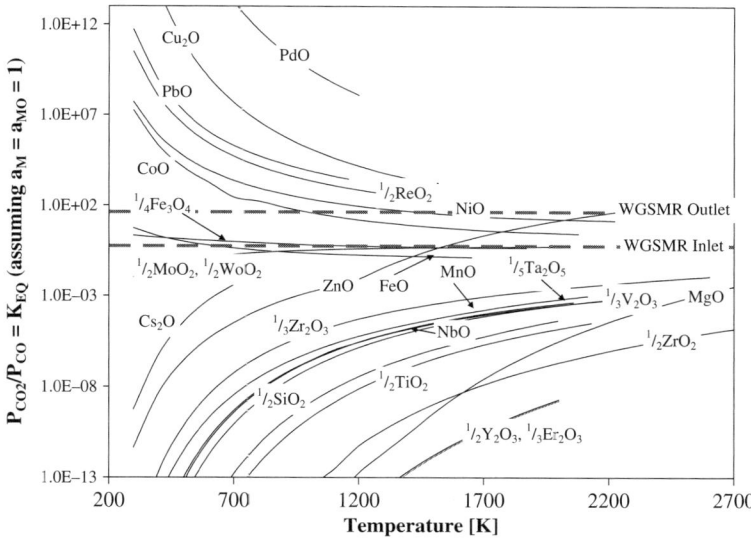

Fig. 10.15 Predicted dissociation pressures of various metal oxides based on the thermodynamic data of Barin [148] and Taylor [146]

Using the aforementioned methodology, the feasibility of pure metal oxidation and/or sulfidation can be estimated based on the gaseous concentrations simulated using COMSOL® model. Also shown in Figs. 10.15 and 10.16 are lines indicating the assumed best and worst case scenario for corrosion; gasifier inlet and outlet composition respectively.

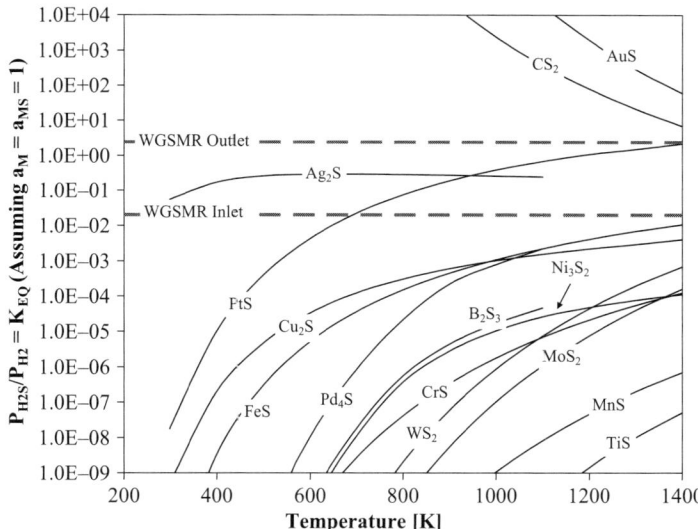

Fig. 10.16 Predicted dissociation pressures of various metal sulfides based on the thermodynamic data of Barin [148] and Taylor [146]

Inspection of the thermodynamic properties of the various corrosion products and gaseous compositions encountered at the WGSMR inlet and outlet, illustrated in Figs. 10.15 and 10.16, suggest that no pure metal is likely to be resistant to both oxidation and sulfidation under the assumed conditions.

Other strategies must be employed to successfully operate a WGSMR in the presence of typical gaseous environments. From a process vantage point, research should continue on the development of methods for contaminant removal. Limiting the extent of conversion in the WGSMR could also be considered as an approach to control corrosion.

Efforts should also continue on the design of membranes that can tolerate the associated gaseous environments. Additionally, the development of alloys that could potentially be corrosion resistant due to activity changes, coating materials that may be "unaffected" by the gaseous environment, and/or the use of corrosion products and associated kinetics to protect the membrane materials could be methods to overcome corrosion issues.

Summary

Coal gasification has been identified as a means of addressing the national issues of fuel independence, energy availability and reliability, and the environment. The abundance of coal in the United States in conjunction with flexibility in the use of syngas, which can be converted to electricity, hydrogen, and/or liquid fuels, is considered a promising near- to mid-term component in the transition to a renewable energy society. Coal gasification processes coupled with the successful identification of a hydrogen membrane that can be integrated into a WGSMR has the potential to increase process efficiency and decrease hydrogen production costs, while producing a "sequestration ready" stream of CO_2.

Significant progress has been achieved in membrane fabrication methods over the past 15 years, including thin film deposition on highly permeable substrates and ultra-thin foil development. However, a significant technical barrier for the implementation of hydrogen membranes in the gasification process is their chemical and mechanical susceptibility to syngas constituents. It has been demonstrated that as little as 0.3 ML of surface coverage by adsorbed gas species can completely inhibit hydrogen adsorption and dissociation.

Additionally, the corrosive nature of the syngas stream can have significant impacts on the mechanical and chemical stability of the membrane. Researchers have shown that a pure palladium membrane can develop a corrosion scale as thick as 35 microns after only 5 days of operation in the presence of $0.1\%H_2S$-H_2, making thin film palladium membrane technologies impractical in such an environment [91].

In conclusion, hydrogen membranes can have significant impact on the overall process efficiency of gasification by implementation as a separation device or an integrated WGSMR. However, chemical degradation from syngas constituents and contaminates prohibits the widespread application of hydrogen membrane technologies to fossil fuel conversion processes. Therefore, both computational and experimental research efforts should focus on the development and identification of membrane materials and configurations that lend themselves to the successful integration of hydrogen membrane technologies and the gasification process.

References

1. Higman C, van der Burgt M. Gasification. Gulf publishing 2003.
2. Marano J, Ciferno J. Membrane selection and placement for optimal co2 capture from igcc power plants. Fifth annual conference on carbon capture & sequestration, Alexandira, VA, 2006.
3. Deville H, Troost L Sur la. Permeabilite du Fer a Haute Temperature. Comptes Rendus Hebdomadaires des Seances de L'Academie des Sciences. 1863;57:965.
4. Deville H Note sur le Passage des Gaz au Travers des Corps Solides Homogenes. Comptes Rendus Hebdomadaires des Seances de L'Academie des Sciences. 1864;59:102.
5. Graham T. Adsorption and separation of gases by colloid septa. Philos Trans R Soc. 1866;156:399.
6. Balovnev YA. Diffusion of hydrogen in palladium at low pressures. Russian J Phys Chem. 1969;43:1382.
7. Davis WD. Diffusion of hydrogen through palladium. general electric company. Schenectady. 1954.
8. Holleck GL Diffusion and solubility of hydrogen in palladium and palladium-silver alloys. J Phys Chem. 1970;74(3):503.
9. Katsuta H, Farraro RJ, McLellan RB. The Diffusivity of hydrogen in palladium. Acta Metallurgica 1979;27:1111.
10. Koffler SA, Hudson JB, Ansell GS. Hydrogen permeation through alpha-palladium. Trans Metall Soc AIME. 1969;245:1735.
11. Toda G. Rate of permeation and diffusion coefficient of hydrogen through palladium. J Res Inst Catal. 1958;6:13–19.
12. Grashoff GJ, Pilkington CE, Corti CW. The purification of hydrogen: a review of the technology emphasizing the current status of palladium membrane diffusion. Report 1983;27(4):157–169.
13. Paglieri SWD. Innovations in palladium membrane research. Sep Purif Methods. 2002;31(1):1–169.
14. Steward SA. Review of hydrogen isotope permeability through materials. Livermore, 1983; 1–21.
15. Buxbaum RE. Composite metal membrane for hydrogen extraction. US Patent 5215729, 1999.
16. Athayde AL, Baker RW, Nguyen P. Metal composite membranes for hydrogen separation. J Memb Sci. 1994;94(1):299.
17. Collins JP, Way JD. Preparation and characterization of a composite palladium-ceramic membrane. Ind Eng Chem Res. 1993;32:3006.
18. Edlund DJ, McCarthy J. The relationship between intermetallic diffusion and flux decline in composite-metal membranes: implications for achieving long membrane lifetime. J Memb Sci 1995;107(1–2):147.
19. Fernandes NE, Fisher SM, Poshusta JC, Vlachos DG, Tsapatsis M, Watkins JJ. Reactive deposition of metal thik(TE: Please clarify if it is 'Thik' or 'Thick') films within porous supports from supercritical fluids. Chem Mater 2001;13:2023.

20. Hou K, Hughes R. Preparation of thin and highly stable Pd/Ag composite membranes and simulative analysis of transfer resistance for hydrogen separation. J Memb Sci. 2003;214(1):43.

21. Hsu C, Buxbaum RE. Electroless and immersion plating of palladium on zirconium. J Electrochem Soc. 1985;132(10):2419.

22. Huang L, Chen CS, He ZD, Peng DK, Meng GY. Palladium membranes supported on porous ceramics prepared by chemical vapor deposition. Thin Solid Films. 1997;302(1–2):98.

23. Itoh N, Tomura N, Tsuji T, Hongo M. Deposition of palladium inside straight mesopores of anodic alumina tube and its hydrogen permeability. Microporous Mesoporous Mater. 2000;39(1–2):103.

24. Jemaa N, Shu J, Kaliaguine S, Grandjean BPA. Thin palladium film formation on shot peening modified porous stainless steel substrates. Ind Eng Chem Res. 1996;35:973.

25. Juda W, Krueger CW, Bombard TR. Diffusion-bonded palladium-copper alloy framed membrane for pure hydrogen generators and the like and method of preparing the same. US Patent 5904754, 1999.

26. Juda W, Krueger CW, Bombard TR. Method of producing thin palladium-copper and the like, palladium alloy membranes by solid-solid metallic interdiffusion, and improved membrane. US Patent 6238465, 2001.

27. Jun CS, Lee KH. Palladium and palladium alloy composite membranes prepared by metalorganic chemical vapor deposition method (cold-wall). J Memb Sci. 2000;176(1):121.

28. Keuler JN, Lorenzen L. Developing a heating procedure to optimise hydrogen permeance through Pd-Ag membranes of thickness less than 2.2 [mu]m. J Memb Sci. 2002;195(2):203.

29. Keuler JN, Lorenzen L, Maichon S. Preparing and testing Pd films of thickness 1–2 micrometer with high selectivity and high hydrogen permeance. Sep Sci Technol. 2002;37(2):379.

30. Keuler JN, Lorenzen L, Sanderson RD, Prozesky V, Przybylowicz WJ. Characterization of electroless plated palladium-silver alloy membranes. Thin Solid Films. 1999;347(1–2):91.

31. Kishimoto S, Inoue M, Yoshida N. Solution of hydrogen in thin palladium films. J Chem Soc Farady Trans. 1986;82:2175.

32. Krueger C, Juda W, Bombard T. Opportunities for high temperature metal hydrogen separation membranes; hy9 corporation: Woburn.

33. Krueger CW. Method of improving and optimizing the hydrogen permeability of a palladium-copper membrane and novel membranes manufactured thereby. US Patent 6372363, 2002.

34. Kusakabe K, Yokoyama S, Morooka S, Hayashi JI, Nagata H. Development of supported thin palladium membrane and application to enhancement of propane aromatization on gasilicate catalyst. Chem Eng Sci. 1996;51(11):3027.

35. Kusakabe K, Takahashi M, Maeda H, Morooka S. Preparation of thin palladium membranes by a novel method based on photolithography and electrolysis. J Chem Eng Jpn. 2001;34(5):703.

36. Li A, Liang W, Hughes R. Characterisation and permeation of palladium/stainless steel composite membranes. J Memb Sci. 1998;149(2):259.

37. Li A, Liang W, Hughes R. Fabrication of defect-free Pd/[alpha]-Al2O3 composite membranes for hydrogen separation. Thin Solid Films. 1999;350(1–2):106.

38. Li A, Liang W, Hughes R. Repair of a Pd/[alpha]-Al2O3 composite membrane containing defects. Sep Purif Technol. 1999;15(2):113.

39. Li A, Liang W, Hughes R, Fabrication of Dense Palladium Composite Membranes for Hydrogen Separation. Catal Today. 2000;56(1–3):45.

40. Mardilovich PP, She Y, Ma YH, Rei MH. Defect-free palladium membranes on porous stainless-steel support. AIChE J. 1998;44(2):310.

41. McCool B, Xomeritakis G, Lin YS. Composition control and hydrogen permeation characteristics of sputter deposited palladium-silver membranes. J Memb Sci. 1999;161:67–76.

42. Moss TS, Peachey NM, Snow RC, Dye RC. Multilayer metal membranes for hydrogen separation. Int J Hydrogen Energy. 1998;23(2):99.

43. Nam SE, Lee SH, Lee KH. Preparation of a palladium alloy composite membrane supported in a porous stainless steel by vacuum electrodeposition. J Memb Sci. 1999;153(2):163.
44. Nam SE, Lee KH A study on the palladium/nickel composite membrane by vacuum electrodeposition. J Memb Sci. 2000;170(1):91.
45. O'Brien J, Hughes R, Hisek J. Pd/Ag membranes on porous alumina substrates by unbalanced magnetron sputtering. Surf Coatings Technol. 2001;253:142–4.
46. Pan XL, Stroh N, Brunner H, Xiong GX, Sheng SS. Pd/Ceramic hollow fibers for H2 separation. Sep Purif Technol. 2003;32(1–3):265.
47. Peachey NM, Snow RC, Dye RC. Composite Pd/Ta metal membranes for hydrogen separation. J Memb Sci. 1996;111(1):123.
48. Pick MA, Davenport JW, Strongin M, Dienes GJ. Enhancement of hydrogen uptake rates for Nb and Ta by thin surface overlays. Phys Rev Lett. 1979;43(4):286.
49. Pick MA, Greene MG, Strongin M. Uptake rates for hydrogen by niobium and tantalum: effect of thin metallic overlayers. J Less Common-Met. 1980;73:89.
50. Roa F, Way JD, McCormick RL. Preparation of micron scale, Pd-Cu composite membranes for H$_2$ separations. ACS. Golden: Colorado School of Mines; 2001.
51. Shu J, Grandjean BPA, Kaliaguine S, Methane steam reforming in asymmetric Pd- and Pd-Ag/porous SS membrane reactors. Appl Catal A Gen. 1994;119(2):305.
52. So J-H, Yang S-M, Bin Park S. Preparation of silica-alumina composite membranes for hydrogen separation by multi-step pore modifications. J Memb Sci. 1998;147(2):147.
53. Souleimanova RS, Mukasyan AS, Varma A. Study of structure formation during electroless plating of thin metal-composite membranes. Chem Eng Sci. 1999;54(15–16):3369.
54. Strokes CL, Buxbaum RE. Analysis of palladium coatings to remove hydrogen isotopes from zirconium fuel rods in canada deuterium uranium-pressurized heavy water reactors: thermal and neutron diffusion effects. Nucl Technol. 1992;98:207.
55. Uemiya S, Kude Y, Sugino K, Sato N, Matsuda T, Kikuchi E. A palladium/porous-glass composite membrane for hydrogen separation. Chem Lett. 1988;17:1687–1690.
56. Uemiya S. State-of-the-art of supported metal membranes for gas separation. Sep Purif Methods. 1999;28(1):51.
57. Uemiya S, Sato N, Ando H, Kude Y, Matsuda T, Kikuchi E. Separation of hydrogen through palladium thin film supported on a porous glass tube. J Memb Sci. 1991;56(3):303.
58. Uemiya S, Kato W, Uyama A, Kajiwara M, Kojima T, Kikuchi E. Separation of hydrogen from gas mixtures using supported platinum-group metal membranes. Sep Purif Technol. 2001;22–23:309.
59. Wu L-Q, Xu N, Shi J. Preparation of a palladium composite membrane by an improved electroless plating technique. Ind Eng Chem Res. 2000;39:342.
60. Yan S, Maeda H, Kusakabe K, Morooka S. Thin palladium membrane formed in support pores by metal-organic chemical vapor deposition method and application to hydrogen separation. Ind Eng Chem Res. 1994;33:616.
61. Yeung KL, Varma A. Novel preparation techniques for thin metal-ceramic composite membranes. AIChE J. 1995;41(9):2131.
62. Yeung KL, Sebastian JM, Varma A. Novel preparation of Pd/vycor composite membranes. Catal Today. 1995;25(3–4):231.
63. Yeung KL, Christiansen SC, Varma A. Palladium composite membranes by electroless plating technique: relationships between plating kinetics, film microstructure and membrane performance. J Memb Sci. 1999;159(1–2):107.
64. Zhao HB, Pflanz K, Gu JH, Li AW, Stroh N, Brunner H, et al. Preparation of palladium composite membranes by modified electroless plating procedure. J Memb Sci. 1998;142(2):147.
65. Shu j, Grandjean B, Kaliaguine S. Catalytic palladium-based membrane reactors. Appl Catal A Gen. 1991;119:1036–60.
66. Knapton AG. Palladium alloys for hydrogen diffusion membranes. Platinum Met Rev. 1977;21:44.
67. McKinley DL, Nitro W. Metal alloy for hydrogen separation and purification. US Patent 3350845, 1967.

68. McKinley DL, Nitro W. Method for hydrogen separation and purification. US Patent 3439474, 1969.
69. Farr JPG, Harris IR. British Patent 1,346,422, 1974.
70. Fort D, Farr JPG, Harris IR. Less-Comon Met. 1975;39:293.
71. Edlund H. A membrane process for hot gas clean-up and decomposition of H2S to elemental sulfur. Phase II Final Report, DE-FG03-91ER81228, 1996.
72. Kulprathipanja A, Alptekin GO, Falconer JL. Way JD Pd and Pd-Cu membranes: inhibition of H_2 permeation be H_2S. J Memb Sci. 2005;254:49–62.
73. Hoang HT, Tong HD, Gielens FC, Jansen HV, Elwenspoek MC. Fabrication and characterization of dual sputtered Pd-Cu alloy films for hydrogen separation membranes. Mater Lett. 2004;58(3–4):525.
74. Roa F, Block MJ, Way JD. The influence of alloy composition on the H2 flux of composite Pd-Cu membranes. Desalination. 2002;147(1–3):411.
75. Way JD. Palladium/Copper alloy composite membranes for high temperature hydrogen separation from coal-derived gas streams. Annual Progress Report, DE-FG26-99FT40585, 2004.
76. Way JD, McCormick RL, Roa F, Paglieri SN. Palladium alloy composite membranes for high temperature hydrogen separation from coal-derived gas streams. 2002.
77. Bryden KJ, Ying JY. Nanostructured palladium-iron membranes for hydrogen separation and membrane hydrogenation reactions. J Memb Sci 2002;203(1–2):29.
78. Itoh N, Xu WC, Sathe AM. Capability of permeate hydrogen through palladium-based membranes for acetylene hydrogenation. Ind Eng Chem Res. 1993;32:2614.
79. Darling AS. Hydrogen separation by diffusion through palladium alloy membranes. symposium on the less common means of separation, institution of chemical engineers. 1963.
80. Ali JK, Newson EJ, Rippin DWT. Deactivation and regeneration of Pd-Ag membranes for dehydrogenation reactions. J Memb Sci. 1994;89(1–2):171.
81. Tosti S, Bettinali L, Violante V. Rolled thin Pd and Pd-Ag membranes for hydrogen separation and production. Int J Hydrogen Energy 2000;25:319.
82. Tosti S, Adrover A, Basile A, Camilli V, Chiappetta G, Violante V. Characterization of thin wall Pd-Ag rolled membranes. Int J Hydrogen Energy 2003;28:105.
83. Fort D, Harris IR. Less-Common Met. 1975;41:313.
84. Hughes DT, Harris IR. Less-Common Met. 1978;61:9.
85. Zetkin AS, Kagan GE, Varaksin AN, Levin ES. Diffusion and Penetrability of Deuterium in the Alloy Pd-53 at. % Cu. Sov Phys Solid State. 1992;34(1):83.
86. Zetkin AS, Kagan GY, Levin YS. Influence of structural transformations on the diffusion parameters of deuterium in palladium-copper alloys. Phys Met Metall. 1987;64(5):130.
87. Piper J. Diffusion of hydrogen in copper-palladium alloys. J Appl Phys. 1966;37(2):715.
88. Levin ES, Zetkin AS, Kagan GE. Solubility of deuterium in Pd-Cu alloys. Russ J Phys Chem 1992;66(2):465.
89. Flanagan TB, Chisdes DM. Solubility of hydrogen(1 atm, 298 K) in some copper/palladium alloys. Solid State Commun. 1975;16(5):529.
90. Howard BH, Killmeyer RP, Rothenberger KS, Cugini AV, Morreale BD, Enick RM, et al. Hydrogen permeance of palladium-copper alloy membranes over a wide range of temperatures and pressures. J Memb Sci. 2004;241(2):207.
91. Morreale B. The influence of H2S on palladium and palladium-copper alloy membranes. PhD Dissertation, University of Pittsburgh, 2006.
92. Kuranov AA, Berseneva FN, Sasinova RA, Laptevskiy AS. Ordering and the mechanical properties of palladium-copper alloys. Phys Met Metall. 1983;56(3):167.
93. Burch R, Buss RG. Pressure-composition isotherms in the palladium-copper-hydrogen system. Solid State Commun. 1974;15(2):407.
94. Sammells AF, Mundschau MV. Nonporous Inorganic Membranes for Chemical Processing. Weinhein, Germany: Wiley-VCH, 2006.

95. Boes N, Zuchner H. Diffusion of hydrogen and deuterium in Ta, Nb, and V. Physica Status Solidi (a). 1973;17:K111.
96. Veleckis E, Edwards RK. Thermodynamic properties in the systems vanadium-hydrogen, niobium-hydrogen, and tantalum-hydrogen. J Phys Chem. 1969;73(3):683.
97. Simonovic BR, Mentus S, Susic MV. Kinetics of tantalum hydriding: the effect of palladization. Int J Hydrogen Energy. 2000;25:1069.
98. Buxbaum RE, Subramanian R, Park JH, Smith DL. Hydrogen transport and embrittlement for palladium coated vanadium-chromium-titanium alloys. J Nucl Mater. 1996; 233–237(Part 1):510.
99. Paglieri SN, Birdsell SA, Snow RC, Springer RW, Smith FM. Development of palladium composite membranes for hydrogen separation. AIChE – Advances in Gas Separation Membranes and Applications II. Los Alamos: Los Alamos National Laboratory; 2003.
100. Buxbaum RE, Hsu PC. Measurement of diffusive and surface transport resistances for deuterium in palladium-coated zirconium membranes. J Nucl Mater. 1992;189(2):183.
101. Buxbaum RE, Kinney AB. Hydrogen transport through tubular membranes of palladium-coated tantalum and niobium. Ind Eng Chem Res. 1996;35:530.
102. Buxbaum RE, Marker TL. Hydrogen transport through non-porous membranes of palladium-coated niobium, tantalum and vanadium. J Memb Sci. 1993;85(1):29.
103. Makrides AC, Newton MA, Wright H, Jewett DN. US Patent 3,350,846, 1964.
104. Buxbaum RE. The use of zirconium-palladium windows for the separation of tritium from the liquid metal breeder-blanket of a fusion reactor. Sep Sci Technol. 1983;8(12–13):1251.
105. Buxbaum RE, Hsu PC. Method for plating palladium. US Patent 5149420, 1992.
106. Simonovic BR, Mentus S, Dimitrijevic R, Susic MV. Multiple hydriding/dehydriding of $Zr_{1.02}Ni_{0.98}$ alloy. Int J Hydrogen Energy. 1999;24:449.
107. Perng TP, Altstetter CJ. On the effective hydrogen permeability in metastable beta titanium alloy, niobium and 2.25cr-1 mo ferritic steel. Metall Trans. 1986;17A:2086.
108. Rothenberger KS, Howard BH, Killmeyer RP, Cugini AV, Enick RM, Bustamante F, et al. Evaluation of tantalum-based materials for hydrogen separation at elevated temperatures and pressures. J Memb Sci. 2003;218(1–2):19.
109. Edlund DPW. Catalytic platinum-based membrane reactor for removal of h2s from natural gas streams. J Memb Sci. 1994;94:111–9.
110. Edlund D, Friesen D, Johnson B, Pledger W. Hydrogen-permeable metal membranes for high-temperature gas separations. Gas Sep Purif. 1994;8(3):131.
111. Edlund DJ, Pledger WA. Thermolysis of hydrogen sulfide in a metal-membrane reactor. J Memb Sci. 1993;77(2–3):255.
112. Philpott J, Coupland DR. Metal membranes for hydrogen diffusion and catalysis. Hydrogen Effects Catal. 1988;679.
113. Hsieh HP. Inorganic membranes for separation and reaction. New York: Elsevier, 2006.
114. Hurlbert RC, Konecny JO. Diffusion of hydrogen through palladium. J Chem Phys. 1961;34(2):655.
115. Edlund D. Hydrogen-permeable metal membrane and method for producing the same. US Patent 6152995, 2000.
116. Ma YH, Mardilovich PP, She Y. Hydrogen gas-extraction module and method of fabrication. US Patent 6152987, 2000.
117. Ma, Akis, Ayturk, Guazzone, Engwall, Mardilovich. Characterization of intermetallic diffusion barrier and alloy formation for pd/cu and pd/ag porous stainless steel composite membranes. Ind Eng Chem Res. 2004;43:2936–45.
118. Thoen PM, Roa F, Way JD. High flux palladium-copper composite membranes for hydrogen separations. Desalination. 2006;193:224–9.
119. Guazzone, Engwall, Ma. Effects of surface activity, defects and mass transfer on hydrogen permeance and n-value in composite palladium-porous stainless steel membranes. Catal Today. 2006;118:24–31.

120. Amandusson H, Ekedahl LG, Dannetun H. The effect of CO and O_2 on hydrogen permeation through a palladium membrane, Appl Surf Sci. 2000;153:259–267.
121. Roa F, Way JD. The effect of air exposure on palladium-copper composite membranes. Appl Surf Sci. 2005;240:85–104.
122. Collins JP, Way JD. Catalytic decomposition of ammonia in a membrane reactor. J Memb Sci. 1994;96(3):259.
123. Li A, Liang W, Hughes R. The effect of carbon monoxide and steam on the hydrogen permeability of a Pd/stainless steel membrane. J Memb Sci. 2000;165(1):135.
124. Lalauze R, Gillard P, Pijolat C. Hydrogen permeation through a thin film of palladium: influence of surface impurities. Sens Actuators. 1988;14(3):243.
125. Paglieri SN, Foo KY, Way JD, Collins JP, Harper-Nixon DL. A new preparation technique for Pd/alumina membranes with enhanced high-temperature stability. Ind Eng Chem Res. 1999;38(5):1925–1936.
126. Gao H, Lin YS, Li Y, Zhang B. Chemical stability and its improvement of palladium-based metallic membranes. Ind Eng Chem Res. 2004;43:6920.
127. Antoniazzi AB, Haasz AA, Auciello O, Stangeby PC. Atomic, ionic and molecular hydrogen permeation facility with in situ auger surface analysis. J Nucl Mater. 1984;670:128–9.
128. Basile A, Criscuoli A, Santella F, Drioli E. Membrane reactor for water gas shift reaction. Gas Separation & Purification. 1999;10(4):243–254.
129. Paturzo L, Basile A. Kinetics, Catalysis, and reaction engineering. Ind Eng Chem Res. 2002;41:1703.
130. Iyoha O, Howard B, Morreale B, Killmeyer R, Enick R. The effects of H_2O, CO and CO_2 on the H_2 permeance and surface charateristics of 1 mm thick $Pd_{80wt\%}Cu$. Topics in Catalysis, 2008;49(1–2):97–107.
131. Wang D, Flanagan TB, Shanahan KL. J Alloys Compd. 2004;372:158.
132. Sakamoto F, Chen FL, Kinari Y, Sakamoto Y. Int J Hydrogen Energy. 1996;210:1024.
133. Campbell CT. $H_2S/Cu(111)$: A model study of sulfur poisoning of water-gas shift catalysis. Surf Sci. 1987;183:100.
134. Burke MLMRJ. Hydrogen on Pd(100)-S: The effect of sulfur on precursor mediated adsorption and desorption. Surf Sci. 1990;237(1–3):1–19.
135. Forbes JG, Gellman AJ, Dunphy JC, Salmeron M. Imaging of sulfur overlayer structures on the Pd(111) surface. Surf Sci. 1992;279:68–78.
136. Vazquez A, Pedraza F, Fuentes S. Influence of sulfidation on the morphology and hydrodesulfurization activity of palladium particles on silica. J Mol Catal. 1992;75(1):63.
137. Feuerriegel U, Klose W, Sloboshanin S, Goebel H, Schaefer JA. Deactivation of a palladium-supported alumina catalyst by hydrogen sulfide during the oxidation of methane. Langmuir. 1994;10(10):3567.
138. Wilke M, Scheffler M. Poisoning of Pd(100) for the dssociations(TE: Please clarify if it is 'dissociations'?) of H_2: a theoretical study of co-adsorption of hydrogen and sulphur. Surf Sci Lett. 1995;329:L605–10.
139. Gravil PA, Toulhoat H. Hydrogen, sulphur and chlorine coadsorption on pd(111): a theoretical study of poisoning and promotion. Surf Sci. 1999;430(1–3):176.
140. Alfonso D, Cugini A, Sholl D. Density functional theory studies of sulfur binding on pd, cu and ag and their alloys. Surf Sci. 2003;546:12.
141. Alfonso D. First-principle study of sulfur overlayers on pd(111) surface. Surf Sci. 2005;596:229–41.
142. Alfonso D, Cugini A, Sorescu D. Adsorption and decomposition of h2s on pd(111) surface: a first principle study. Catal Today. 2005, 99, 315.
143. Kajiwara M, Uemiya S, Kojima T. Stability and hydrogen permeation behavior of supported platinum membranes in presence of hydrogen sulfide. Int J Hydrogen Energy. 1999;24:839.
144. Edlund D. A membrane reactor for H_2S decomposition. Advanced coal-fired power systems. Bend: Bend Research, Inc; 1996.
145. Mundschau M, Xie X. Dense inorganic membrane for production of hydrogen from methane and coal with carbon dioxide sequestration. Catal Today. 2006;118(1–2):12–23.

146. Taylor JR. Phase relationships and thermodynamic properties of the pd-s system. Metall Trans B. 1985;16B:143–8.
147. Niwa K, Yokokawa T, Isoya T. Equilibria in the PdS-H2-Pd4S-H2S and Pd4S-H2-Pd-H2S systems. Bull 14th Annual Meeting Chem Soc Jpn. 1962;35–9:1543–5.
148. Barin I, Sauert F, Schultze-Rhonhof E, Sheng W. Thermodynamic properties of inorganic substances. Weinheim, Federal Republic of Germany. New York: VCH. 1993.
149. Zdrazil F. Comparative study of activity and selectivity of transition metal sulfides in parallel hydrodechlorination of dichlorobenzene and hydrodesulfurization of methylthiophene. J Catal. 1997;167:286–95.
150. Pecoraro TA, Chianelli RR. Hydrodesulfurization catalysts by transition metal sulfides. Appl Surf Sci. 1981;67:430–45.
151. Vasudevan PT, Fierro JLG. A review of deep hydrodesulfurization catalysis. Cat Rev Sci Eng. 1996;28–2:161–88.

Chapter 11
Un-supported Palladium Alloy Membranes for the Production of Hydrogen

Bruce R. Lanning, Omar Ishteiwy, J. Douglas Way, David Edlund, and Kent Coulter

Thin self-supported permeable membranes of palladium alloys such as $Pd_{60}Cu_{40}$ have many applications in which hydrogen separation is required. Magnetron sputtering onto selected flexible or non-flexible substrates, for example using a water-soluble release agent or backing, has allowed lift-off and production of high quality films of controllable alloy composition down to $3 \, \mu m$ in thickness. The targeted manufacturing process is a low-cost reel-to-reel process that will theoretically meet the targets established by DOE. The results from a recently completed project and some of the remaining problems are discussed.

Introduction

Coal gasification and fuel cells are two of our nation's most promising technologies for the efficient production of clean electricity. At the heart of these technologies is hydrogen, the most simple, ubiquitous element known in the universe. Unfortunately, the ability to produce pure hydrogen has been a particular challenge that has impeded progress in both areas and will only become a more significant issue in the years ahead. Hydrogen is costly to produce or to separate from gas mixtures such as reactor effluent or waste streams due to the high capital and energy expenditures associated with compression, heat exchange, cryogenic distillation and pressure swing adsorption (PSA).

An affordable, tough and selective hydrogen separating membrane, on the other hand, could significantly reduce these costs and ultimately replace traditional unit operations or be integrated into an existing process to recover hydrogen. Polymer membranes, representing one type of commercially available membranes, currently compete with the other technologies to reduce the hydrogen/carbon monoxide ratio in synthesis gas (syngas), or to recover hydrogen from the purge of off-gas streams in ammonia or petrochemical plants. Polymer membranes are economical in some

B.R. Lanning (✉)
Southwest Research Institute, TX 78228, USA
ITN Energy Systems, Inc., 8130 Shaffer Parkway, Littleton, CO 80127-4107, USA
e-mail: blanning@itnes.com

A.C. Bose (ed.), *Inorganic Membranes for Energy and Environmental Applications*,
DOI 10.1007/978-0-387-34526-0_11, © Springer Science+Business Media, LLC 2009

applications, although the higher temperatures of most chemical reactions and many waste gas and reforming (i.e., coal gasification/natural gas) streams preclude their use, at least without process modifications such as cooling prior to introduction into the module. In general, membrane systems, such as the polymer type, require lower capital investment although their main liability is that recompression of the permeated hydrogen is usually required.

Considerable research in the area of inorganic membranes for hydrogen gas separation for purification at high temperatures has taken place in recent years, much of which has been supported by DOE. Of the two general classes of high-temperature membranes available (ceramic and metal), microporous ceramic membranes have been developed and commercialized to a greater extent for gas separation. Such materials, however, pose key challenges from several perspectives. Typically, the ceramics must exhibit an extremely fine, highly controlled pore size that can be difficult to fabricate over large areas. In addition, the mechanical integrity of thin ceramic membranes can come into question given the harsh environment associated with coal gasification while some of the ceramics being considered are rather exotic and expensive to make. Hence, the ability to manufacture ceramic membranes at low cost and in a continuous process has yet to be fully established.

Metal membranes, however, appear to have significant advantages over ceramic and polymer membranes in terms of manufacturability, lifetime (durability) and ease of sealing for the former and higher operating temperatures and selectivity for the latter. Of the metal membranes, self-supporting, dense palladium alloy membranes have been shown to exhibit extremely high hydrogen perm-selectivity and are able to produce high purity hydrogen feed streams needed for recycling. The outcome leads to saving on downstream separation requirements, equipment size and energy usage.

An excellent review of palladium membrane research has been published by Paglieri and Way [1], covering all aspects of performance, fabrication and applications. The present contribution is intended to enlarge upon the prospects for vacuum deposition and, in particular, magnetron sputtering for membrane production.

Palladium as a Membrane Material

Pd-based membranes operate by dissociative adsorption of molecular hydrogen on membrane surfaces.

$$H_2 \text{ (gas)} \rightleftharpoons 2H \text{ (surface)}$$

which comes about because of the powerful catalytic properties of palladium. Given the equilibrium between H_2 molecules in the gas phase and hydrogen atoms within the surface (related by the respective rate constants for adsorption and dissociation), the concentrations of atomic hydrogen just within the metal is proportional to the square root of hydrogen partial pressure according to Sieverts' law.

The dissociated atomic hydrogen diffuses through the membranes and is recombined into molecular hydrogen that desorbs on the other side of the membrane.

The hydrogen permeation process is influenced by the surface topography, the purity of the metal and its defect structure (e.g., grain boundaries and dislocations). Within the metal, the hydrogen occupies octahedral interstitial sites. At high hydrogen concentrations, above 20°C, the α phase of palladium hydride exists, and one of the problems associated with pure palladium as a membrane is hydrogen embrittlement and the distortion of the metal by repeated adsorption/desorption cycles of exposure.

Palladium Alloys

A solution to this problem is to alloy the palladium, for example with silver, copper or ruthenium. Such alloys have higher hardness and tensile strength than Pd. Not only is $Pd_{60}Cu_{40}$ appreciably cheaper than palladium, but it has increased permeability and sulfur tolerance. The hydrogen permeance of a range of Pd–Cu alloys over a wide range of temperatures has been studied by Howard et al. [2], and the increased permeability is attributed to the formation of a bcc structure. It has been suggested that the electronic effects are the cause of this behavior [3]. According to Roa et al. [4] the highest hydrogen permeability is obtained at about 60 wt% Pd measured at 350°C, while McKinley [5] recommends 42 wt% Cu and 58 wt% Pd to obtain a single phase β structure.

It is an important advantage of Pd–Cu alloys that they are resistant to the poisoning effects of sulfur at high temperatures, perhaps caused by the formation of impermeable sulfur compounds. Progress at the National Energy Technology Laboratory (NETL) on S-tolerant hydrogen membranes has been reviewed by Killmeyer et al. [6]. Carbon, for example originating from pump oil, is another deleterious impurity that must be avoided (1).

Economic Constraints

Palladium is an expensive metal and this imposes limits on the thickness of material that can be used for hydrogen purification in competition with other industrial methods. Emonts et al. estimated that films less than about 5 μm in thickness need to be used in a fuel-cell methanol reformer [7], while Criscuoli et al. [8] concluded that 20 μm is an upper limit for membranes to be economically competitive. These economic estimates overlook the possibility of recycling the palladium or palladium alloy. This becomes a very real possibility in the use of free-standing membranes rather than composite structures with other metals or ceramics. Recycling prospects probably increase the thickness constraint to something between 5 μm and 8 μm, a value that is also consistent with factors such as limitations on the volume of space occupied by a multiple membrane assembly.

In the past, thin metal membranes have been fabricated by rolling between precision rollers, but potential for pinhole defects limits this method for Pd-alloy membranes to a current state of the art of 25 μm (0.001 in.). Difficulty in controlling deformation across the length of the press rolls used in forming the membranes limits the practical width of the membranes to approximately 10 cm. Other methods, such as traditional 'thick-film' coating techniques, have been used to fabricate self-supporting membranes. Coating methods, such as electroplating and electroless deposition, have been demonstrated, but have significant concerns with contamination from organic carbon, and the ability to keep a controlled and consistent bath chemistry over multiple cycles and large areas.

Even though palladium utilization has been increased by the techniques presented above for self-supporting membranes, further reductions in thickness are needed to make this technology economical. As an example of an economic driver, for a residential hydrogen purifier or reformer, a series of more than a dozen membranes are required to purify a sufficient volumetric flow rate. In addition to the cost of the palladium (approximately one ounce for a stationary fuel cell reformer), the cost of labor and materials to assemble multiple membrane elements is high. Hence, a method to fabricate thinner, large-area palladium-alloy membranes in a continuous or even semi-continuous manner would represent a significant breakthrough in the development and commercialization of hydrogen purifiers for fuel cells and coal gasification.

Vacuum Deposited Palladium Membranes

Vacuum deposition methods are used in a wide variety of industries including semiconductors, machine tools, razor blade manufacture and packaging. At the manufacturing level, vacuum-based processes can be cost-effective. For example, an important subclass of vacuum deposition processes is vacuum roll coating, the treatment of large flexible substrates often hundreds of meters in length and up to several meters wide. The economics and scale-up issues of roll coating have been well established for a number of markets. Using these methods, flexible polymeric materials and metal foils have been coated with a variety of metal and ceramic materials for use in capacitors, magnetic media, thin film batteries and food packaging. These products are all subject to economic constraints and the competitiveness with other coating methods.

One form of physical vapor deposition widely used is electron beam evaporation in which the intense focused power of an energetic electron beam heats a crucible of the source material in vacuum and it evaporates on to the work piece. This is less suitable for alloy deposition because differing vapor pressures of the constituents alter coating composition by something akin to distillation. Much better, therefore, is magnetron sputtering in which ions from a dense plasma strike the surface of a target and release atoms kinetically by collision. The composition of the deposited coating remains close to that of the source material after an initial period of equilibration. The impinging atoms have much greater energies than those in the case of thermal evaporation and this increases ad-atom mobilities on the work-piece surface to produce denser, more pin-hole free films, especially at high deposition rates (several nm/sec).

Moss and Dye [9] and Peachey et al. [10] used sputtering to deposit thin (100 nm) films of palladium on both faces of 40 μm thick foils of tantalum or of vanadium, after first removing native oxide by argon ion sputtering. This was in order to remove the diffusion barriers presented by metal oxides. The palladium provides the necessary catalytic activity for dissociative chemisorption of hydrogen. The resulting permeabilities were 20 times greater than that of a 40 μm Pd foil, and the H_2/He ideal separation factor was very high, at 50,000. The membrane was stable up to 300°C but at higher temperatures diffusion of palladium into the substrate foil decreased the permeability. This is generally the case with composite metallic membranes and is a strong reason for the preparation of self-supporting palladium alloy membranes of monolithic composition.

In summary, a vacuum-based process must be capable of enhancing catalyst activity at the palladium surface, perhaps by atomic displacement, and of optimizing diffusion pathways, perhaps by formation of small grains and many grain boundaries. Finally, the film nucleation and growth process must favor production of a thin but pin-hole free film and avoidance of island formation known as Volmer-Weber growth.

Self-Supporting Palladium Alloy Membranes

The novel feature of our approach is to prepare freestanding, thin membranes by vacuum deposition on to a suitable temporary substrate that can easily and cleanly be removed. Important requirements for the substrate material are that it be very smooth and free of contaminants, pinholes and surface defects. The materials should also have decent thermal stability and be inexpensive (relative to the membrane), reusable and/or recyclable. Based on these requirements, two initial approaches were envisioned: (1) deposition of the membrane onto a polymeric substrate, which can be subsequently chemically dissolved or (2) deposition onto a rigid substrate with and without pre-treatment with a thin release coating that may be water soluble.

In the first case, high-quality polystyrene (PS) or polyvinyl alcohol (PVA) film 0.002–0.005″ thick was used. Pd-alloy films were deposited by vacuum deposition methods (discussed in further detail below), initially onto individual sheets approximately 10 in.^{2} each, then onto larger sheets up to 75 in.^{2} in area, and ultimately onto continuous rolls up two 12″ wide and several feet long. The metallized polymer films were then cut to size (as needed) and mounted between a set of high transmission metal mesh. At this point, appropriate sealing attachments could also be incorporated. Finally, the polymer film is removed by exposure (e.g., immersion) to a suitable solvent. It is essential that any hydrocarbon residue be removed in order to limit carbon contamination of the membrane during operation at elevated temperatures.

Experimental

Two different approaches to the problem of releasing deposited alloy films have been assessed. One is to use a polymeric substrate that can be dissolved in a suitable solvent, while the other is to choose a solid substrate to which the deposited film adheres very poorly so that it is easily lifted off. Yet another procedure is to coat a polymeric film such as Kapton™ with a water-soluble release agent such as sodium iodide or barium chloride. There are many other such possibilities.

In the first experiments, palladium copper alloy was deposited by magnetron sputtering from a target with the composition Pd 60%/Cu 40% (by weight) in an argon plasma, on to flexible film of polyvinyl alcohol (PVA, trade name 'Solublon') and on to polystyrene. The alloy films were from 0.5 to 3 microns thick and had a minimal intrinsic stress. This was evident from the fact that the films had no tendency to curl up.

In other experiments, thermal evaporation from separate sources of palladium and copper was carried out using electron beam to vaporize material. Pd–Cu alloy films were formed on polymer backing materials in the 12″ wide drum or web-coating system shown in Fig. 11.1. Deposition rates of palladium and copper were independently controlled using crystal quartz monitor or electron impact emission spectroscopy (EIES). After having established uniformity profiles and elemental distribution across the deposition zone, a design of experiments (DOE) approach was conducted to correlate processing parameters (i.e., deposition/web feed rates, drum temperature, polymer pre-treatment, etc.) with final properties (response) of the Pd–Cu alloy films (i.e., composition, defects, strain). Finally, in other experiments magnetron sputtering was used to fabricate free-standing Pd–Cu alloy films using rigid silicon and glass backing materials, i.e., vacuum processing parameters were optimized to promote dense film formation with minimal adhesion to backing material.

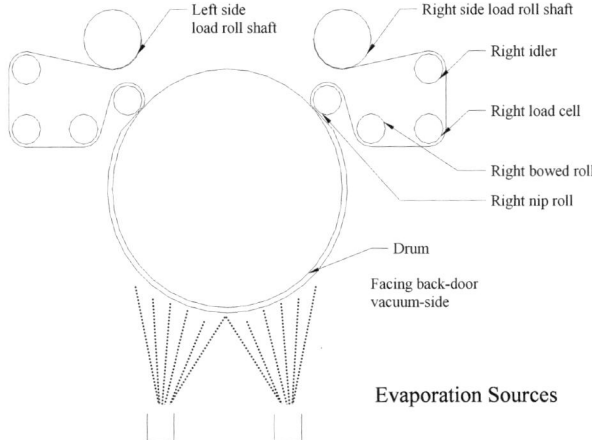

Fig. 11.1 Schematic of web roll coater with evaporation sources

Plastic Backing Removal

Pd–Cu alloy coated plastic samples were first cut into discs 75 mm in diameter, sandwiched together with a porous metal disc (Monel) between two microscreens and then clamped (refer to schematic in Fig. 11.2). To remove the plastic backing material, the samples are then lowered horizontally with the coated polymer discs up, Monel mesh down, into the appropriate solvent; hot water (60–80°C) for Solublon and chloroform (room temperature) for polystyrene. Polymer dissolution (removal rate) was evaluated as a function of temperature and time. Nominal times were 30 s for the Solublon and 600 s for the polystyrene. Upon dissolution of the polymer backing material, samples were removed from the solvent, carefully disassembled and then dried.

H_2 *Permeation Testing*

A membrane foil is first sandwiched between two circular supports, such as alumina paper, and then sealed with either a Kalrez O-ring (max. use to 315°C) or Grafoil packing material (allowing a 650°C upper use temperature in oxygen-free environments) in the 25 mm Millipore membrane cell. The membrane is then checked with helium to confirm a tight seal and that the membrane is defect (pinhole) free. Subsequently, the membrane is heated to operating temperature to begin permeation testing.

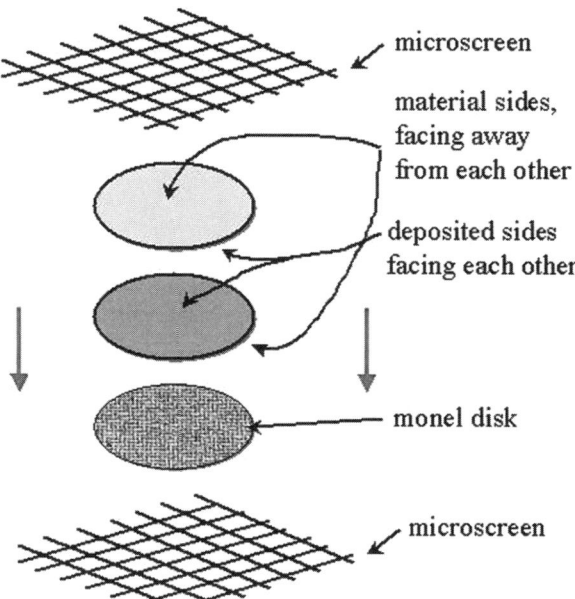

Fig. 11.2 Schematic of sandwich configuration

Pd–Cu Alloy Deposition on Plastic

Although procedures have been developed and routinely employed to deposit and release Pd–Cu films, 2–5 μm-thick and >75 in.[2] in area from PS and PVA polymer backing materials (refer to Fig. 11.3 showing appearance of 6″ by 6″ film deposited on the polymeric film), we observe appreciable through-thickness defects (pinholes) for film thicknesses less than 6 μm. Though a small percentage of the defects are related to the deposition process (and therefore potentially eliminated), the majority of the defects are due to characteristics of the sacrificial substrate.

Some defects appear to be associated with the tendency for polymeric films to become electrostatically charged and hence to attract fine air-borne particulates. These impede the formation of a continuous film of a few microns thickness. Other problems are caused by a low density of nucleation points for metal films to grow and spread across the surface, a process that requires some wettability.

Two treatments methods were investigated to promote spreading of the Pd–Cu films on the polymer backing materials, that is, Ar/O_2 plasma and deposition of a precursor layer such as SiO_x with a more favorable surface energy. Although argon plasma treatment (with 5% O_2) was shown to decrease water contact angles (increase film wettability) on all plastics tested (PS, PVA, Kapton and PET), we were not able to completely eliminate defects in films less than 5 μm. Based upon the success of deposition on oxidized silicon wafers, an evaporated layer of silica (SiO_x) was applied to plasma-treated polymer substrates. Unlike the case for rigid silicon substrates, the SiO_x-coated polymer substrates did not yield films with a lower defect density. Therefore, irrespective of the surface treatment options tested (plasma or seed layer), we were not able to produce defect-free Pd–Cu films on flexible polymer substrates over large areas at thicknesses <6 μm (it is important to note however that defect-free films have been produced at thicknesses >6 μm).

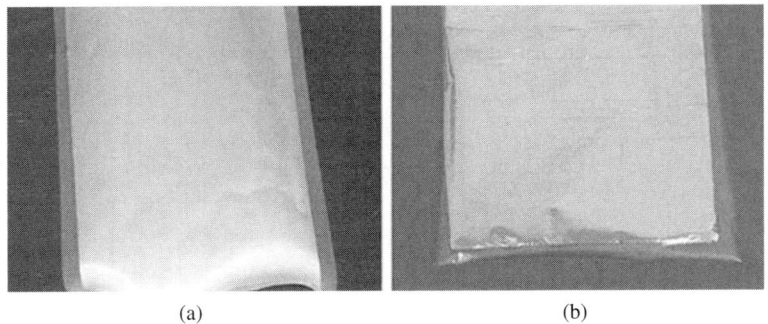

(a)   (b)

Fig. 11.3 Pd–Cu alloy films on (**a**) polystyrene and (**b**) polyvinyl alcohol films

Pd–Cu Alloy Deposition on Silicon

The initial alloy coatings deposited on to polymers showed a tendency to exhibit pinhole defects most of which have since been attributed to the presence of dust particles. Polymer films have a propensity to accumulate an electrostatic charge that attracts fine particles. As an alternative, interim method to address the issue of defects in the Pd–Cu alloy membrane films, films were deposited onto smooth, silicon wafers. Particulate and other contaminants can be more readily controlled (i.e., minimized) on a silicon surface, in comparison to plastic, and is considerably smoother than plastic. In experiments using magnetron sputtering, we were able to produce defect-free Pd–Cu films on 12 in. diameter silicon and 4 in. square glass plates. A released film, 4 μm thick, is shown in Fig. 11.4. Films as thin as 0.7 μm have been produced in this way.

Although we have established procedures for the formation of Pd–Cu alloy films over large areas (> 75 in.^{2}) on sacrificial polymer backing materials, formation of Pd–Cu alloy materials on silicon wafers has helped us to address the issue of defect formation (i.e., pinholes). In general, the key factors that affect formation of a thin, dense, defect-free, Pd–Cu alloy film are surface energy, roughness and oxygen/moisture content of the backing material. By using thermally oxidized silicon wafers, we have been able to reduce the surface roughness while at the same time, control surface chemistry and, more specifically, oxygen activity. Correspondingly, using vacuum processing conditions that have been optimized to minimize intrinsic film stress, we have produced pinhole-free Pd–Cu alloy films over large areas at thicknesses below 5 μm.

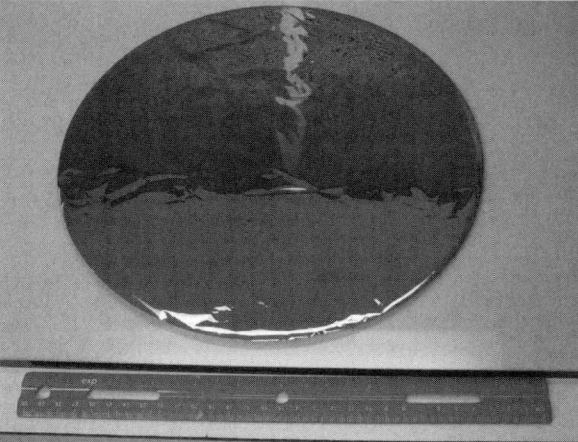

Fig. 11.4 Free-standing Pd–Cu Foils, 4 μm-thick; (**a**) from rigid, 12″ silicon wafer

Annealing

As has been previously reported, as-deposited membranes are usually in the alpha phase and transform to a beta or mixed alpha and beta phase upon heating to 400°C [2]. Because these membranes are very thin it may not be possible to assemble the membranes into a module while still in the fcc phase as the change in lattice parameter associated with the phase transformation can result in a slight contraction that could place significant stress on the membrane and cause it to rupture. Hence we began a series of annealing experiments to convert the membrane to the desired phase prior to assembly in a module. The membrane is held between two pyrex glass plates with simple binder clips that have been stripped of paint prior to use. Stainless steel lock wire is used to maintain a separation between the plates to limit air entrapment and allow the membrane to expand or contract more freely without curling. After mounting, the membrane samples were placed in a tube furnace purged with a continuous flow of Ar annealed at 450°C for 12 h. XRD showed that the as-received membranes were in the pure alpha phase while the annealed samples were purely beta (Fig. 11.5). SEM images indicated that the membrane was unchanged during the annealing process.

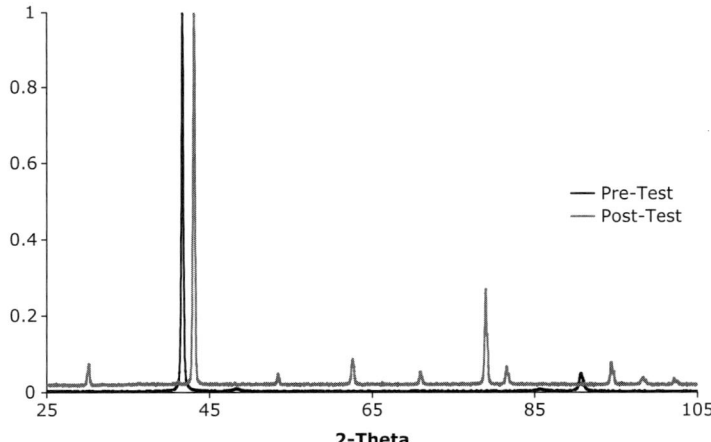

Fig. 11.5 XRD spectrum of 051206#1 before and after testing

Economics of Vacuum Deposited Separation Membranes

Of the elements that comprise a hydrogen purification module, the membrane is presumed to be by far the most significant cost contributor. Analogous to a computer's central microprocessor, the membrane lies at the heart of the purification system and is the key element defining system performance. The DOE has set aggressive performance and cost targets for several membrane properties in 2005 and 2010 including flux, cost per square foot, hydrogen purity and differential pressure.

For Pd alloy membrane fabrication, the materials cost is expected to be higher relative to other fixed costs. Another key element impacting the cost is material throughput. If a greater square footage of material can be produced over a given time, the labor and capital equipment costs can be substantially reduced. In fact, we believe that it is highly likely a semiconductor process tool can be adapted to produce full-size Pd alloy membranes up to 12 in. in diameter. The membrane throughput for such a system can be estimated to first order using the following empirical equation:

$$P = 0.785 \, N \, D/d$$

where P is the throughput (ft^2/min), N is the number of process stations, D is the deposition rate (nm/min) and d is the target membrane thickness in nm. A typical system can allow up to five 12-in. wafers to be processed simultaneously and deposition rates of 600 nm/min are possible. So for a membrane thickness of 4 microns (4000 nm), 0.59 sq ft/1 min of membrane can be produced. Obviously this estimate neglects the contributions of cycle time and downtime due to maintenance although these may be included by altering the above equation in the following manner:

$$P = 0.785 \, N \, D \, U \, (1{-}C)/d$$

where U is the average uptime percentage and C is the percentage of time in each cycle where material is not being deposited. Assuming 75% uptime and 25% coating cycle idle time, then 0.33 sq ft/min of membrane will be produced on average.

The next issue to consider is raw materials cost. The cost of palladium greatly exceeds that of copper so we can reasonably omit the latter from consideration. Sputter deposition processes are highly efficient with more than 95% of sputtered material typically deposited on the support material in a production system. The balance of the material can be recovered as scrap and recycled. The other cost consideration is the fabrication of the PdCu alloy target. This is usually done by taking powders of each material in appropriate quantity to make the desired alloy composition then hot pressing the power in vacuum or inert atmosphere to make a plate typically 0.5 in. thick. For this calculation we add an additional 25% to the cost of palladium for target manufacturing and material recovery. Based on this as well as the composition, density and thickness of the membrane and market price of Pd, we can calculate the membrane raw materials cost per square foot using the following empirical equation:

$$R = 1.2 \times 10^{-2} \, P \, W \, (W + 3) \, T$$

where R is the raw materials cost per square foot, P is the market price ($/oz), W is the weight percentage and T is the membrane thickness in microns. Hence a 4 micron thick, 60 wt% Pd alloy membrane with a market price of $330/oz, will have approximately $35/ft^2 of Pd in it. Combine this information with the projected material throughput, we can arrive at a total cost for manufacturing based on the following factors

$$T = (F + L + E)/(P * S * 1.75 \times 10^5) + R$$

where T is the total cost per square foot, F is the annual equipment deprecia-
tion, L is the fully burdened annual labor costs, E is the annual cost of utilities and
maintenance, P is the throughput per minute and S is the number of 8 h shifts per
day. If we assume a \$1.5M piece of equipment with level amortization over 3 years,
4 full time personnel (3 technicians and 1 engineer) working a total of two shifts at
\$0.50M/yr, and \$0.20M/yr in utilities and maintenance, and the above estimates for
productivity and raw materials, we get a total cost of \$45.40 per square foot. Even
if we have significantly underestimated the throughput, equipment or labor costs,
this cost estimate is still more than an order of magnitude lower than the DOE 2010
target, which gives us great confidence that the process will be cost effective.

Hydrogen Permeation Testing

The first successful membrane to be tested was a 12.7 μm-thick foil with a composition slightly off the ideal $Pd_{60}Cu_{40}$ (i.e., slightly higher palladium weight fraction).
The membrane was heated to 250°C, and the H_2 permeability at this temperature
was determined to be 3.8×10^{-5} cm^3(STP) $\cdot$ cm/cm$^2 \cdot$ s $\cdot$ cm Hg$^{0.5}$ (for comparison, the permeability of a $Pd_{60}Cu_{40}$ foil at 250°C from the patent literature [USP
3,439474] is 9.2×10^{-5} cm^3[STP] $\cdot$ cm/cm$^2 \cdot$ s $\cdot$ cm Hg$^{0.5}$). This is good agreement
given that the palladium composition of the foil sample is higher than 60 weight per
cent and that the H_2 permeability declines sharply for higher Pd contents.

Figure 11.6 shows molar flux versus driving force for this membrane at 250°C.
The membrane was then heated up to 300°C. The flux declined to approximately
one-third of the value at 250°C. One possible reason for this could be carbon contamination from the O-ring. Upon removing the membrane from the cell, the O-ring
appeared to have degraded. Another explanation could be that the membrane did
not undergo a phase change to the higher permeability FCC β-phase, and that it
may have moved into the mixed α and β phase.

As shown in the phase diagram of Fig. 11.7, if the composition of the Pd–Cu
membrane is greater than the pure β boundary and within the α/β phase (at an
approximate palladium concentration [in weight percent] of 61%), then a two phase
structure can exist and thereby reduce the efficiency of the membrane. For a constant
composition, the membrane can transform from the single phase β to the two phase
$\alpha + \beta$ structure by merely heating up the membrane and thereby crossing the phase
boundary.

XRD analysis was performed on the treated 12.7 μm-thick membrane discussed
above as well as on a piece of as-received foil. The treatment consisted of: (1) exposure to H_2 at 250°C for 24 h, (2) 300°C for 72 h, (3) 250°C for 24 h and finally (4)
air quench to room temperature. The XRD pattern is shown in Fig. 11.8. Analysis
indicates that the foil is in the β phase.

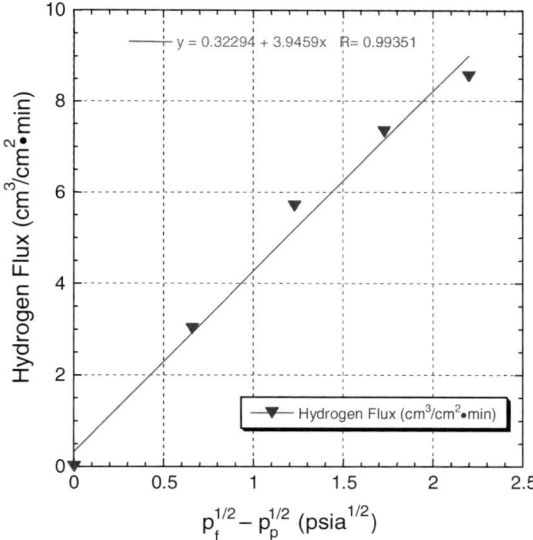

Fig. 11.6 Pure hydrogen flux data at 250°C for 13 μm foil prepared by IBAD on silicon wafer support. Feed pressures range from 5 to 20 psig. Atmospheric pressure is 12 psia in Golden, CO

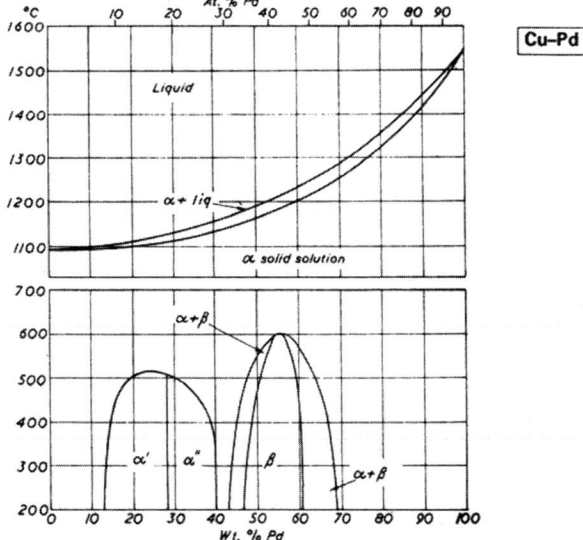

Fig. 11.7 Pd–Cu phase diagram from [11]

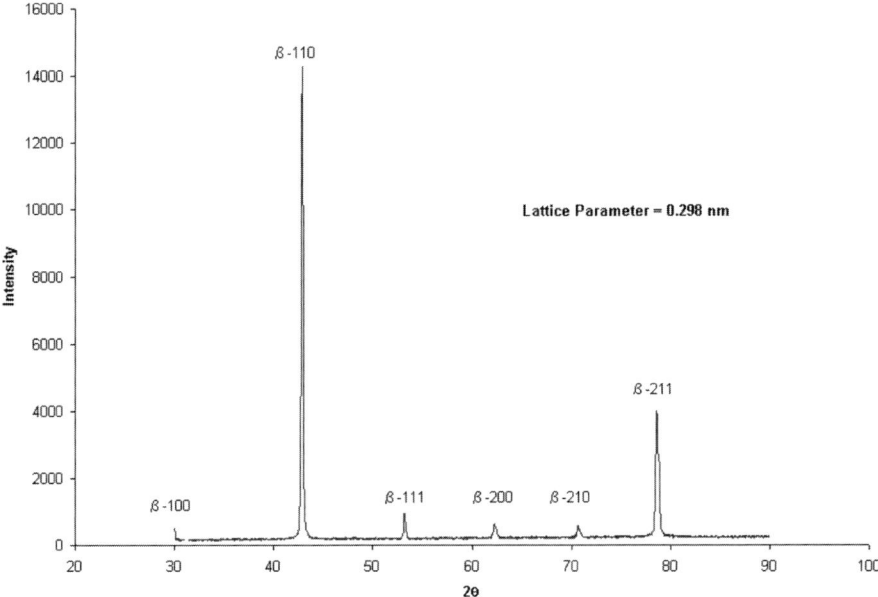

Fig. 11.8 XRD pattern of heat treated PdCu foil

An additional membrane, thickness $\sim 9\,\mu m$, displayed a pure hydrogen flux of $16.8\,cm^3(STP)/cm^2 \cdot min$; this is 2.2 times greater than that of the thicker membrane presented above at the same temperature and driving force. This is shown in Fig. 11.9 along with the results from other membranes. Further tests are ongoing with this membrane.

The H_2 permeability of the $9\,\mu m$-thick membrane above is $5.1 \times 10^{-5}\,cm^3(STP) \cdot cm/cm^2 \cdot s \cdot cm\,Hg^{0.5}$. Correcting the permeability value to $350°C$ using the data in the McKinley patent (USP 3,439,474), [5] we obtain a value of $7.4 \times 10^{-5}\,cm^3(STP) \cdot cm/cm^2 \cdot s \cdot cm\,Hg^{0.5}$. This value compares very well to the permeability reported by McKinley for a 62.5% Pd membrane of $7.9 \times 10^{-5}\,cm^3(STP) \cdot cm/cm^2 \cdot s \cdot cm\,Hg^{0.5}$. Another comparison is that the hydrogen permeability of the $9\,\mu m$ membrane is 56% of the value at the same temperature reported by McKinley for a PdCu alloy membrane with the optimum 60 wt.% Pd composition. We believe that the discrepancies are due to differences in the Pd composition of our samples and the cold-rolled foils used by McKinley and Edlund [12].

Another membrane $6\,\mu m$ thick was tested, although exhibiting a small helium leak that was attributed to Knudsen diffusion (i.e., pure hydrogen flux vs. driving force measurements showed a pressure dependence of ΔP^n where n was approximately 0.6, indicating that there is likely a pinhole in the membrane). This membrane was sealed in the cell using a Kalrez® O-ring. All of the following hydrogen fluxes were therefore corrected for the observed leak. The membrane was heated and then held at $257°C$ for 12h under helium. The hydrogen flow was then initiated and

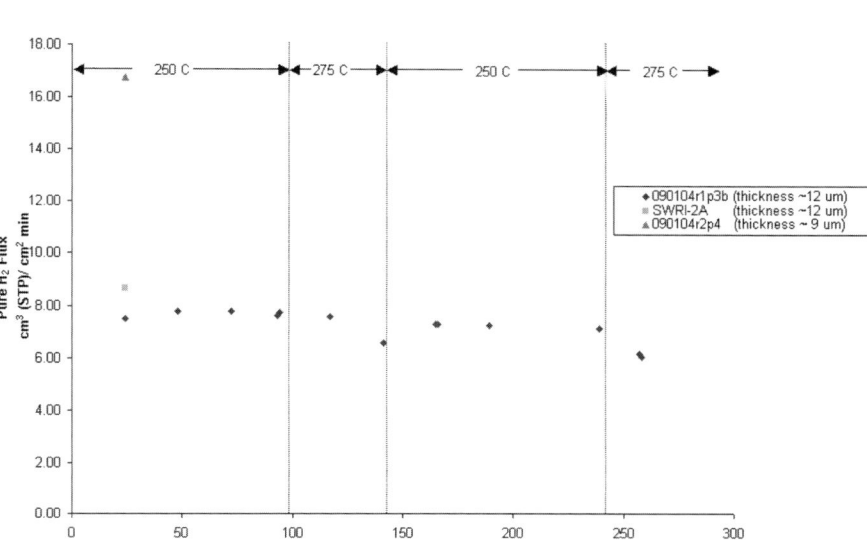

Fig. 11.9 Pure hydrogen flux for a 20 psig feed pressure for various membranes

the pure hydrogen flux at 20 psig was measured. The flux at these conditions slowly decreased from about $28\,cm^3/cm^2$ min to about $20\,cm^3/cm^2$ min over a period of 5 days. At this point, the flow to the membrane was switched from hydrogen to helium and left overnight with a feed pressure of 0 psig. After remaining under helium for approximately 15 h, the membrane was tested again for pure hydrogen permeation at 20 psig. The pure hydrogen flux jumped to $36\,cm^3(STP)/cm^2$ min. The pure hydrogen flux was measured again at $36\,cm^3/cm^2$ min the following day after the membrane was exposed to helium again overnight. This flux corresponds to a pure hydrogen permeability of $7.4 \times 10^{-5}\,cm^3(STP) \cdot cm/cm^2 \cdot s \cdot cm\,Hg^{0.5}$ at 250°C. This value is within 20% of the pure hydrogen permeability at 250°C reported in the McKinley patent [5].

Remaining Issues

In this ongoing research program, there are several issues that need to be addressed. One, mentioned briefly above, is the fact that the Pd–Cu alloy composition is critical in the as-prepared and annealed samples. Figure 11.10 shows hydrogen permeability as a function of the palladium content (in weight per cent) at 350°C, and it is evident that there is a critical composition close to 60 wt.% Pd.

Analysis of our films shows that they are slightly palladium-rich with a composition near Pd 63.3 wt.%, Cu 36.7 wt.%. A similar problem was encountered by

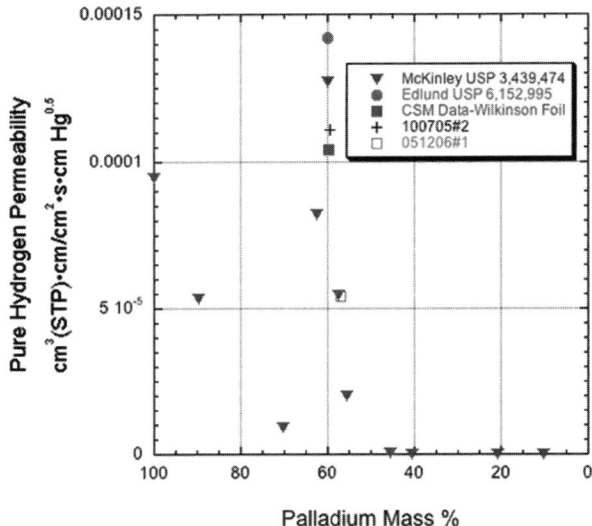

Fig. 11.10 H$_2$ permeability of PdCu foils at 400°C

Athayde et al. [13] in the sputtering of Pd-Ag alloys: their films were also palladium-rich relative to the sputtering target. This is something that can be adjusted by addition of copper to the target.

Another issue is that of achieving good seals without damaging the thin membranes. The pressure from O-ring seals is considerable and can exert lateral stress on thin films. There are various solutions to this problem that remain to be evaluated.

Other flexible substrates than polystyrene and PVA may have better properties in terms of coating nucleation and wettability and so allow the continuous production of membranes by a reel-to-reel process.

Conclusions

Self-supporting Pd–Cu alloy membranes have been produced with thicknesses down to about 3 μm. The polymer substrates (PS and PVA) evaluated so far require thicker coatings, above about 7 μm, in order to be pinhole-free.

Good hydrogen permeability rates have been measured and self-supporting membranes are free from the problems due to metallic interdiffusion in composite structures, and so should exhibit a long life at temperatures above 300°C.

It has been shown to be feasible to produce pore-free membranes below 5 μm in thickness that should consequently be competitive with other methods for hydrogen separation in energy applications.

Acknowledgments   The authors gratefully acknowledge support of this work from the US Department of Energy (National Energy Technology Lab) through cooperative agreement No. DE-FC26-03NT41849.

References

1. Paglieri SN, Way JD. Innovations in palladium membrane research. Sep Purif Methods. 2002;31(1):1–169.
2. Howard BH, Killmeyer RP, Rothenberger KS, Cugini AV, Morreale BD, Enick RM, et al. Hydrogen permeance of Pd–Cu alloy membranes over a wide range of temperatures and pressures. J Memb Sci. 2004;241:207–18.
3. Vert ZL, Drozdova NV. Solubility of hydrogen in alloys of the palladium-copper system. J Appl Chem USSR. 1990;63(11):2365.
4. Roa, F, Block MJ, Way JD. The influence of alloy composition on the H_2 flux of composite Pd–Cu membranes. Desalination. 2002;147:411–6.
5. McKinley DL. Metal alloy for hydrogen separation and purification. US Patent 3,350,845, 1967.
6. Killmeyer RP. NETL progress on S-tolerant hydrogen membranes. Fuel Cell Bull. 2004;2:3.
7. Emonts B, Hansen JB, Jorgensen SL, H̄lein B, Peters R. Compact methanol reformer test for fuel-cell powered light-duty vehicles. J Power Sources. 1998;7:288.
8. Criscuoli A, Basile A, Drioli E, Loiacono O. A economic feasibility study for water gas shift membrane reactor. J Memb Sci. 2001;181:21.
9. Moss TS, Peachey NM, Snow RC, Dye RC. Multilayer metal membranes for hydrogen separation. Int J Hydrogen Energy. 1998;23(2):99.
10. Peachey NM, Snow RC, Dye RC. Composite Pd/Ta metal membranes for hydrogen separation. J Memb Sci. 1996;111:123.
11. Smithells CJ. In: Brades EA, editor. Smithells metals reference book. 6th ed. London: Butterworth-Heinemann; 1983.
12. Edlund DJ. Hydrogen permeable membrane and method for producing same. US Patent 6,152,995, 2000.
13. Athayde AL, Baker RW, Nguyen P. Metal composite membranes for hydrogen separation. J Memb Sci. 1994;94:299.

Chapter 12
Palladium-Copper and Palladium-Gold Alloy Composite Membranes for Hydrogen Separations

Fernando Roa, Paul M. Thoen, Sabina K. Gade, J. Douglas Way, Sarah DeVoss, and Gokhan Alptekin

Electroless plating was used to fabricate PdCu and PdAu alloy composite membranes using tubular Al_2O_3 and stainless steel microfilters to produce high temperature H_2 separation membranes. The composite membranes were annealed and tested at temperatures ranging from 350 to 400°C, at high feed pressures ($\leq$250 psig) using pure gases and gas mixtures containing H_2, carbon monoxide (CO), carbon dioxide (CO_2), H_2O and H_2S, to determine the effects these parameters had on the H_2 permeation rate, selectivity and recovery. Reformulating the palladium electroless plating solution dramatically increased the H_2 permeability of the metal membranes and improved their stability in gas mixtures containing CO, CO_2 and steam. No flux reduction was observed for a PdAu membrane for the water-gas shift (WGS) mixture compared to a pure H_2 feed gas at the same 25 psig partial pressure difference and 400°C. A typical pure H_2 flux for a PdAu membrane was 245 $SCFH/ft^2 = 0.93$ mol/m^2.s for a 100 psig H_2 feed gas pressure at 400°C. This flux exceeds the 2,010 DOE Fossil Energy pure hydrogen flux target. The H_2/N_2 pure gas selectivity of this PdAu membrane was 1,000 at a partial pressure difference of 100 psi. However, inhibition of the H_2 flux was observed for similar WGS experiments with PdCu composite membranes. H_2S caused a strong inhibition of the H_2 flux of the Pd-Cu composite membranes, which is accentuated at levels of 100 ppm or higher. Adding 5 ppmv to the WGS feed mixture reduced the hydrogen flux by about 70%, but the inhibition due to H_2S was reversible. At H_2S levels above 100 ppm, the membrane suffered some physical degradation and its performances was severely affected. The use of sweep gases improved the hydrogen flux and recovery of a Pd-Cu composite membrane.

J.D. Way (✉)
Colorado School of Mines, Golden, CO 80401, USA
e-mail: dway@mines.edu

A.C. Bose (ed.), *Inorganic Membranes for Energy and Environmental Applications*,
DOI 10.1007/978-0-387-34526-0_12, © Springer Science+Business Media, LLC 2009

Introduction

The US Department of Energy's FutureGen program aims to demonstrate the technical and economical feasibility of a coal gasification plant to produce power with near zero emissions, including the emissions of carbon dioxide (CO_2). In this approach, coal is first gasified to produce a carbon monoxide (CO)-rich synthesis gas. As shown in Fig. 12.1, the impurities in the syngas are removed and the CO content is reduced by converting the CO to CO_2 and H_2 through the WGS reaction. Finally, the hydrogen is separated from other compounds, mainly CO, CO_2 and water. Although no coal-based facilities have been constructed that produce both hydrogen and electric power, system level studies indicate that the efficiency of the coal-to-hydrogen plant could be enhanced if the WGS and H_2 separation were combined into a single step [1]. A key feature of this process intensification improvement is that a WGS membrane reactor would produce both a high pressure CO_2 product stream and a high-purity hydrogen stream for power generation. This would greatly facilitate the economics of carbon sequestration as this CO_2 rich stream has value for enhanced oil recovery as well as carbon credits in the European Union.

A successful hydrogen separation membrane for this application should provide robust performance, high hydrogen throughput, high selectivity and recovery, resistance to inhibition from impurities and long life at low cost. Palladium (Pd) alloy membranes have all of these attributes. Palladium and its alloys, as well as nickel, platinum and the metals in Groups III–V of the Periodic Table are all permeable to hydrogen. Hydrogen-permeable metal membranes made of palladium and its alloys are the most widely studied due to their high hydrogen permeability, their chemical

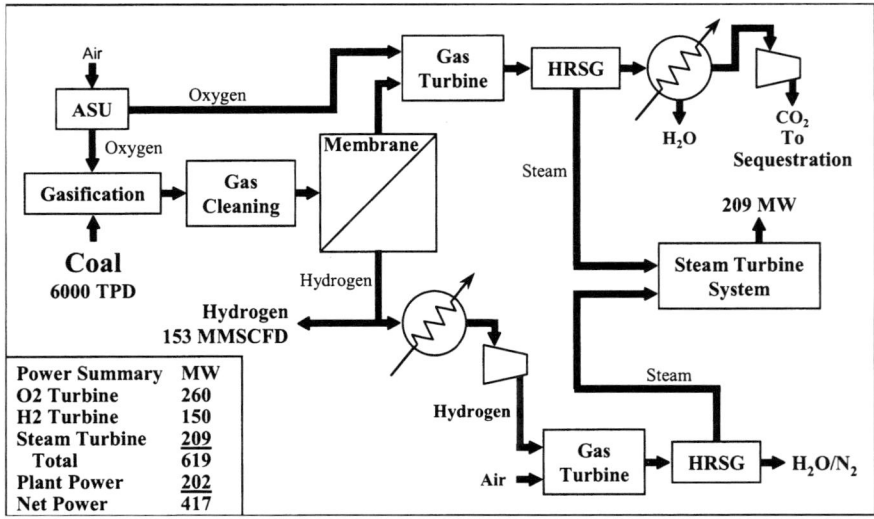

Fig. 12.1 Schematic of a coal-to-hydrogen co-generation plant [1]

compatibility with many hydrocarbon containing gas streams, and their theoretically infinite hydrogen selectivity.

Many palladium alloys such as $Pd_{73}Ag_{27}$, $Pd_{95}Au_5$, and $Pd_{60}Cu_{40}$ possess higher hydrogen permeability than pure palladium. In his pioneering work in the 1960s, McKinley and coworkers [2, 3] reported that binary alloys of Pd with Au and Cu had pure hydrogen permeabilities greater than Pd that were unaffected by thermal cycling, and had some resistance to sulfur poisoning by hydrogen sulfide. This is shown graphically in Fig. 12.2. In particular, the inhibition or reduction of the pure hydrogen flux due to exposure to 4 ppm hydrogen sulfide through the 40mass% Au alloy was the least compared to pure Pd, PdAg, and PdCu alloys. Our recent research efforts have focused on developing sulfur resistant, composite metal membranes consisting of a relatively thin palladium or palladium-alloy coating on porous ceramic or metal supports. Such a composite membrane can achieve high flux because the palladium film is very thin, which also minimizes materials costs.

The sulfur resistance of PdCu foil membranes was further investigated by researchers at the DOE NETL laboratory [4]. They reported the best sulfur resistance, essentially no inhibition, with a 20mass% Cu in Pd binary alloy having an FCC crystal structure. However, as shown in Fig. 12.3, this $Pd_{80}Cu_{20}$ alloy has 20% of the H_2 permeability of pure Pd and about 2 times less than 40% Au.

In previous publications from our group [5–8], we have discussed many issues surrounding the fabrication and testing of Pd and Pd-Cu membranes. For the present

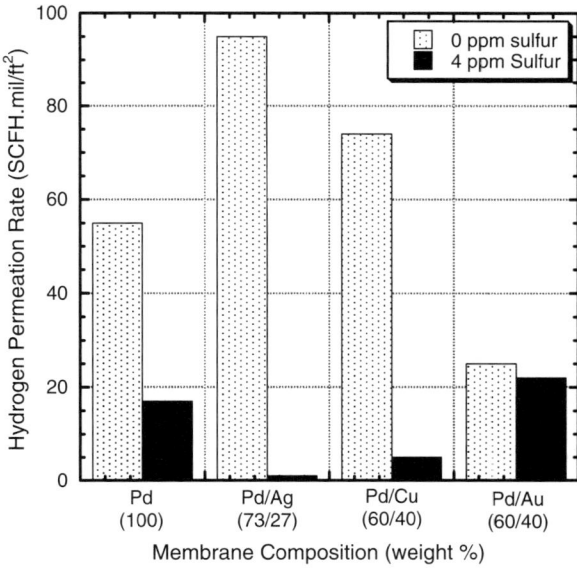

Fig. 12.2 H_2 permeation rates with and without 4 ppm H_2S at 350°C and 75 psig for pure Pd and Pd alloys from McKinley [3]

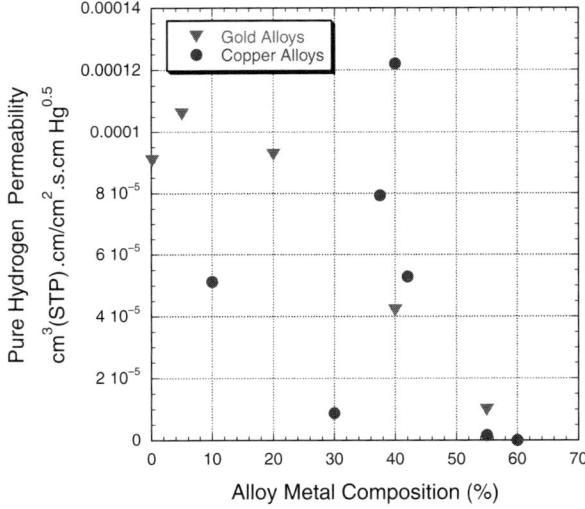

Fig. 12.3 The effect of bulk alloy composition on pure H_2 permeability of foil membranes at 350°C from McKinley [3]

work, we tested the permeation properties of Pd, Pd-Cu, and Pd-Au membranes supported on tubular, porous ceramic and stainless steel substrates. Feed gases consisted of pure H_2 and N_2 as well as mixtures containing H_2S, H_2O, CO, and CO_2 to simulate separation of hydrogen from an equilibrium WGS mixture. We also investigated the effect that sweep gases had on membrane performance. These conditions emulate those expected to be observed in real-world applications. By testing the membranes under these conditions, we hoped we could learn more about their behavior and possibly to improve their performance.

The DOE Fossil Energy cost and performance targets for hydrogen selective membranes are summarized in Table 12.1 below.

Table 12.1 DOE hydrogen membrane performance goals

Performance Criteria	2007 Target	2010 Target	2015 Target
Flux[a]	100	200	300
Operating Temperature (°C)	400–700	300–600	250–500
Sulfur Tolerance	Yes	Yes	Yes
Membrane Cost ($/ft^2)	150	100	<100
WGS Activity	Yes	Yes	Yes
Operating Pressure Capability (psi)	100	400	800–1,000
CO Tolerance	Yes	Yes	Yes
Hydrogen Purity (%)	95	99.5	99.99
Stability/Durability (years)	3	7	>10

a) SCFH/ft^2 @ 100 psi ΔP H_2 @ 50 psia perm side pressure.

Experimental

Materials

Composite Pd alloy membranes were fabricated by sequential deposition of palladium and copper or gold from electroless plating baths onto symmetric 0.2 μm cut-off α-alumina tubes (GTC-200) fabricated by CoorsTek, formerly Golden Technologies. The symmetric (constant pore size) GTC supports were chosen over the higher flux Pall Exekia T1-70 filters to maximize strength and safety working at high feed and transmembrane gas pressures.

More recently, composite membranes were prepared using Pall AccuSep© microfilters, which are zirconia-coated, porous stainless steel, tubular membranes [9]. The primary challenges in preparing Pd alloy composite membranes on stainless steel supports are twofold. First, the mean pore size of the porous metal material is typically about 2 μm requiring thicker Pd alloy films to bridge these larger pores [10]. Second, a diffusion barrier layer is usually required to prevent intermetallic diffusion between the Pd alloy film and the stainless steel support. A strategy used by Pall Corporation to mitigate both of these issues is to deposit an oxide coating on the stainless steel filter as shown in Fig. 12.4. This creates a diffusion barrier and reduces the surface roughness of the substrate.

Gases used in this study were nominally 99.999% pure (UHP grade) and were used without further purification.

Membrane Preparation

Prior to palladium plating, the surface of the membrane support is first seeded with Pd particles. The seeding procedure involves impregnation of the ceramic support with a chloroform solution of palladium acetate, followed by calcination and reduction in flowing hydrogen or aqueous hydrazine solutions [5].

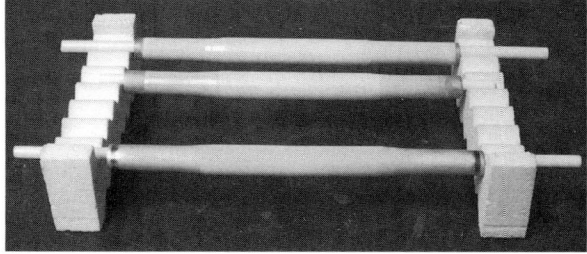

Fig. 12.4 Two Pall AccuSep ZrO_2 coated porous stainless steel filters at top. Pd coated AccuSep filter at bottom. These substrates have approximately 30 cm^2 of active area

Table 12.2 Summary of membranes fabricated and tested

Sample #	Support	Thickness (μm)	Alloy Composition (mass %)	H_2 Flux at 20 psig, 400°C (mol/m^2.s)	EDTA in Pd Film?
GTC-6	GTC-200	7	$Pd_{94}Cu_6$	0.026*	Yes
GTC-30	GTC-200	6	$Pd_{94}Cu_6$	0.11	No
GTC-31	GTC-200	7	$Pd_{87}Au_{13}$	0.22	No
Pall-77	AccuSep®	9	$Pd_{85}Au_{15}$	0.2	No

*Tested at 350°C

Pd and Cu or Au electroless plating baths were then used in sequence to deposit films ranging from 6 to 9 microns in thickness. Conditions and bath compositions for Pd and Cu electroless plating can be found in our previous work [11].

A description of our electroless Au plating process was given by Foo [12]. We believe that the mechanism of Au plating is a combination of displacement plating, caused by the more positive reduction potential of Au^{3+}, along with electroless plating of Au using hydrazine as the reducing agent.

Pd electroless baths commonly contain ethylene diamine tetra acetic acid (EDTA), a complexing agent that increases the stability of the plating bath [6]. Recently, we have determined that the use of EDTA can leave up to 6mass% carbon in the Pd membranes and this carbon reduces permeability and can react with CO_2 present in the feed gas, contributing to membrane instability [13, 14]. The majority of the membranes discussed in this article were prepared without using EDTA.

Table 12.2 summarizes the properties of membranes investigated in this work.

High Temperature Permeation Tests

The membrane to be tested was loaded into a stainless steel module, which in turn was mounted in a tube furnace. Figure 12.5 shows a sketch of the module and a photograph of a typical membrane. To avoid embrittlement, the membranes were heated under helium and no H_2 was introduced until the membrane reached 350°C. Annealing the two metals was achieved during the initial single gas permeability tests. Transmembrane pressure differentials varied from a 5 psig to 250 psig (up to 18 bar), while operating temperatures varied from 350 to 400°C. Permeate pressure was local atmospheric pressure (12 psia = 820 mbar), except for the test using a sweep gas.

Characterization

Scanning electron microscopy (SEM) was utilized to determine film thickness; X-ray diffraction (XRD) and energy dispersive X-ray spectroscopy (EDAX) were used to study crystal structure and determine Pd alloy composition.

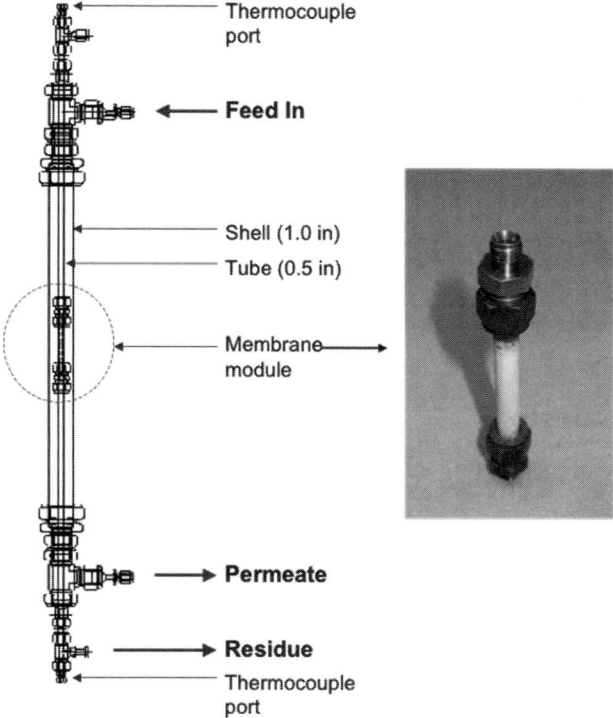

Thermocouple
port

Feed In

Shell (1.0 in)
Tube (0.5 in)

Membrane
module

Permeate

Residue
Thermocouple
port

Fig. 12.5 Photograph of a typical membrane and the high temperature testing module

Results and Discussion

Pure Hydrogen Permeation

The pure H_2 flux of a $Pd_{94}Cu_6$ alloy film plated onto a GTC $0.2\,\mu m$ symmetric support (membrane GTC-30) is shown in Fig. 12.6 as a function of feed pressure over the range of 10–100 psig. Please note that the flux of membrane GTC-30 is about 110 SCFH/ft^2 at 400°C for a 100 psig feed pressure. Therefore this composite membrane, fabricated on an inexpensive, symmetric support easily exceeds the 2007 DOE Fossil Energy hydrogen flux target. The ideal H_2/N_2 selectivity for this membrane ranged from 227 at 80 psi to 407 at 20 psi. The calculated "n-value" or pressure exponent for GTC-30 is 0.6, which is slightly higher than the theoretical value of 0.5, which is observed when diffusion of hydrogen atoms through the bulk Pd film is limiting hydrogen permeation. The n-value of 0.6 may indicate some hydrogen mass transfer resistance of the symmetric ceramic support. Two additional data sets are given in Fig. 12.6 showing that the pure hydrogen flux declined after mixed gas permeation tests with a WGS equilibrium mixture and that subsequent oxidation of the membrane restored about half of the flux reduction due to the WGS tests. We have previously observed that oxidation can improve the flux of composite

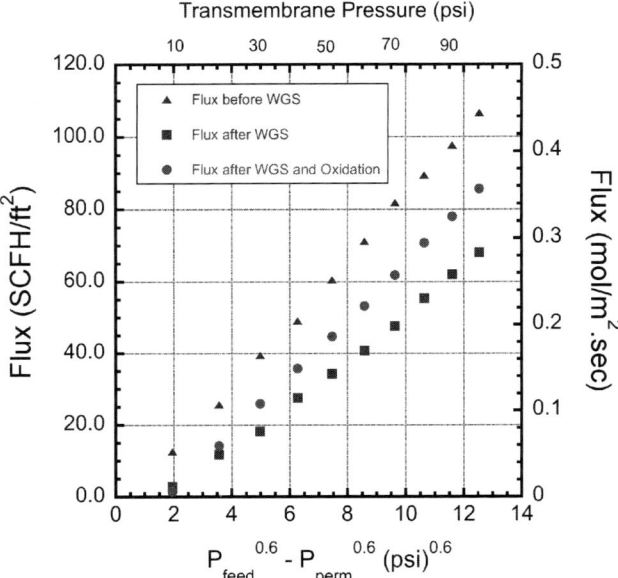

Fig. 12.6 The influence of feed pressure on pure H_2 flux for $Pd_{94}Cu_6$ membrane GTC-30 at 400°C. $SCFH/ft^2$ refers to $ft^3(STP)/ft^2 \bullet h$

palladium alloy membranes [15]. The fact that oxidation increases the flux after testing the membrane with a feed mixture containing CO and CO_2 suggests that carbon (coke) may be forming on the PdCu membrane surface [13].

Pd-Au alloy membranes have also been deposited on symmetric, 0.2 μm, alumina GTC supports. As demonstrated in Fig. 12.7, the hydrogen flux at 400°C and a partial pressure difference of 100 psi is 245 $SCFH/ft^2$, which is over twice the DOE Fossil Energy 2007 flux target and higher than the 2010 target. The H_2/N_2 pure gas selectivity of membrane GTC-31 is about 1,000 at a partial pressure difference of 100 psi. These data demonstrate that by reducing the membrane thickness and reducing carbon contamination in the Pd film, we have improved the H_2 flux by about a factor of five compared to similar membranes made using the same symmetric, ceramic support from GTC (CoorsTek) [5]. The pure hydrogen flux for the $Pd_{87}Au_{13}$ membrane is about twice that of the $Pd_{94}Cu_6$ membrane at similar feed pressure. This is roughly the same ratio estimated from the data of McKinley [3] shown in Fig. 12.3.

Gas Mixture Experiments

CO/CO$_2$ Effects

In order to investigate the impact that CO and CO_2 have on hydrogen permeation through Pd-Cu membranes, membrane GTC-6, shown in Fig. 12.8, was fed a

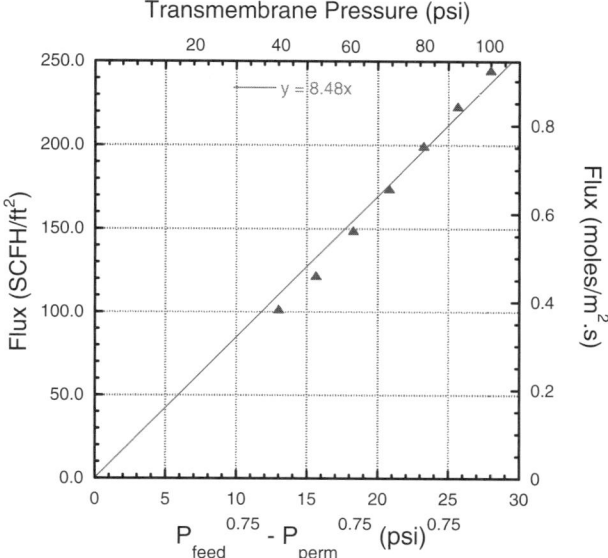

Fig. 12.7 The influence of feed pressure on pure H_2 flux for $Pd_{87}Au_{13}$ membrane GTC-31 at 400°C. Least squares fit of data is shown to aid in estimation of flux at other driving forces

Fig. 12.8 Scanning electron microscope (SEM) image of PdCu membrane #GTC-6 used for $CO/CO_2/H_2S$ tests. The apparent or visible thickness is 7 μm

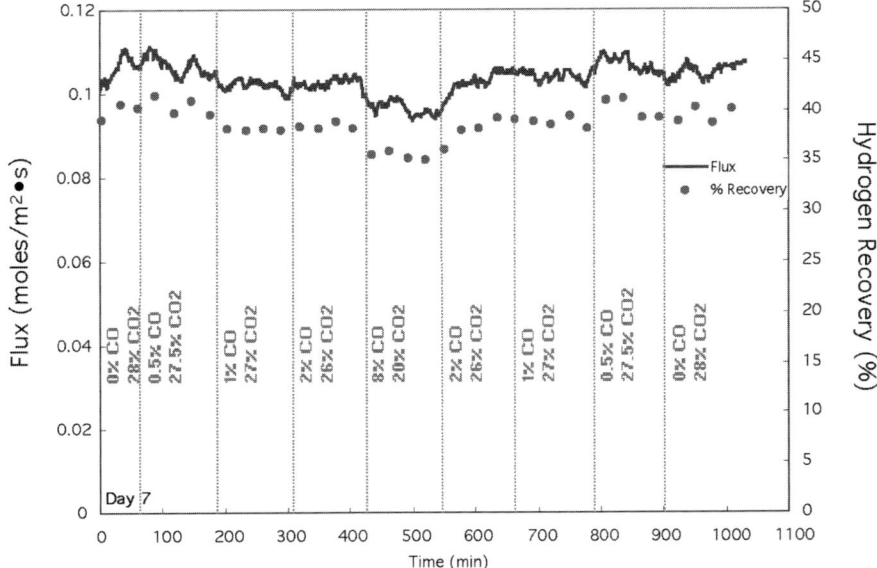

Fig. 12.9 The impact of the CO/CO$_2$ levels on the H$_2$ flux for Pd-Cu composite membrane GTC-6. Balance was 51% H$_2$ and 21% water, at 262 psia, 350°C

gaseous mixture containing 51% H$_2$, 21% water and varying concentrations of CO and CO$_2$. The sum of the CO and CO$_2$ compositions was held constant at 28mole%. Figure 12.9 shows the effect of varying the CO/CO$_2$ concentrations on the H$_2$ flux at 350°C and 250 psig feed pressure. Shown also in Figure 12.9 is the hydrogen recovery, which was maintained at about 40%. At the highest CO concentration of 8mole%, the reduction in the hydrogen flux was about 17%. The copper alloy concentration was 10% that may have increased the resistance of the membrane to inhibition or coking by CO.

Chabot and coworkers studied the effects of gas phase impurities such as CO, CO$_2$, and CH$_4$ on the hydrogen flux through a PdAg alloy foil membrane [16] over a range of gas compositions and temperatures. The degree of inhibition (hydrogen flux reduction) they reported for 9.5mole% CO in H$_2$ feed gas mixture was similar to what we observed. Inhibition was a minimum between 350 and 400°C. Above 400°C, CO methanation was also observed.

Similar degrees of inhibition due to CO exposure were reported by Hughes and coworkers [17] for a Pd composite membrane prepared by electroless plating. At 380°C and 50 psig feed pressure, a CO concentration of 12mole% caused a reduction in the H$_2$ flux of about 20%.

WGS Mixture

The effect of changing the feed gas mixture to PdAu membrane GTC-31 from H$_2$/N$_2$ to the WGS mixture (51% H$_2$, 21% water, 26% CO$_2$, and 2% CO) at

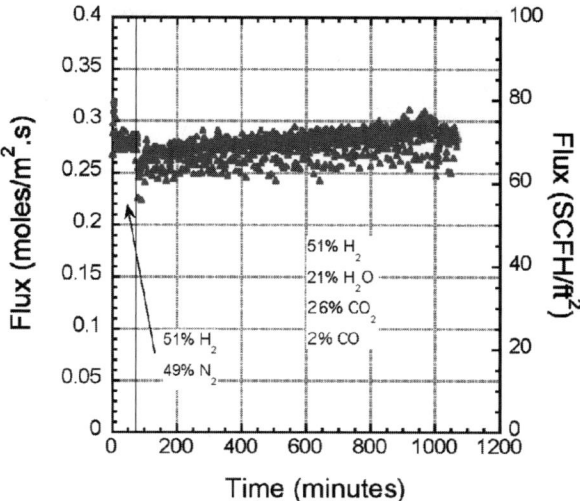

Fig. 12.10 The influence of the feed gas mixture on the flux of $Pd_{81}Au_{19}$ membrane GTC-31 at constant H_2 partial pressure at 400°C and 72 psia. The H_2 partial pressure difference is 25 psi

constant hydrogen partial pressure on the flux is shown in Fig. 12.10. The total feed pressure was 72 psia and the temperature was 400°C. The permeate flux was about 70 SCFH/ft² or 0.26 mol/m².s at a hydrogen partial pressure difference of approximately 25 psi. This was essentially the same flux measured for pure hydrogen similar temperature and H_2 partial pressure difference. For example, as shown in Fig. 12.7, the pure H_2 flux measured for GTC-31 when the feed pressure was 25 psig was 70 SCFH/ft² = 0.26 mol/m².s. The driving force corresponding to the mixture composition was 8.5 $psi^{0.75}$. The pure gas flux was calculated using the least-squares fit equation in Fig. 12.7. The interesting result is that no inhibition or reduction of the hydrogen flux was observed when CO, CO_2, and steam are present in the feed gas for the PdAu membrane.

SEM and EDAX data for PdAu membrane GTC-31 obtained after permeation testing are given in Figs. 12.11 and 12.12 below. The composition of this membrane is 87mass% Pd and 13mass% Au.

An additional PdAu alloy membrane Pall-77 was fabricated and tested with the WGS gas mixture at 400°C. As shown in Fig. 12.13, a ~9 micron thick PdAu membrane on a Pall AccuSep© stainless steel support (Pall-77) was fed a WGS mixture and produced an H_2 permeate stream containing only 5 ppm CO and a H_2 permeate purity of 99.9% using graphite seals. The flux of this membrane is modest, only 61 SCFH/ft² for a hydrogen partial pressure difference of 25 psi. Assuming the support resistances are similar, we would expect a slightly lower flux compared to GTC-31 given that Pall-77 has larger membrane thickness and this is what was observed.

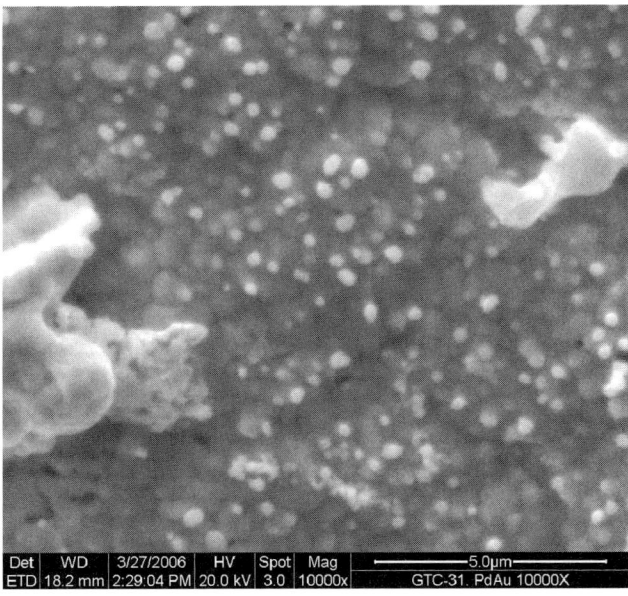

Fig. 12.11 SEM image of PdAu membrane GTC-31 after several hundred hours of pure and mixed gas permeation testing

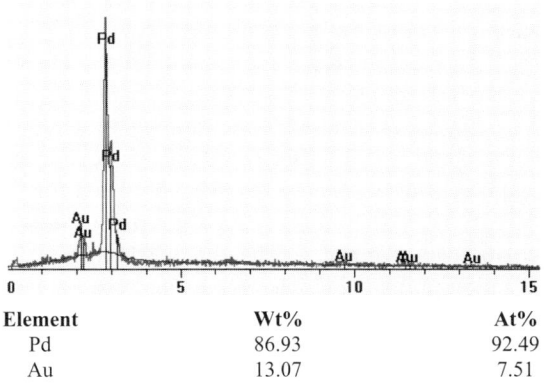

Element	Wt%	At%
Pd	86.93	92.49
Au	13.07	7.51

Fig. 12.12 EDAX spectrum and alloy composition of $Pd_{87}Au_{13}$ membrane GTC-31

WGS Mixture Including H_2S

Figure 12.14 presents permeation and exit gas concentration data corresponding to a permeation experiment for membrane GTC-6 fed the same WGS mixture at 262 psia and 350°C in which H_2S was added at different concentrations from 0 to 250 ppm to study its effect on the permeation behavior of the membrane.

H_2S causes a strong inhibition on the H_2 flux at all concentrations with the effects of this inhibition being greatest at higher H_2S levels. For example, in the presence of 10 ppm H_2S, the H_2 flux was reduced by approximately 50%. The H_2 flux is not

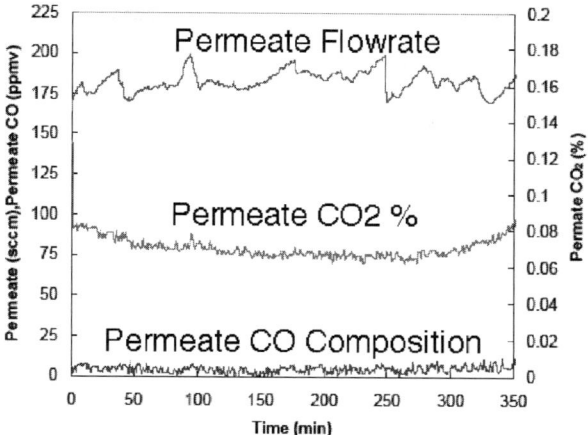

Fig. 12.13 The influence of the feed gas mixture on the permeate flow rate and composition of $Pd_{81}Au_{19}$ membrane Pall-77 for the WGS feed mixture at 400°C and 72 psia. The H_2 partial pressure difference is 25 psi. Feed composition = 51% H_2, 26% CO_2, 2% CO, and 21% H_2O. The hydrogen permeate flow corresponds to a flux of 61 SCFH/ft^2 or 0.23 moles/m^2 s. The CO composition in the permeate averaged 5 ppmv

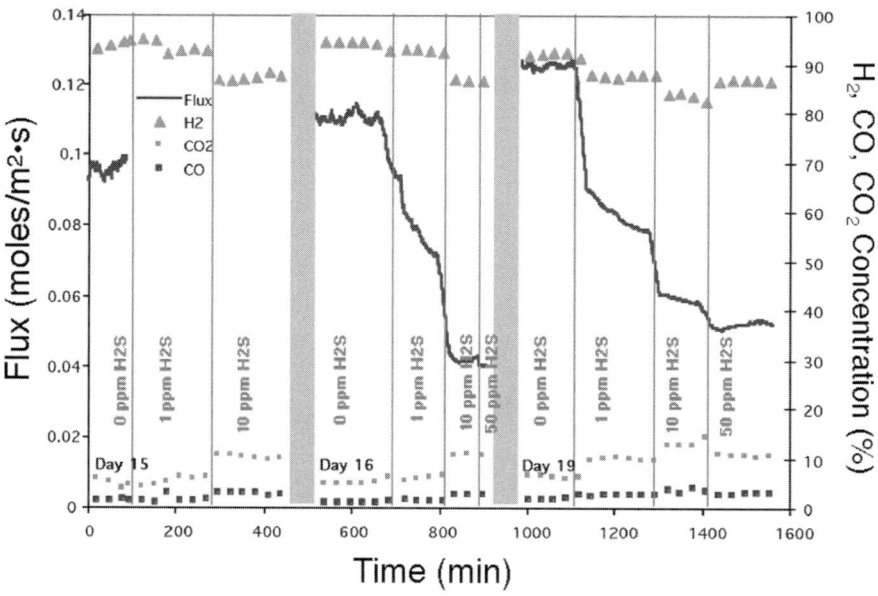

Fig. 12.14 The effect of H_2S inlet concentration on the permeate stream gas concentration and flux of membrane GTC-6. Feed consisting of 51% H_2, 21% H_2O, 26% CO_2, and 2% CO at 262 psia and 350°C

shown in Fig. 12.14 from 100 to 400 min due to equipment malfunction, but the experiment continued. The solid grey bars at about 500 min and 950 min indicate a period when the membrane was exposed to pure H_2. Note that the flux was recovered each time to the levels prior to H$_2$S exposure. At approximately 1,600 min, the membrane was exposed to 250 ppm H$_2$S and the H$_2$ flux increased continuously until the H$_2$S concentration was reduced again to 0 ppm when the H$_2$ flux was 0.18 mole/m^2 s. We interpreted this H$_2$ flux increase as membrane failure due to pore formation.

McKinley [2, 3] presented data showing that Pd alloy membranes containing Au and Cu have excellent resistance to H$_2$S in mixtures with H$_2$. More recently, Edlund [18] and Morreale and coworkers [4] have shown that exposure to 1,000 ppm of H$_2$S in H$_2$ did not significantly change the H$_2$ flux of 50–100 μm thick PdCu foil membranes at temperatures and/or compositions corresponding to the fcc phase [19]. The study by Morreale and coworkers used a feed gas mixture containing 1,000 ppm H$_2$S in pure hydrogen. This high partial pressure of hydrogen could favor the dissociation of palladium sulfide to form H$_2$S. An additional difference in the present study is that instead of a binary mixture of only H$_2$ and H$_2$S, our feed mixture contained steam that we have observed to have a much greater poisoning effect [20].

The experiment described in Fig. 12.14 was repeated for Pd$_{94}$Cu$_6$ membrane GTC-30 and the permeation data are shown in Fig. 12.15. It is interesting to examine how the permeate flux changes in the presence of the WGS feed mixture even without H$_2$S. The flux drops by about 33% when the feed gas is changed from H$_2$/N$_2$ to the WGS mixture. This is in sharp contrast to what was observed for the PdAu membrane, shown in Fig. 12.10, where there was no change in the flux for the WGS feed gas. Introduction of 5 ppm H$_2$S into the WGS mixture reduces the flux by another 37%. However the inhibition caused by H$_2$S in the feed gas is completely

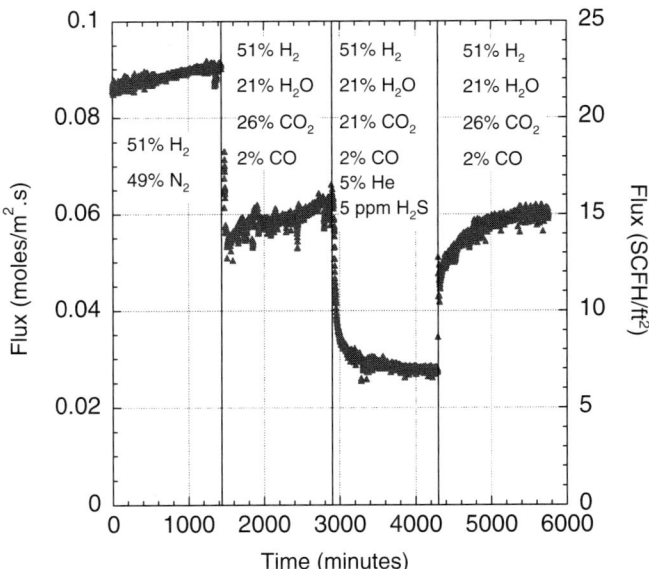

Fig. 12.15 The influence of the feed gas composition on the flux of Pd$_{94}$Cu$_6$ membrane GTC-30 at 72 psia feed pressure and 400°C

Fig. 12.16 SEM image of PdCu membrane GTC-30 after several hundred hours of pure and mixed gas permeation testing

reversible. The final feed gas composition in Fig. 12.12 is the WGS mixture alone and the flux returned to the same value of 0.06 mole/m² s = 15 SCFH/ft².

In addition, there may have been some carbon deposition on the membrane. After the series of WGS tests, membrane GTC-30 was again tested with pure hydrogen, shown in Fig. 12.6. The pure H_2 flux was 50% lower after WGS testing. An oxidation treatment partially restored the pure hydrogen flux.

A scanning electron microscope image of the surface of membrane GTC-30 after several hundred hours of pure and mixed gas permeation testing is shown in Fig. 12.16. The EDAX spectrum and tabulated composition data for this membrane is shown in Fig. 12.17. The composition of GTC-30 is 94mass% Pd and 6mass% Cu.

Sweep Gas Effects

As shown in Fig. 12.18, membrane #GTC-6 was fed a gas mixture consisting of 51% H_2, 26% CO_2, 21% H_2O, 2% CO, and 1 ppm H_2S at 350°C and 17 barg. Two conditions were run, a helium sweep and a helium and steam sweep. We observed that the composition of the permeate was affected little by the introduction of the sweep gases with the H_2 fraction remaining fairly constant over the period of the testing at ~91%. The effect of the different sweep gases on the total flux of the membrane was a bit more pronounced. The use of helium and helium/steam sweep gases improved, although marginally, the performance of the membrane with the first condition producing an H_2 flux increase of about 5% and the latter of 9%.

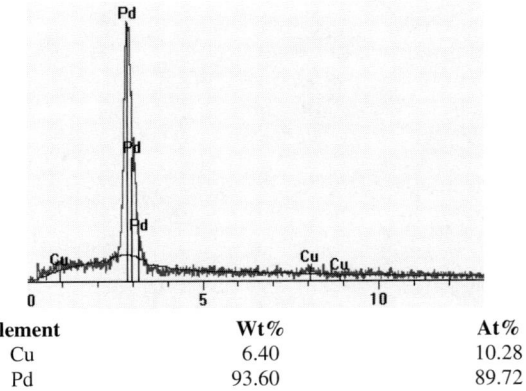

Element	Wt%	At%
Cu	6.40	10.28
Pd	93.60	89.72

Fig. 12.17 EDAX spectrum and alloy composition of PdCu membrane GTC-30

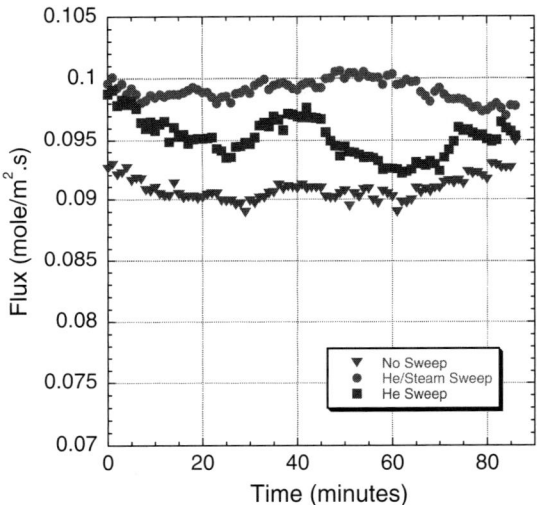

Fig. 12.18 The effect of different permeate sweep gas conditions on the flux for membrane #6 at 350°C. The feed pressure was 17 barg while the permeate pressure was 3.45 barg. The top curve is the He/steam sweep, the middle data set is for the He sweep, and the lowest flux is the case with no sweep gas

This observation is surprising given literature reports that steam adsorbs strongly on Pd membranes [17] and also causes inhibition of the H_2 flux when present in the feed gas stream at lower temperatures. The enhancement due to the steam sweep stream could be due to a cleaning effect where the steam reacts with CO or carbon present on the permeate side of the membrane to form CO_2 as we have observed that the Pd-Cu membrane is a catalyst for the WGS reaction. Carbon dioxide is less strongly bound to the surface of the membrane than CO that would free additional sites for the formation of molecular hydrogen. The He sweep could have a similar effect of encouraging desorption of CO from the permeate side of the membrane simply by reducing the CO concentration in the permeate gas stream.

Conclusions

PdAu and PdCu composite membranes supported on 0.2 micron porous alumina tubes and Pall AccuSep® stainless steel substrates were fabricated using electroless plating and tested with a variety of feed gases including pure H_2, a WGS equilibrium mixture and the WGS mixture with H_2S. Reformulating the palladium electroless plating solution by removing EDTA dramatically increased the permeability of the metal membranes and improved their stability in gas mixtures containing CO, CO_2, and steam.

For a PdAu membrane (GTC-31), no flux reduction was observed for the WGS mixture compared to a pure H_2 feed gas at the same 25 psig partial pressure difference and 400°C. A typical pure H_2 flux was 245 SCFH/ft^2 = 0.93 mol/m^2 s for a 100 psig H_2 feed gas pressure at 400°C. This flux exceeds the 2010 DOE Fossil Energy pure hydrogen flux target. The H_2/N_2 pure gas selectivity of membrane GTC-31 was about 1,000 at a partial pressure difference of 100 psi.

An additional PdAu membrane, supported on a Pall AccuSep® support achieved a 99.9% pure H_2 stream was produced from the WGS mixture.

Similar WGS experiments were performed with a PdCu composite membrane. At the same driving force, the hydrogen flux was roughly 50% of that measured for the PdAu GTC-31 membrane. Exposure to the WGS feed mixture reduced the hydrogen flux by 33%. Adding 5 ppmv to the WGS feed mixture reduced the hydrogen flux by about 70%, but the inhibition due to H_2S was reversible.

The inhibition caused by CO/CO_2 gases on a 7 μm thick Pd-Cu composite membrane (GTC-6) was modest (≤17%) over a wide range of concentrations at

Table 12.3 DOE hydrogen membrane performance goals and current status of CSM membrane technology

Performance Criteria	2007	2010	CSM Status
Flux (SCFH/ft^2)[a]	100	200	206 (PdAu)
Operating Temperature (°C)	400–700	300–600	350–450
Sulfur Tolerance	Yes	Yes	Yes
WGS Activity	Yes	Yes	Yes (PdCu)
Cost ($/ft^2)	150	100	TBD
Operating Pressure Capability (psi)	100	400	300
CO Tolerance	Yes	Yes	Yes[b]
Hydrogen Purity (%)	95	99.5	99.93[c]
Stability/Durability (years)	3	7	3 months pure H_2, 2 weeks PdAu

a) Calculated @ 150 psia feed H_2 and 50 psia perm side from data in Fig. 12.7.
b) Tested with 2mol% CO, see Figs. 12.9–12.13
c) From mixture test with WGS feed mixture, see Fig. 12.11. CO permeate concentration less than 5 ppmv.

350°C. H_2S caused a strong inhibition of the H_2 flux of the same Pd-Cu composite membrane, which is accentuated at levels of 100 ppm or higher. The membrane was exposed to 50 ppm H_2S 3 times without permanent damage. At H_2S levels above 100 ppm, the membrane suffered some physical degradation and its performances were severely affected. The use of sweep gases improved the hydrogen flux and recovery of a Pd-Cu composite membrane.

Table 12.3 summarizes the performance parameters for our Pd alloy composite membranes and compares them to the DOE Fossil Energy targets from Table 12.1.

Acknowledgments The authors gratefully acknowledge support of this work from the U.S. Department of Energy through Grants DE-FG2699-FT40585 and DE-FG26-03NT41792 from the NETL University Coal Research Program and Grant DE-FG03-93ER14363 from the DOE Office of Science, Division of Basic Energy Sciences.

References

1. Gray D, Tomlinson G. Hydrogen from Coal. MTR 2002–31, Mitretek, 2002.
2. McKinley DL. Metal alloy for hydrogen separation and Purification, US Patent 3,350,845, 1967.
3. McKinley, D.L. Method for hydrogen separation and purification. US Patent 3,439,474, 1969.
4. Morreale BD, Ciocco MV, Howard BH, Killmeyer RP, Cugini AV, Enick RM. Effect of hydrogen-sulfide on the hydrogen permeance of palladium-copper alloys at elevated temperatures. J Membr Sci. 2004;241:219–24.
5. Paglieri SN, Foo KY, Way JD, Collins JP, Harper-Nixon DL. A new preparation technique for Pd/alumina membranes with enhanced high temperature stability. Ind Eng Chem Res. 1999;38(5):1925–36.
6. Collins JP, Way JD. Preparation and characterization of a composite palladium-ceramic membrane. Ind Eng Chem Res. 1993;32(12):3006–13.
7. Roa F, Block MJ, Way JD. The influence of alloy composition on the H_2 flux of composite Pd–Cu membranes. Desalination. 2002;147:411–6.
8. Roa F, Way JD. The influence of alloy composition and membrane fabrication on the pressure dependence of the hydrogen flux of palladium-copper membranes. Ind Eng Chem Res. 2003;42(23):5827–35.
9. http://www.pall.com/pdf/Accusep_Membrane.pdf.
10. Ames RL, Bluhm EA, Way JD, Bunge AL, Abney KD, Schreiber SB. Physical characterization of 0.5 μm cut-off sintered stainless steel membranes. J Membr Sci. 2003;213:13–23.
11. Roa F, Way JD, McCormick RL, Paglieri SN. Preparation and characterization of Pd-Cu composite membranes for H_2 separation. Chem Eng J. 2003;93(1):11–22.
12. Foo KY. Hydrogen separation in palladium ceramic membranes and palladium-gold ceramic membranes. M.S. Thesis in Chemical Engineering & Petroleum Refining, Colorado School of Mines: Golden, Colorado, 1995.
13. Kulprathipanja A, Alptekin GO, Falconer JL, Way JD. Effects of water gas shift gases on Pd-Cu alloy membrane surface morphology and separation properties. Ind Eng Chem Res. 2004;43(15):4188–98.
14. Thoen PM, Roa F, Way JD. High flux palladium–copper composite membranes for hydrogen separations. Desalination. 2006;193:224–229.
15. Roa F, Way JD. The effect of air exposure on palladium–copper composite membranes. Appl Surf Sci. 2005;240:85–104.

16. Chabot J, Lecomte J, Grumet C, Sannier J. Fuel clean-up system: poisoning of palladium–silver membranes by gaseous impurities. Fusion Tech. 1988;14:614–8.
17. Li A, Hughes R. The effect of carbon monoxide and steam on the hydrogen permeability of a Pd/stainless steel membrane. J Membr Sci. 2000;165:135–41.
18. Edlund D. A membrane reactor for H_2S decomposition. Advanced Coal-Fired Power Systems '96 Review Meeting, 1996, Morgantown, WV.
19. Piper J. Diffusion of hydrogen in copper–palladium alloys. J Appl Phys. 1966;37(2):715–21.
20. Kulprathipanja A, Alptekin GO, Falconer JL, Way JD. Pd and Pd–Cu membranes: inhibition of H_2 permeation by H_2S. J Membr Sci. 2005;254:49–62.

Chapter 13
Composite Pd and Pd/Alloy Membranes

Yi Hua Ma

This chapter deals with the synthesis and characterization of composite Pd and Pd/alloy membranes for hydrogen separation and reaction applications. The electroless plating of Pd on porous stainless steel with an intermetallic diffusion barrier is emphasized. Membranes with the intermetallic diffusion barrier prepared by the in-situ oxidation coupled with bi-metal multilayer (BMML) deposition have been shown to be stable under reaction conditions for over 6,000 h at 500°C. Techniques for the membrane characterization include macroscopic flux measurements by shell-and-tube arrangement and microscopic surface and microstructure analysis by XRD, SEM, TEM, XPS and AFM. Factors affecting hydrogen flux and long-term stability of the membrane include membrane thickness, types of supports and mass transfer resistances. Interdisciplinary approach and industry university collaboration are emphasized. This article concludes with an optimistic view on the prospect of a near future commercialization of composite Pd and Pd/alloy porous metallic support membranes for large scale industrial hydrogen production processes.

Introduction

Palladium is one of the six transition metals that are also known as the "platinum metals" and has high hydrogen solubility to make it attractive as a hydrogenation catalyst and hydrogen separation membrane. The early works of the palladium as a hydrogen separation membrane is mostly concentrated on palladium foils. An excellent review on hydrogen permeation in palladium and palladium hydrogen systems was presented in detail by Lewis [1].

In the past decade, because of increased demands in the hydrogen production for petroleum and petrochemical industry and potentially large increases in hydrogen needed for the hydrogen economy, there have been considerable advances in developing processes for hydrogen separations and productions. Of particular importance

Y. Hua Ma (✉)
Center for Inorganic Membrane Studies, Department of Chemical Engineering, Worcester Polytechnic Institute, Worcester, MA 01609, USA
yhma@WPI.EDU

A.C. Bose (ed.), *Inorganic Membranes for Energy and Environmental Applications*,
DOI 10.1007/978-0-387-34526-0_13, © Springer Science+Business Media, LLC 2009

is the substantial increase in research and development of the hydrogen separation and production by membranes and membrane reactors at high temperatures and pressures.

This paper provides discussions on the hydrogen permeation through Pd membranes, methods of the formation of thin Pd and Pd/alloy films on porous supports with special emphasis on electroless plating, porous metallic supports, formation of intermetallic diffusion barrier and long term membrane stability at high temperatures.

Pd-Hydrogen System: Thermodynamics and Hydrogen Permeation

Figure 13.1 shows that the pressure–concentration relation (p–C isotherm) for the hydrogen–palladium system exhibits an immiscibility gap region between the α and β phase with both phases having an FCC crystalline structure at temperatures below $\sim 300°C$ [2, 3]. Since the lattice parameter of the β phase is larger than that of the α phase, the nucleation of the β phase from the α phase leads to hydrogen embrittlement. Although loading hydrogen in palladium at temperatures above 300°C also causes changes in the lattice parameter, these changes are gradual and do not involve phase changes. Therefore, unless very low pressures are used [4, 5], it is necessary to operate a pure palladium membrane at temperatures above 300°C to avoid the hydrogen embrittlement. In order to minimize the hydrogen embrittlement problem, alloying Pd with other metals such as Ag or Cu will lower the critical temperature for the α to β phase transformation.

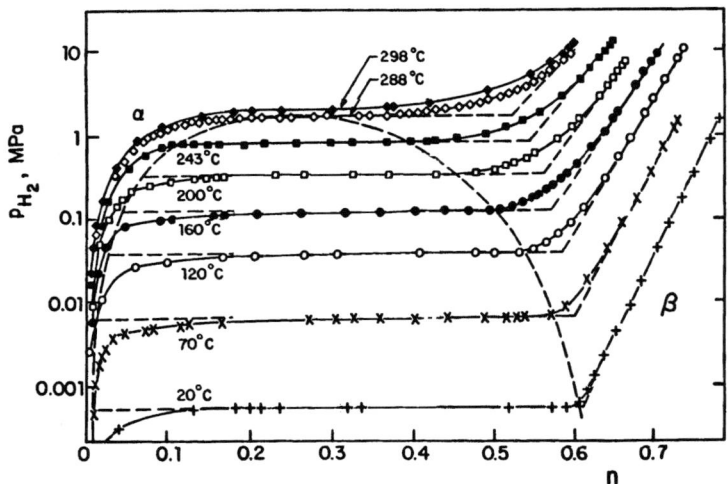

Fig. 13.1 Isotherms for Pd-H system where $n =$ H/Pd

Although there may be a number of steps for hydrogen permeation through a composite Pd membrane [6], the major steps are: (i) adsorption and dissociation of hydrogen molecules to hydrogen atoms at the membrane surface of the high pressure side, (ii) diffusion of hydrogen atom through the bulk of the Pd layer, and (iii) re-association of hydrogen atoms and desorption of hydrogen molecules at the membrane surface of the low pressure side. Under most normal operating conditions, the rate of the hydrogen permeation is generally controlled by the hydrogen diffusion through the bulk of the Pd layer.

Assuming that the rate of hydrogen permeation is controlled by the hydrogen diffusion through the Pd layer and no phase transformation between α and β occurs, the hydrogen flux J can be expressed by the Fick's first law,

$$J = -D\frac{\Delta C}{\Delta X} = D\frac{(C_{HP} - C_{LP})}{\delta} \tag{13.1}$$

For low H/Pd concentrations, $C = kp^{1/2}$ is a good approximation outside the immiscibility region, Eq. 13.1 then reduces to

$$J = kD\frac{(p_{HP}^{1/2} - p_{LP}^{1/2})}{\delta} \tag{13.2}$$

where J is the hydrogen flux, m^3/m^2 s, or kmole/m^2 s, k the Henry's law constant defined above, kmole/m^3 Pa$^{1/2}$, D the diffusivity, m^2/s and δ the thickness of the membrane m. Subscripts HP and LP designate the high and low pressure side, respectively. Both k and D are functions of temperature and kD is the permeability. Since it is often difficult to accurately determine the membrane thickness δ, $P = kD/\delta$ (kmole/m^2 s Pa$^{1/2}$) defined as the permeance is frequently used. Eq. 13.2 is commonly known as the Sieverts' law. However there have been cases reported in the literature where the exponent of the pressure is not 1/2 due to factors such as the non-linearity of the $p^{1/2}$-C isotherm, surface reaction and existence of significant mass transfer resistances in addition to that of diffusion through the Pd bulk [6]. Therefore, Eq. 13.2 can be expressed in a more general form as

$$J = P(p_{HP}^{n} - p_{LP}^{n}) \tag{13.3}$$

where $1 \geq n \geq 1/2$. For the Sieverts' law, n = 1/2.

Deposition of Pd on Porous Substrates

Some of the methods using more sophisticated equipment for the deposition of Pd or Pd/alloy on porous substrates include, among others [7], the metal-organic chemical vapor deposition (MOCVD) [8, 9], and electron beam evaporation and ion-beam sputtering [10]. On the other hand, electroless plating uses relatively simple equipment. The techniques using sophisticated equipment are quite powerful and especially useful for the deposition of alloy films. However, their major drawbacks are low area of the prepared membranes, difficult for large scale production and/or high cost of the necessary equipment. Therefore the practical applicability of these techniques may require further investigation. The electroless plating, on the other hand, is attractive and widely used because of the possibility of uniform deposition on complex shapes and large substrate areas, hardness of the deposited film, and very simple equipment [11].

Electroless Deposition

An electroless palladium plating bath generally includes: a palladium ion source (e.g. $PdCl_2$, $Pd[NH_3]_4Cl_2$, $Pd[NH_3]_4[NO_3]_2$, and $Pd[NH_3]_4Br_2$), a complexant (e.g. ethylenediamine tetra acetic acid [EDTA] and ethylenediamine [EDA]), a reducing agent (e.g. hydrazine [$NH_2 - NH_2$], sodium hypophosphite [$NaH_2PO_2 \cdot H_2O$], and trimethylamine borane), stabilizers and accelerators. The electroless plating of Pd on a porous stainless support normally consists of pre-treatment of the porous stainless steel (PSS) support, surface activation and plating [7]. Pre-treatment of the PSS support consists of: (i) complete removal of foreign contaminants (grease, oil, dirt, and corrosion products), (ii) activation of the support surface with palladium nuclei, which during the electroless plating initiates an autocatalytic process of the reduction of a metastable Pd salt complex on the target surface, and (iii) electroless plating with a plating solution consisting of $(Pd(NH_3)_4)Cl_2H_2O$, Na_2EDTA, NH_4OH, and H_2NNH_2 at pH around 10.4 and temperature of 60°C. The Pd deposition occurs through the following autocatalytic reaction:

$$2Pd(NH_3)_4Cl_2 + H_2NNH_2 + 4NH_4OH \rightarrow 2Pd^0 + N_2 + 8NH_3 + 4NH_4Cl + H_2O$$

or

$$2Pd^{2+} + H_2NNH_2 + 4OH^- \rightarrow 2Pd^0 + N_2 + 4H_2O$$

Pre-seeded palladium nuclei at the activation stage reduce the induction period of the autocatalytic process at the beginning of the deposition. The composition of a typical Pd plating solution is shown in Table 13.1 [7].

The thickness of the plated Pd layer can be determined gravimetrically from the plated amount and the density of Pd ($12.0\,g/cm^3$). It can also be estimated directly from the SEM picture of the cross-section of the membrane. However, the latter

Table 13.1 Composition of the electroless palladium-plating bath

$Pd(NH_3)_4Cl_2 \cdot H_2O$, g/l	4.0
NH_4OH (28%), ml/l	198
Na_2EDTA, g/l	40.1
H_2NNH_2 (1 M), ml/l	5.6–7.6
p^H	~10.4
Temperature, °C	60
$V_{solution}/S_{plating\,area}$, cm^3/cm^2	~3.5

method is destructive and generally not preferred. For the formation of the Pd/alloy membranes, sequential plating followed by thermal treatment ("coating and diffusion treatment") has been used to obtain homogeneous alloy films [12]. Although it is more difficult to obtain appropriate solution compositions, co-plating can also be used for making composite Pd/alloy membranes. In addition to lowering the critical temperature, some of the added advantages of Pd/alloy membranes include higher hydrogen permeance (e.g. Pd/Ag [13, 14], Pd/Au [15], and sulfur tolerant Pd/Cu [16]). The formation of homogeneous alloys by high temperature heat treatment, however, remains a serious challenge, which is especially true for using porous metal substrates due to the possibility of the intermetallic diffusion between the components in the substrate and the membrane layer.

Porous Substrates

Both porous ceramic and dense metallic membranes are suitable for hydrogen separation and reaction applications. Porous ceramic membranes have the main advantages of high hydrogen flux and good chemical stability but suffer from low separation factors because of the small size of the hydrogen molecule. On the other hand, dense metallic membranes, in particular palladium and palladium/alloy, have high separation factors (in theory, infinite) but, in general, have low hydrogen fluxes. One way to minimize this problem is to support Pd or Pd/alloy thin films on either porous ceramic or metallic supports to provide the necessary mechanical strength and at the same time to increase the hydrogen flux, thereby reducing the palladium cost.

The main advantages of porous ceramic supports include small and uniform pore size distribution. Both of these make it easier to plate thinner membranes on porous ceramic supports. The major drawbacks for porous ceramic supports are fragile, susceptible to cracking and difficulty in sealing ceramic to metal parts for process integration. On the other hand, the advantages of porous metal supports include easy module fabrication and process integration. However, the main disadvantages of porous metal supports include large pore size, wide pore size distribution and intermetallic diffusion between the chemical elements in the support and the deposited film causing the deterioration of the hydrogen permeability.

As indicated in the above discussion, Pd and Pd/alloy membranes deposited on a porous metal support can easily be integrated into the process and are preferred for industrial applications especially for large scale process applications. However, in order to have a long term stability on permeabilities and selectivity, it is necessary to minimize the intermetallic diffusion between the chemical elements in the support and the deposited film by having a barrier layer between the deposited film and the support. A variety of methods have been used to produce the barrier layer including a thin aluminum oxide layer sandwiched between the vanadium foil and the palladium foil (Pd/γ-alumina/V) [17], sputtering titanium in a nitrogen atmosphere to form a TiN layer [18], applying nickel powder followed by sintering combined with the application of a thin γ-alumina layer by the sol gel method [19] and a controlled in situ oxidation of the porous metal at temperatures greater than 400°C to produce an in situ oxide barrier layer [20]. The thickness of the formed oxide layer from the in situ oxidation increased but with negligible effects on the pore size as the oxidation temperature was increased [21]. The oxide layer was predominantly formed on both the exterior surface and in the pores and pore mouths near the surface, resulting in no significant increases in the mass transfer resistance in the support. The membrane prepared with the oxide barrier layer has been shown to be stable in the temperature range of 350 and 450°C for over 6,000 h [22]. The temperature range for the membrane stability can be increased by the bi-metal multi-layer (BMML) deposition technique, which involved the formation of a porous Pd-Ag composite layer by consecutive deposition of Pd and Ag layers with no intermediate surface activation [23, 24]. The BMML formed an extremely effective intermetallic diffusion barrier and several membranes prepared by this technique have been stable under hydrogen permeance conditions for over 500 h at temperatures exceeding 500°C [23]. In addition, one of the membranes was stable for 6,000 h under steam reforming conditions [25]. Further improvements of the in situ oxide barrier layer to increase the thermal stability of the membranes are covered in several additional US patent and patent applications [26–28].

Membrane Characterization

Although the tubular geometry is normally preferred in the chemical industry, both tubular and flat plate geometries have been used for making composite Pd and Pd/alloy membranes. Methods used for the membrane characterization include, among others, macroscopic permeation flux measurements and microscopic surface and microstructure analysis by various techniques such as X-ray Diffraction (XRD), Scanning Electron Microscopy (SEM), Transmission Electron Microscopy (TEM), X-ray Photoelectron Spectroscopy (XPS) and Atomic Force Microscopy (AFM).

Permeation Flux Measurements

For tubular membranes, the permeation flux measurements are generally carried out by a shell-and-tube arrangement as shown in Fig. 13.2 [7]. Since the membrane is plated on the outside of the porous substrate, the feed gas flows upward through the outside of the membrane (shell side), and the permeate gas is collected on the tube side. A purge gas can also be used on the tube side if desired. The upstream pressure is monitored by a capacitance pressure transducer and the permeate side pressure is kept atmospheric. The gas permeation rate (the volumetric flow rate) is measured in the permeate side at atmospheric pressure and room temperature. For mixture experiments, the permeate and retentate can be analyzed by an online gas chromatograph equipped with appropriate detectors, such as Thermal Conductivity Detector (TCD), Flame Ionization Detector (FID) or others, depending on the specific gas mixture

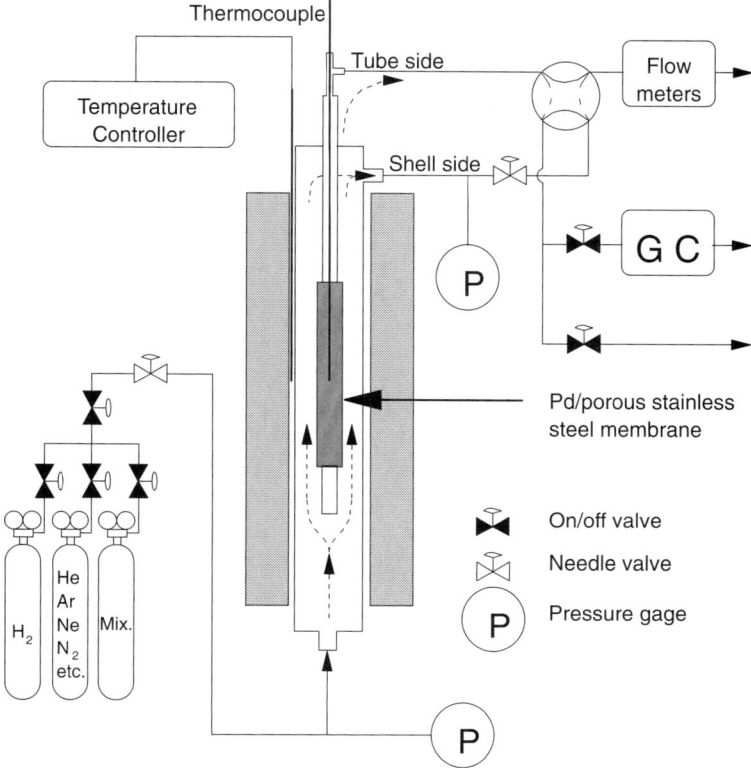

Fig. 13.2 Schematic diagram of the experimental set-up for permeation measurements

used in the experiments. A typical Sieverts' plot for a composite pure Pd/stainless steel membrane at 4 temperatures is shown in Fig. 13.3 [29]. The slope of the line gives the permeance of the membrane at each temperature, from which the activation energy for hydrogen permeation can be determined from the slope of the Ahrrenius plot shown in Fig. 13.4. The relatively low activation energy (E) is due, in part, to the mass transfer resistances in the porous stainless steel support. It should be cautioned that although the n value may vary from membrane to membrane and the Ahrrenius plot based on different n values will give different fitted values for the activation energy, only the activation energy value obtained from Sieverts' plot has theoretical meaning associated with the hydrogen diffusion through the bulk of the palladium layer [30].

Microscopic Surface and Microstructure Analysis

XRD, SEM, TEM, and AFM are extremely useful techniques for understanding the surface morphology, phase identification and microstructure of the thin Pd and Pd/alloy layer. The formation of an alloy phase can be obtained from X-ray spectra

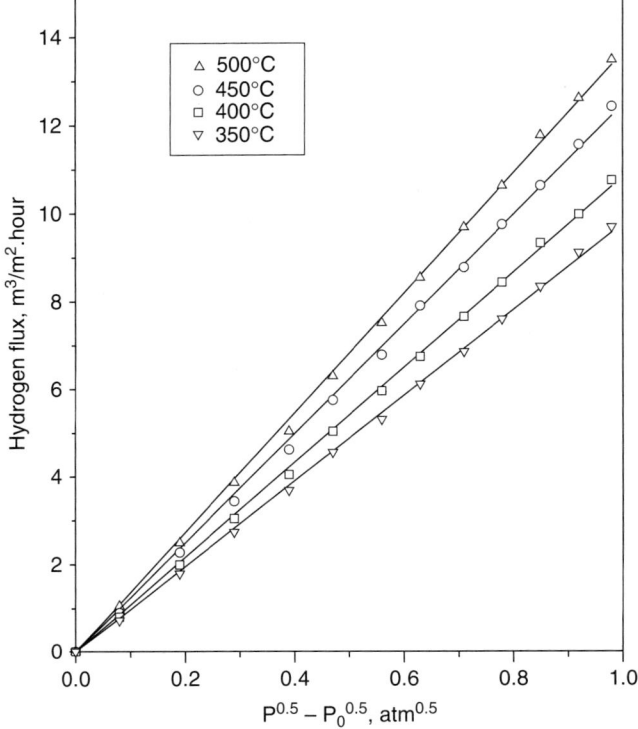

Fig. 13.3 Hydrogen flux as a function of the difference in the square roots of the shell (P) and tube (P_o) side pressures (Sieverts' law) for Pd membrane on the PSS support with media grade 0.2 μm

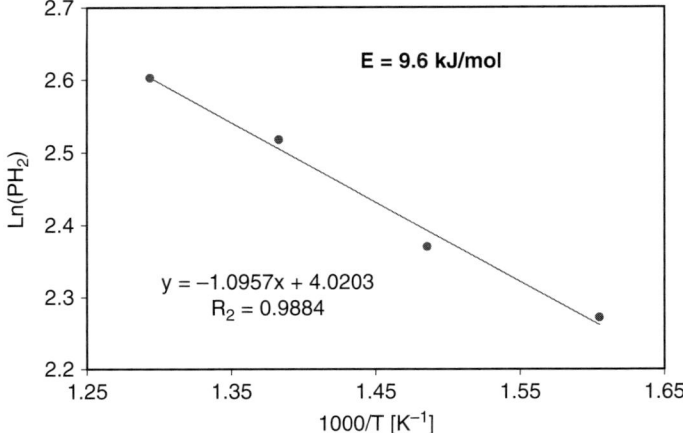

Fig. 13.4 Arrhenius relation between the hydrogen permeance and temperature

although it is difficult to determine the homogeneity of the formed alloy phase through the entire membrane layer. The homogeneity of the alloy film, however, may be determined by using SEM equipped with EDS to analyze the cross-sectional composition of each element. Figure 13.5 shows Pd/Cu alloy film on a porous stainless steel support after the heat treatment at 600°C for 10 h under helium atmosphere [21]. Figure 13.6 shows the composition profiles of different elements along the direction of the arrow in Fig. 10.5 by EDS line scan [21] demonstrating the relative homogeneity of the formed Pd/Cu layer.

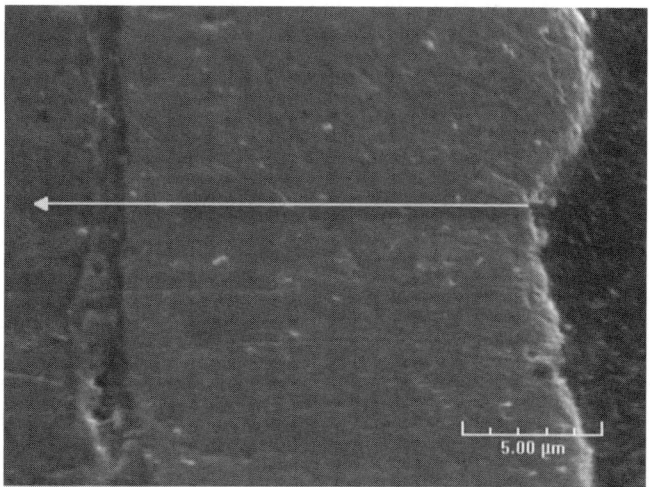

Fig. 13.5 Pd/Cu film on PSS support that was oxidized at 800°C for 10 h. The annealing was carried out at 600°C for 10 h under helium (3KX)

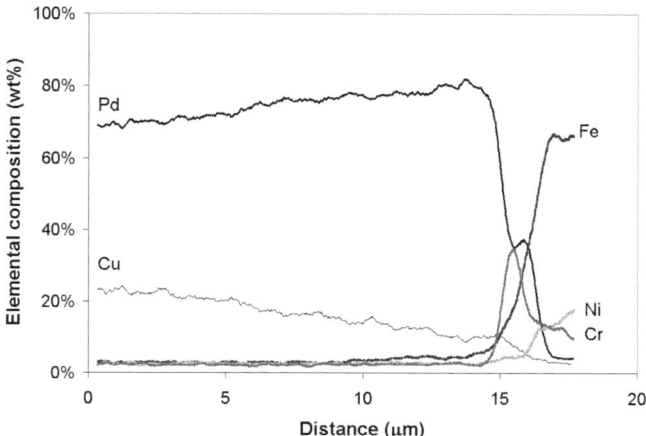

Fig. 13.6 Compositions profiles of different elements along the arrow in Fig. 13.5 by EDS line scan

AFM can be used to study the surface morphology of a membrane. Varma's group [31] appeared to be the first one to use AFM to measure the surface roughness to characterize the differences between conventional electroless plating and electroless plating assisted by osmotic pressure of a thin palladium layer supported on Vycor glass surface. AFM was also used to characterize the oxidized Pd surface [32, 33].

The microstructure of Pd and Pd/alloy membranes can be examined by SEM and TEM. Since the size of Pd grains is in the range of nanometers, TEM will be more appropriate for the examination of the change of grain sizes as a function of thermal treatment and long term uses at high temperatures. A powerful technique to study the alloy formation as a function of temperature and time is the High Temperature X-ray Diffractometer (HTXRD) [34]. Other surface techniques, such as XPS and Electron Probe Microanalyzer (EPMA), are also useful for the membrane composition analysis.

Factors Affecting Membrane Flux and Long-Term Stability

There are a number of factors affecting the quality of the membrane and its long-term stability. Only some of the more important ones are discussed here to illustrate the complexity of the problem. Depending on the specific applications, factors important to one application may not be critical to the others. Few studies dealt with the membrane's long-term stability. Most long-term stability studies reported in the literature are in tens, with few in hundreds, of hours. However, in order for the membrane to be viable for industrial applications, long-term stability tests should be carried out in thousands of hours. The 2007 stability/durability target set by the US DOE is 3 years while the ultimate goal for 2015 is greater than 10 years [35].

The long-term thermal stability between 350 and 450°C for a composite Pd porous stainless steel membrane with an oxide intermetallic diffusion barrier layer is shown in Fig. 13.7. As shown in the figure, the membrane showed no sign of deterioration after over 6,000 h of continuous testing.

Although thermal cycling stability may not be an important factor for large scale industrial applications, it may be crucial for other applications such as fuel cells for mobile applications. There are practically no thermal cycling studies reported in the literature. When performing thermal cycling studies, the cycle temperatures, the rate of temperature changes and the number of cycles are important factors to consider. Although the first two factors may depend on specific applications, the number of cycles should, in general, be in thousands of cycles in order to provide meaningful data for practical applications.

Membrane Thickness

One of the most important factors affecting the hydrogen flux is the thickness of the membrane. Thin layers can generally be deposited on substrates with a relatively smooth surface and small pores with a uniform or narrow pore size distribution [36].

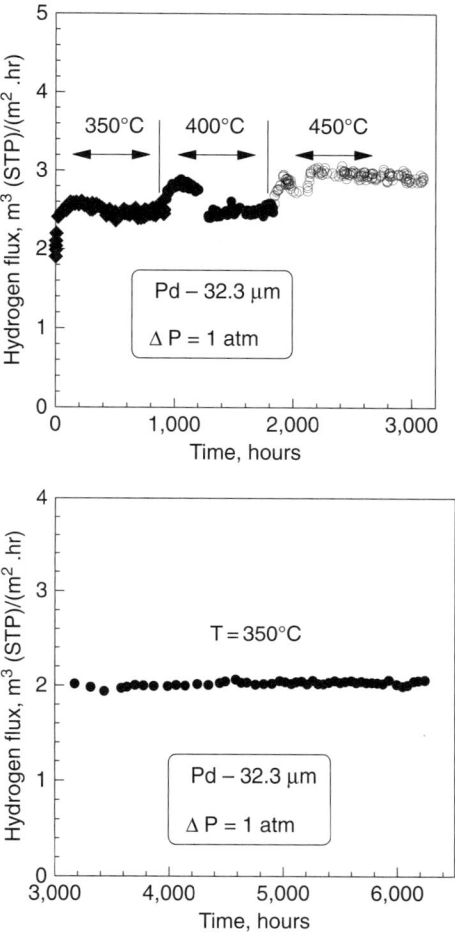

Fig. 13.7 Long-term thermal stability of composite palladium-porous stainless steel membrane

Since the hydrogen flux is inversely proportional to the membrane thickness, thin membranes give high hydrogen fluxes, assuming the permeation process is controlled by the hydrogen diffusion through the bulk of the Pd or Pd/alloy layer. In addition, a two-fold reduction of the membrane thickness simultaneously increases the hydrogen flux by a factor of two. The net economic benefit is the reduction of the membrane material cost by a factor of four.

The selectivity of a membrane may, among other factors, be also affected by the membrane thickness. Thin membranes may give high hydrogen fluxes with a lower selectivity. For applications in the electronic industry and fuel cells, high purity hydrogen is required but is not necessarily essential to applications in refineries. Therefore, performing an optimization study will be needed in order to maximize the benefits to obtain the required membrane thickness.

Table 13.2 Coefficients of thermal expansion for some metal and ceramic material

Material	Coefficient of Thermal Expansion 10^{-5} m/m K
Palladium	1.2
Austenitic stainless steel	1.6–1.8
Copper	1.4–1.8
Silver	2.0
Nickel and its alloys	1.2–1.7
Alumina ceramics	0.6–0.7
Zirconia	0.6
Gold	1.4

Support Type

Because of the cost and difficulty of obtaining metal powders with small and uniform particles, porous metallic substrates have, in general, larger pores with a wider pore size distribution than those of their ceramic counterpart. In addition, intermetallic diffusion, which is not a concern for ceramic substrates, between the chemical elements in the substrate and the deposited layer will cause the permeance to deteriorate. On the other hand, metallic substrates offer invaluable advantages of ease of module fabrication and process integration that is especially important for large scale industrial applications but not afforded by ceramic supports.

The difference in coefficients of thermal expansion between the substrate and the deposited film is another important factor to consider. Table 13.2 lists the coefficients of thermal expansion for some materials relevant to this discussion. It is clear that these materials cover a wide range of values of the coefficients of thermal expansion. Since process applications of the composite Pd and Pd/alloy membranes involve high temperatures and temperature cycling in certain cases, the matching of the values between the support and the membrane layer is one of the critical factors for obtaining a leak-stable membrane.

Mass Transfer Resistances

There has been some discussion on the permeation process being controlled by surface reaction for thin membranes in the literature. In this case, the reduction of the membrane thickness will not increase the hydrogen permeation fluxes. The same phenomena will occur when the permeation process is controlled by one or more of the mass transfer resistances, including the formation of a boundary layer between the bulk gas phase and the membrane surface and flow through the support pores by Knudsen diffusion and viscous flow. There are only a few studies in the literature addressing the existence of these resistances. The reduction of these resistances is critical for the membrane to achieve its maximum permeation flux.

Concluding Remark

Since Graham [37] first observed that palladium possessed the unusually high hydrogen permeation flux and good hydrogen solubility, the Pd-hydrogen system has been studied by various investigators for over 100 years. The use of palladium membranes for hydrogen separations has been investigated for close to 50 years with the granting of what appeared to be the first Pd membrane patent to De Rosset [38] of Universal Oil Products Company in 1958 and subsequent granting of patents to McKinley [39], McBride et al. [40] and McKinley [41, 42], all of Union Carbide Corporation. However, no significant large-scale industrial applications have resulted from these inventions. Considerable progress has been made in recent years through the renewed interests in using Pd membranes for high temperature separation and reactor applications in energy related fields. It is likely that large-scale industrial applications of Pd and Pd/alloy may result from these intense research activities in the foreseeable future.

The perception that palladium is too expensive to be economically feasible for large-scale applications is misleading. Since it is possible to reduce the thickness of a composite membrane to less than 10 μm, the quantity of palladium used is so small that its cost becomes an insignificant fraction of the total cost. On the other hand, the cost of the support may play a considerably more important role in making composite membranes economically viable for large scale industrial applications. When large quantities of supports are needed their cost will most likely come down, thereby making the process more competitive to the conventional steam reforming processes. It should be emphasized that to get the maximum economical and operational benefits, Pd and Pd/alloy membrane reactors should be used to combine reaction and separation in a single unit operation.

Although the recent surge in Pd and Pd/alloy research has produced some rather encouraging results, many technical challenges, such as making even thinner membranes with long term stability of permeance and selectivity, still need to be addressed. It is imperative that interdisciplinary approaches be employed to tackle these technical challenges. The collaboration between industry and academic should be encouraged to bring technologies to market in a speedy and timely manner. Such interdisciplinary approaches and industry/university collaboration can make the commercialization of Pd and Pd/alloy membranes for large-scale chemical engineering process applications a reality in the near future.

References

1. Lewis FA. The palladium hydrogen system. London: Academic Press; 1967.
2. Wicke E, Nernst GH. Ber Bunsenges Physik Chem. 1964;68:224–35.
3. Frieske H, Wicke E. Ber Bunsenges Physik Chem. 1973;77:48–52.
4. Jewett DN, Makrides AC. Trans Faraday Soc. 1965;61:932–9.
5. Amandusson H, Ekedahl LG, Dannetun H. J Memb Sci. 2001;193:35–47.
6. Ward TL, Dao TJ. Memb Sci. 1999;153:211–31.
7. Mardilovich PP, She Y, Ma YH, Rei MH. AIChE J. 1998;44:310–22.

8. Yan S, Maeda H, Kusakabe K, Morooka S. Ind Eng Chem Res. 1994;33:616–22.

9. Morooka S, Yan SY, Kudskabe K. Sep Sci Tech. 1995;30:2877–889.

10. Peachey NM, Snow RC, Dye RC. J Memb Sci. 1996;111:123–33.

11. Ma YH, Mardilovich IP, Engwall EE. Annals of the New York Academy of Sciences. In: Li NN, Drioli E, Ho WS, Lipscomb GG, editors. Advanced membrane technology. Vol. 984. New York Academy of Sciences, New York, NY; 2003. pp. 346–360.

12. Uemiya S, Matsuda T, Kikuchi EJ. Memb Sci. 1991;56:315–25.

13. Shu J, Grandjean BPA, Ghali E, Kaliaguine SJ. Memb Sci. 1993;77:181–95.

14. Cheng YS, Yeung KL. J Memb Sci. 1999;158:127–41.

15. Gryaznov V. Sep Purif Methods. 2000;29(2):171–87.

16. Roa F, Way JD. Appl Surf Sci. 2005;240:85–104.

17. Edlund DJ, McCarthy JJ. Memb Sci. 1995;107:147–53.

18. Shu J, Adnot A, Grandjean BPA, Kaliaguine S. Thin Solid Films. 1996;286:72–9.

19. Nam SE, Lee KH. J Memb Sci. 2001;192:177–85.

20. Ma Y, Mardilovich PP, She Y. US Patent 6,152,987, 2000.

21. Ma YH, Akis BC, Ayturk ME, Guazzone F, Engwall EE, Mardilovich IP. I & EC, Res. 2004;43:2936–45.

22. Ma YH, Mardilovich PP, She Y. Proceedings, ICIM$_6$, 1998, pp. 246–9.

23. Ayturk ME, Mardilovich IP, Engwall EE, Ma YH. Presented at the AIChE 2004 Annual Meeting, Austin, TX, 7–12 Nov 2004.

24. Ma YH, Mardilovich IP, Engwall EE. US Patent 7,175,694, 2007.

25. Matzakos A. Presentation at the 2006 NHA Annual Hydrogen Conference, Long Beach, CA, 15 March 2006.

26. Ma YH, Mardilovich IP, Engwall EE. US Patent 7,172,644, 2007.

27. Ma YH, Mardilovich IP, Engwall EE. US Patent 7,225,726, 2007.

28. Ma YH, Mardilovich IP, Engwall EE. US Patent 7,390,536, 2008.

29. Mardilovich IP, Engwall EE, Ma YH. Desalination. 2002;144:85–9.

30. Guazzone F, Engwall EE, Ma YH. Catal Today. 2006;118(1–2):24–31.

31. Souleimanova RS, Mukasyan AS, Varma A. Chem Eng Sci. 1999;54:3369–77.

32. Aggarwal S, Monga AP, Perusse R, Ramesh R, Ballarotto V, Williams ED, et al. Science. 2000;287:2235–7.

33. Roa F, Way JD. Appl Surf Sci. 2005;240(1–4):85–104.

34. Ayturk ME. PhD dissertation, Department of Chemical Engineering, Worcester Polytechnic Institute, Worcester, MA 01609, USA, 2007.

35. Hydrogen from Coal RD&D Plan, Office of Fossil Energy, Sep 2007 – DRAFT.

36. Ma YH, Mardilovich IP, Mardilovich PP. Preprints-ACS. Div Petro Chem. 2001;46:154–6.

37. Graham T. Phil Trans R Soc. 1866;156:415.

38. De Rosset AJ. US Patent 2,824,620, 1958.

39. McKinley DL. US Patent 3,247,648, 1966.

40. McBride RB, Nelson RT, McKinley DL, Hovey RS. US Patent 3,336,730, 1967.

41. McKinley DL. US Patent 3,439,474, 1969.

42. McKinley DL. US Patent 3,350,845, 1967.

Chapter 14
Model-Based Design of Energy Efficient Palladium Membrane Water Gas Shift Fuel Processors for PEM Fuel Cell Power Plants

Mallika Gummalla, Thomas Henry Vanderspurt, Sean Emerson, Ying She, Zissis Dardas, and Benoît Olsommer

An integrated, palladium alloy membrane Water-Gas Shift (WGS) reactor can significantly reduce the size, cost and complexity of a fuel processor for a Polymer Electrolyte Membrane fuel cell power system. Physics-based system modeling that accounts for component performance under transient conditions is essential to derive the full benefit from this technology. Modeling elucidates system designs and operating philosophies that maximize system efficiency and performance. This advanced WGS membrane reactor fuel processor is enabled by high activity noble metal alloy/ceria-based WGS catalysts [1, 2] and porous metal tube supported palladium alloy membranes such as those investigated by Ma and co-workers [3–6].

Introduction

Palladium membranes have been known for years, and the concept of a membrane reactor that would allow the continuous removal of a reaction product from an equilibrium controlled reactor has been discussed and to some degree practiced for several decades. Hydrogen production, via some type of hydrocarbon reforming or catalytic partial oxidation, coupled with the WGS reaction $CO + H_2O \leftrightarrows H_2 + CO_2$ ($\Delta G° = -41$ kJ/mole) is one application for such a membrane reactor.

In the past, there was no need for small-scale (< 1 MW) distributed hydrogen production from hydrocarbons where such reactors would likely be first employed. In any case, unsupported palladium membranes have required too much expensive metal to be economically feasible for all but special purposes. The potential availability of cost effective, high hydrogen permeance, palladium membranes supported on porous, stainless steel, tubular substrates combined with the advent of stable, high volumetric activity WGS catalysts is the driving force behind the system modeling

M. Gummalla (✉)
United Technologies Research Center, 411 Silver Lane, East Hartford, CT 06108, USA
GummalM@UTRC.utrc.com

A.C. Bose (ed.), *Inorganic Membranes for Energy and Environmental Applications*,
DOI 10.1007/978-0-387-34526-0_14, © Springer Science+Business Media, LLC 2009

described in this article. The construction of a fuel cell power plant system model allows the determination of the best overall design parameters to achieve a set of system requirements.

In the work described here, the performance requirements for the fuel processing system (FPS) will be reviewed followed by an overview of a fuel cell power plant. The layout of a particular FPS is considered that incorporates a WGS membrane reactor (WGSMR) with a very active WGS catalyst developed at UTRC [1, 2] and a supported, thin, palladium alloy membrane. Although the WGSMR design is the focus of this particular article, a brief overview of palladium membranes, which make such a study possible, is given for background. For the purposes of this study, it is assumed that a suitable membrane with high durability and adequate performance properties (e.g. permeance) is chosen. Once the system and the behavior of membranes have been introduced, the rest of this article will focus on the modeling results.

Performance Requirements

The industrial requirements on catalyst performance, acceptable reactor size, and dynamic response of the catalytic reactor, arise from the end application and the system level interaction between the catalytic reactor and the remaining components of the device. For example, in fuel cell applications for electrical power generation, processing of the fuel (gasoline, natural gas, diesel, etc.) involves a train of catalytic reactors to produce hydrogen that is used in the fuel cell stack. There are also stringent cost, volume, life, and performance criteria that need to be met for application of fuel cell power plants in automobiles and auxiliary power units. These criteria are the primary drivers for the design specifications of the individual reactors and components. These design specifications, in turn, drive the research to develop high quality catalysts and palladium membranes in the WGSMR. As discussed below, the properties of the membrane have a strong impact on performance and life of the system.

Therefore, fuel cell power plant system level models, detailed component models, and their analysis are needed to determine the performance targets for the WGSMR, subject to the power plant cost, weight and performance targets set by the Department of Energy (DOE), USA, for future automotive applications with an on-board fuel processing system as shown in Table 14.1 [7]. Keeping these criteria in mind, the fuel cell power plant must now be considered.

Table 14.1 DOE's fuel processing targets for an on-board fuel processing system

Requirement	Target
Efficiency	78%
Power density (with insulation)	> 700 W/L
Cost	$25/kW
Start-up time	< 1 min for 33% full power (FP)
10–90% transient response time	≤ 5 s

Power Plant Description

Fuel cell power plants are designed to operate on gasoline, natural gas, or other fuels to produce electricity that can be used to drive electric motors in the automotive, or supply power to the grid, in the case of distributed power generation. Fuel cell power plants are complex systems that include chemical reactors, fuel cell stacks, turbo machinery, heat exchangers, valves, power conditioning, and other electromechanical components. The design of such a system requires consideration of several strongly interacting subsystems that are themselves complex and challenging to design for high performance.

The various sub-systems in the fuel cell power plant are: the fuel-processing subsystem (FPS), the thermal and water management subsystems (TMS, WMS), the cell stack subsystem (CSS), and the power conditioning subsystem (PCS) as shown in Fig. 14.1. In the FPS, the fuel is converted to a hydrogen-rich gas stream that can be used in the fuel cell. The efficiency of the fuel processing system is the ratio of the useful chemical energy in the species utilized downstream to the total chemical energy in the fuel entering the system. The FPS includes a desulfurizer, fuel reformer, WGSMR, and a catalytic burner to burn any unutilized hydrogen and unconverted hydrocarbons to generate heat and steam that is recuperated in the overall power plant.

A conventional FPS, shown in Fig. 14.2, includes a reformer, two WGS reactors, and two Preferential Oxidation (PrOx) reactors, located downstream of the WGS. For PEM fuel cells, it is a necessity to assure $< 10\,ppm$ of CO in the cell stack. These reactors form a considerable fraction of the FPS weight, volume, and cost. Replacing this train by an integrated hydrogen permeation selective membrane on the water gas shift reactor, shown in Fig. 14.3, results in a considerable reduction in the number of components, cost, and volume of the FPS. This will make fuel cell power plants practical and affordable for power generation in a wide range of applications, especially for residential and transportation. Numerous published works [8, 9] in the area of catalytic membrane reactors can be quoted in the experimental [10] and numerical [11, 12] domains.

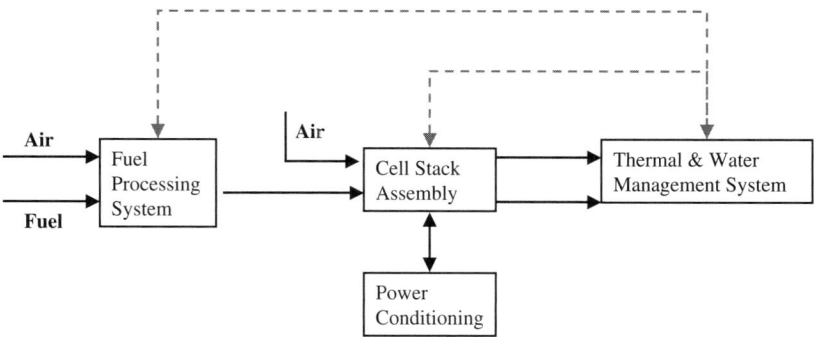

Fig. 14.1 Schematic of the fuel cell power plant showing the various subsystems and the interactions between them

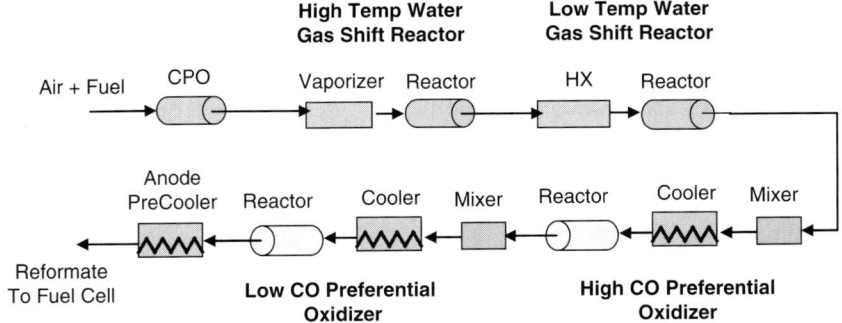

Fig. 14.2 Schematic of a conventional FPS

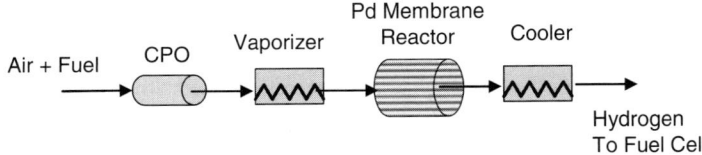

Fig. 14.3 Schematic of simplified fuel processing system with a WGSMR

The CSS converts the chemical energy in the hydrogen fuel to electrical energy. The efficiency of the fuel cell is the gross electrical power generated to the chemical energy supplied to it. The cell stack, subsystem includes the fuel cell stacks and the components that enable hydrogen recycle and so on. In the process of generating electrical power, the CSS also generates heat and water. The cell stack, considered in this system analysis, is based on porous water transport plate technology [13] and operates close to ambient pressure.

The TMS and WMS are greatly integrated in a system with a WGSMR. The heat from the exhaust gases from FPS and CSS is utilized for raising steam. The water generated by the fuel cell is partially recovered in the fuel cell stack and in an energy recovery device (ERD) down stream of the plant exhaust and is converted to steam, which is used in the fuel processing system. For efficient management of the energy and water, system level modeling and analysis is a great resource and is described later.

Palladium Alloys for Long Life and Performance

Producing long, lasting, defect-free, thin palladium membranes for a WGSMR is a major technical challenge. There are materials issues that need to be considered when designing the membranes for a WGSMR. Although the path to solve these problems is not the subject of this article, some of the issues to be considered in the appropriate choice of a membrane are described here. For the modeling work to follow, it is assumed that suitable membrane materials are selected to meet the FPS requirements and that such membranes are capable of being fabricated economically.

It is well known that at temperatures below 295°C, hydrogen dissolves in palladium forming the β-phase hydride. The β-phase hydride has a lower density and higher hydrogen content than the α-phase hydride that is stable at higher temperatures. This means in practice that unalloyed palladium, if exposed to hydrogen pressure at temperatures below 295°C, expands due to the formation of the β-phase hydride and then contracts above 295°C due to the transition to the α-phase hydride. Such volumetric changes due to temperature transitions can result in cracks developing and subsequent membrane failures.

Binary addition elements, having in general Face-Centered-Cubic (FCC) crystal structures, stabilize the α-hydride phase against the β-hydride phase transition, reducing the problem of embrittlement, and also increase hydrogen permeability above that of pure palladium (see Table 14.2).

The equilibrium amount of hydrogen in Pd-Ag alloys increases as the relative amount of silver in the alloy is increased. However, the diffusion coefficient of hydrogen in the alloy decreases with increasing silver content. As a result of these two opposing factors, the hydrogen permeability (the product of the solubility coefficient and the diffusion coefficient) for the Pd-Ag system reaches a maximum of 1.7 times the permeability of pure palladium at 23wt.% Ag at a temperature of 350°C. However, Pd-Ag is rapidly and irreversibly poisoned by sulfur and, even with the best sulfur clean-up technologies, there is a reasonable chance that a process upset or a change in feedstock would expose the membrane to sulfur. Clearly neither Pd nor Pd-Ag membranes are suitable for use with sulfur containing feed gas.

There has been a great interest in the Pd-Cu system due to its reported resistance to sulfur poisoning compared to the Pd-Ag system. Within the Pd-Cu binary system, the 60wt.% Pd-40wt.% Cu ($Pd_{0.47}Cu_{0.53}$ in atomic or mole%) composition was found to exhibit the highest hydrogen permeation [14–16] presumably because of the formation of an ordered, Body-Centered-Cubic (BCC) crystal structure (B2 phase). This phase is an anomaly since both palladium and copper have FCC crystal structures. The stability range of the BCC phase narrows as the temperature increases, and at 60wt% Pd-40wt% Cu is very close to the B2/FCC–B2 solvus line [17] in the operating region of a WGSMR at 350–400°C with possible excursions to 450°C (723 K). Howard et al. [18] tabulates hydrogen permeance values of various Pd-Cu alloys from the literature, presents additional data and discusses the impact

Table 14.2 Improvement in permeability of various binary Pd alloys at 350°C in hydrogen [5]

Alloy Metal	Wt.% for Maximum Permeability	Normalized Permeability (P_{alloy}/P_{Pd})
Y	10	3.8
Ag	23	1.7
Ce	7.7	1.6
Cu	40	1.1
Au	5	1.1
Ru, In	0.5, 6	2.8
Ag, Ru	30, 2	2.2
Ag, Rh	19, 1	2.6
Pure Pd	–	1.0

of crossing the solvus line by changing the temperature on hydrogen permeance. Hydrogen permeance is affected not only by the alloy composition but by the reformate composition as well.

Combined experimentation, atomic scale predictions and thermodynamic modeling were made of H interactions with the Pd-Cu system in the vicinity of the Pd-Cu BCC phase region, in order to establish the necessary fundamental understanding to support the development of Pd membrane alloys with improved thermal and chemical stability. Experimental measurements of H solubility in the $Pd_{0.44}Cu_{0.56}$ B2 alloy were made with a Sievert's type apparatus [19]. This data was assessed along with the existing experimental data describing the binary phase interactions and first principles predicted finite temperature data for hypothetical end-member phases, to develop a thermodynamic description of the ternary Pd–Cu–H system encompassing the PdCu B2 phase [19]. The thermodynamic assessment was used to model H solubility in the Pd-Cu B2 phase in terms of Sievert's constant, the proportionality factor between H solubility and the square root of pressure at a given temperature. First principles ground state and lattice dynamics simulations were used to predict favorable pathways for thermally activated hydrogen diffusion within the B2 lattice. The predicted diffusivity was shown to be in reasonable agreement with previous measurements by Völkl and Alefeld [20] and Piper [21]. The newly derived solubility and diffusivity parameters were evaluated within a mass transfer model to predict the ideal bulk permeability in the absence of other mass transfer contributions [22], which was found to be consistent with the permeability rankings of Ma [5].

Reformate obtained from low to moderate sulfur liquid fuels (< 15–30 ppm S by weight) would contain about 1–2 ppm sulfur species by volume. This sulfur concentration is low enough to be tolerated by the 60wt.% Pd-40wt.% Cu system according to the literature. Kulprathipanja et al. give an excellent discussion of the impact of sulfur [23] and other components of reformate [24], CO, CO_2 and H_2O on self supporting Pd-Cu foil and supported thin film Pd-Cu membranes.

Thin Pd alloy membranes supported on very smooth nano-porous oxide layers deposited on porous stainless steel tubing offer advantages over self-supporting foils. In our view, system protection from catastrophic failure is more significant than the presumed cost advantage. Assuming a tube and shell design that optimizes fabrication cost and heat and mass transfer considerations, the size of the integrated WGSMR is governed primarily by the permeance of the palladium alloy membrane. For a given reformate composition and a given alloy, permeance is inversely proportional to membrane thickness. A mathematical model is used to identify the performance metrics of the membrane and the reactor that meet the FPS requirements of Table 14.1.

Modeling and Analysis of Membrane Reactor Systems

Modeling of fuel cell systems can be carried out at various levels of fidelity, during different stages of the technology development process. The first phase in the technology development process, after concept synthesis, is demonstration of feasibility.

To screen high potential concepts, estimate the maximum attainable system efficiency, and to establish the feasible design space, low fidelity thermodynamic modeling is often handy. To design components and systems that meet high performance
metrics, while lowering the development risk, physics-based detailed component
models [25, 26] of higher fidelity are required. Physics-based modeling, as the
name suggests, incorporates the essential physics that represents the system, and
subsequent analysis of the transient behavior and steady-state performance provide
design guidance to the extent of validity of the assumptions. Such physics-based
system level models, developed for the fuel cell power plant including the membrane
reactor module, are discussed below.

System Level Models of Fuel Cell Power Plants

The fuel cell power plant, shown schematically in Fig. 14.1, consists of several components including the fuel cell stack, chemical reactors for processing the fuel and
burning the effluent gases, heat exchangers, compressors, blowers, expanders, pipes,
valves, and control units. These components are represented as lumped models with
physics essential for conceptual level of analysis. These models can be built and
analyzed in chemical process modeling tools such as gPROMS [27], ASPEN [28],
ABACUSS [29], any other simulation package, or even in conventional software
languages. Each of these tools provides some advantages and disadvantages over
the other, and the choice of the tool depends on the problem at hand and the detail
required for modeling them, ease of use, reusability, and maintainability. For the
current system, gPROMS is used as the simulation tool, due to the ease of use and
the capability to simulate detailed distributed models for steady state and transient
operation. The system model accounts for pressure drop in various components,
the energy used to drive the mechanical components, the energy conversion losses
(electrical to mechanical, chemical to chemical in the FPS, chemical to electrical in
the fuel cell), heat losses from the components, and effective use of thermal energy.
Furthermore, in these models, continuity of mass, momentum, and energy across
interfaces is ensured. A solution of the system level model provides the local pressure, temperature, flow rate, and composition of each of the streams in the power
plant, representing a virtual power plant. The overall electrical efficiency of the
system and efficiencies of the different subsystems are also tracked. Using these
models, the impact of the WGSMR for hydrogen separation has been studied, and
the impact of the reactor design changes on the performance metrics was evaluated.
Since WGSMR is a critical component of the compact fuel processing system, a
higher fidelity model, described below, is used both at the component level and at
the system level.

WGSMR Modeling

A typical WGSMR consists of three main sections: (1) the reformate side with a
WGS catalyst, (2) the palladium membrane, and (3) the permeate side to collect the

separated hydrogen gas. Membrane reactors can be conceived to be in a tubular or planar setup. Even in a tubular setup, few combinations for the location of the WGS catalyst are plausible. Although different designs can be conceptualized with these three essential sections to generate and separate hydrogen from the reformate, only one of the configurations is discussed here. In this configuration, shown schematically in Fig. 14.4, the inner tube is filled with the WGS catalyst where the reformate gas flows, the wall of the inner tube contains a hydrogen selective permeation membrane, and an outer shell is used to collect the hydrogen separated by the membrane.

The WGSMR is modeled [26] using the conservation equations of mass, energy and momentum. The reactor is represented in a one-dimensional distributed domain, where the mass transfer limitations in the radial direction were neglected, capturing the variations in the axial domain, along the length of the reactor. The mass transfer resistance in the radial direction can be neglected in cases where the radial transport time scale in the reformate side and permeate side for hydrogen is considerably less than that through the membrane. The reactor is modeled as three units with mass transfer between them. The most involved model is the inner WGS section, containing supported catalyst, represented with multiphase gas transport, heat and mass transport limitations to the catalyst surface, proprietary WGS reaction kinetics, and appropriate thermodynamics.

For the assumption of negligible external transport of hydrogen to and from a membrane, the hydrogen permeability of the membrane is modeled using Sievert's law, represented as Eq. 14.1 [30]. In this equation, Q_0 is the hydrogen permeability constant, l is the thickness of the membrane, E_a is the activation energy for membrane permeability, R is the universal gas constant and T is the average temperature (in K) of the inner and the outer shells. The flux is the volumetric flow of hydrogen per unit area per unit time. In a palladium membrane, the separation process occurs in a few steps, first with hydrogen dissociating on one side (the high pressure side), then diffusing as atoms through the palladium lattice and finally reassociating to

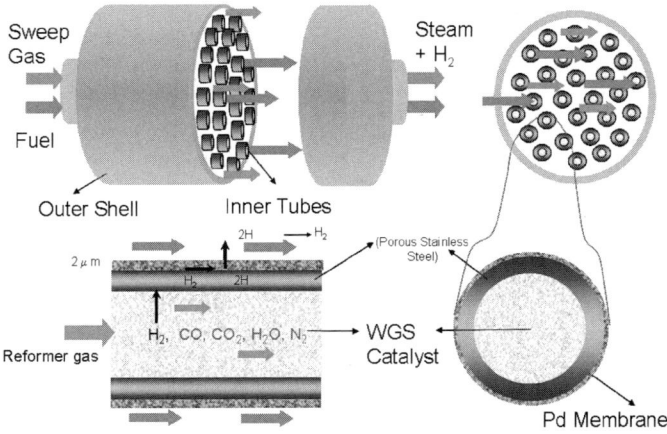

Fig. 14.4 Schematic of the WGSMR showing the internal tubes and flow of reactants

molecular hydrogen at the low pressure side. Since the H diffusion step in the Pd bulk is the rate-limiting step, the flux of hydrogen is proportional to the difference in the square root of the partial pressure of hydrogen between the two sides of the reactor/separator and is inversely proportional to the thickness of the membrane.

$$flux = \frac{Q_0}{l} \cdot \exp(-\frac{E_a}{RT}) \cdot (P_{H_2h}^{0.5} - P_{H_2l}^{0.5}) \qquad (14.1)$$

The hydrogen flux through the membrane can be increased by different operational strategies. For higher flux, greater difference in partial pressure of hydrogen between the reformate side and the permeate side is required. One way is to increase the partial pressure of hydrogen on the reformate side by increasing the pressure and/or mole fraction of hydrogen. The other way is to decrease the pressure on the permeate side by maintaining vacuum or using sweep gas to decrease the mole fraction of hydrogen, thus decreasing the partial pressure there. Each of these methods has positive and negative effects on the system efficiency, and will be discussed below. The analysis of the membrane reactor will be discussed first, followed by the analysis of the fuel processing subsystem and system level results, and last with the impact of the WGSMR design on the system performance metrics.

Reactor Design Analysis

The WGSMR simulations assumed that the reformate gas is supplied to the inner tube filled with water gas shift catalyst, operating at 6 bar. The reformate gas is distributed to several 1-in diameter tubes in a staggered arrangement (all tubes are considered identical). The outer shell geometry is designed to envelope these tubes, with steam as sweep gas. A fuel cell system generates electricity from hydrogen, and in the process generates water. This water is used to raise steam used as an inlet to the outer shell.

Figure 14.5 shows the partial pressure of hydrogen in the inner and outer sections of the reactor along its length, where the reformate and the permeate flow in a cocurrent configuration. It is seen that the driving force for permeation is higher near the inlet and very low at the outlet. The partial pressure of hydrogen in the permeate side is zero at the entrance of the reactor, providing a large driving force for hydrogen separation. Additionally, hydrogen is produced at the front end of the reactor, increasing the partial pressure of hydrogen, which is subsequently removed, causing the peak in the partial pressure of hydrogen at the entrance of the reactor. The CO fraction decreases along the length of the reactor due to conversion to hydrogen, lowering the equilibrium limited reaction rate. Since the hydrogen removed from the reformate side is added to the permeate side, the partial pressure of hydrogen increases along the length of the reactor for the co-flow configuration.

Analysis of the hydrogen permeation rate across the membrane indicated that the driving force is higher near the inlet of the reactor for all conditions simulated. The rate of WGS reaction along the length of the reactor is also indicated in the same

Fig. 14.5 The radial averaged reaction rates and partial pressure of hydrogen along the length of the reactor in a co-flow configuration

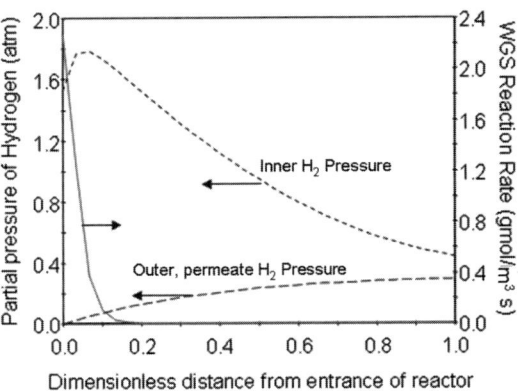

plots. For the operating conditions assumed, the reaction approaches equilibrium within 20% of the reactor length. Note that this factor would strongly depend on the length of the reactor and the recovery fraction of hydrogen. Recovery fraction is defined as the ratio of the hydrogen separated to the sum of hydrogen generated through the WGS reaction and hydrogen that enters the reactor with the reformate. Though the reaction rate is relatively low over the remaining reactor length due to the shifting of the equilibrium, it is a finite value greater than zero. The additional length of the reactor is predominantly utilized for hydrogen separation. It is evident from these simulations that hydrogen separation is the slower process relative to the shift reaction. Therefore, in these systems the remaining 80% of the reactor can be empty or filled with low performing catalyst. Hence, potential cost savings in the catalyst can be obtained without greatly compromising reactor efficiency.

Figure 14.6 shows the effect of hydrogen recovery from the WGSMR for increasing membrane thickness, in cocurrent-flow and in countercurrent-flow configurations, with identical reactor geometry and conditions. Permeation of hydrogen through the membrane is inversely proportional to the membrane thickness, as indicated by Sievert's Law. As seen from Fig. 14.6, to achieve more than 90% hydrogen recovery, a $\sim$2–3 μm thick Pd membrane is required, for an assumed volume of the reactor.

As the thickness of the membrane increases, for identical tube geometry, the size of the reactor increases to achieve the same recovery fraction. Furthermore, since the surface area is proportional to the diameter of the inner tubes and the reactor volume is proportional to the square of the diameter, the overall size of the reactor would scale inversely proportional to the radius of the tubes, for constant surface area of the membrane. This is representative for the cases where the permeance through the membrane determines the size of the reactor, and the hydrogen removal fraction is proportional to the area of the membrane. Other contributing factors that tend to provide better recovery are increasing the number of inner tubes and increasing the flow rate of the outer shell carrier fluid (sweep gas). However, there is a trade between the reactor geometry (determining the overall volume of WGSMR) and the hydrogen recovery that can be achieved.

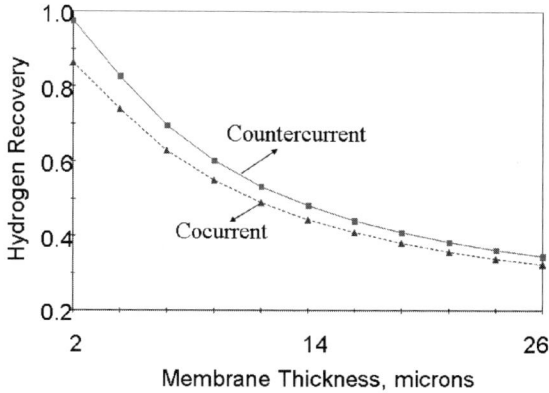

Fig. 14.6 Effect of membrane thickness on hydrogen recovery through the membrane

Typically, the cost of the membrane reactor increases proportional to the volume, arising from the cost of manufacturing, installation, and usage of expensive materials like palladium. When steam is used as the sweep gas, increasing the sweep gas flow rate in the outer shell would increase the thermal energy requirements to sustain the system, thus reducing the overall efficiency. Furthermore, in the case where there is no sweep gas flow, or relatively low flow rate, higher pressure needs to be maintained at the reformate side or vacuum on the permeate side to maintain the same hydrogen recovery fraction. There is a trade off between the thermal and parasitic power required to sustain the power plant reflected in the system efficiency.

The geometric flow of the permeate and the reformate relative to each other also alters the extent of hydrogen recovery for identical operational conditions. Also shown in Fig. 14.6 is the removal fraction for the countercurrent flow case, wherein the sweep gas enters the reactor towards the reformate exit section. At this location, the reformate has a lower fraction of hydrogen and hence lower partial pressure. With pure steam, at this location, the partial pressure of hydrogen on the permeate side is even lower, enabling further removal of hydrogen towards the end of the reformate side. This results in overall better driving force for hydrogen removal and hence better hydrogen recovery, for identical reactor volume and operating conditions. It is important to note that the effect of flow geometry and location dependent driving force for hydrogen separation along the length of the reactor is better represented with distributed models and are ignored in lumped models.

Fuel Processing Subsystem

The fuel processing subsystem primarily consists of a desulfurizer, reformer (operational conditions varied from that of partial oxidation conditions to auto thermal reforming), WGSMR, catalytic burner, heat exchangers and vaporizers. Although higher steam fractions to a reformer are known to yield higher efficiency, a fixed size membrane WGS reactor downstream of the reformer can recover only a fraction of the hydrogen produced. Figure 14.7 shows the reformer efficiency and hydrogen

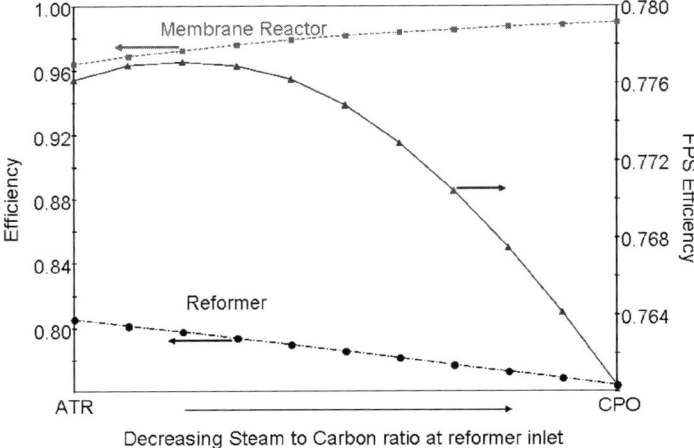

Fig. 14.7 Model based prediction of FPS and reformer efficiency is shown along with hydrogen recovery from the fixed volume membrane reactor

recovery fraction of the membrane reactor, and their product defining the efficiency of the fuel processing system. Higher amounts of hydrogen require larger membrane surface area for removal, and with a fixed size it is difficult to remove a greater fraction of hydrogen without altering the design or operating conditions. Therefore, for maximum efficiency of the fuel processing system optimal operating conditions can be established, while meeting the size requirements. The heating value of the fuel not used in the FPS is used to raise steam, rather than using a larger membrane reactor to separate hydrogen and subsequently use the lower heating value to raise steam. It should be noted that FPS efficiency is not a monotonic function of the hydrogen removal rate.

System

A system level model of the power plant has been implemented and simulations were carried out over a wide range of design parameters with the aim to minimize fuel processing system volume and maximize the overall system efficiency. Although the FPS and CSS efficiencies have been defined earlier, the mechanical efficiency reflects the net power generated by the power plant to the gross power produced by the CSS. The electrical power in excess of the net power is lost during the process of power conditioning, for operation of blowers, compressors and pumps. Since the FPS operates at high pressure, the air compressor and a turbine are operated on the same shaft to minimize the electrical power required to sustain the system, hence improving the mechanical efficiency of the system. With the efficiency definitions provided above, the system efficiency is the product of the FPS, CSS and mechanical efficiencies.

For a specific design change in the system, the virtual fuel cell power plant is simulated to determine its effect on the performance metrics. In a typical simulation, with a 7-L WGSMR (that corresponds to a 4-μm membrane thickness), the CSS efficiency is at 54% (end of life), the FPS efficiency is at 68% with a reformer efficiency of 79% and the mechanical efficiency is at 84%. The overall system efficiency is at 31%. Conventional systems operating on gasoline are projected to provide near 26% efficiency. It should be noted that to decrease the size of the FPS unit to meet multiple objectives of cost and dynamic response, a maximum efficiency system does not translate to a maximum fuel processing efficiency. To maximize the FPS efficiency, either the thermal energy to raise steam or parasitic energy to maintain vacuum needs to be provided. These design changes could lower the mechanical efficiency in the system, affecting the overall system efficiency. Although the analysis discussed so far is based on steady-state system level and component models, the dynamics of the FPS with focus on the membrane reactor to meet the performance targets must also be analyzed.

Start-Up Time and Transient Capability

One of the concerns with bulky reactors is the time taken to start up the reactor. Startup involves bringing the temperature of the essential components to the required level to kick off the reaction kinetics, supplying the required fuel to the reactors, and ensuring durable operation where all the reactors work seamlessly. In conventional systems, a large number of units need to be brought up in temperature before activating the reactions. In the case of a membrane reactor-based FPS, the catalytic partial oxidation (CPO) reformer, vaporizers, and membrane reactor need to be brought up in temperature.

Figure 14.8 shows the response of the membrane reactor for a predetermined startup procedure adopted. The fuel heat of combustion warms up the CPO and its effluent hot gases subsequently warm up the membrane reactor. Since both the reactors are relatively compact and the thermal masses are low, they can be warmed up in less than 30 s. The red line shows the flow of the reformate from the CPO. The CPO being a compact exothermic reactor, responds to fuel flow change in a few seconds and is capable of providing the required flow rate to the WGSMR within 20 s after initiating the startup protocol. The steam is not necessarily available at this point for the most efficient hydrogen removal from the membrane reactor. However, even with no steam as sweep gas, the simulation shows that 50% of the hydrogen required for full power can be provided by the compact membrane reactor in ≈ 30 s of startup. Assuming that the steam can be generated within 1 min, 100% power can be provided seconds after that demonstrating the capability to meet the startup time requirements.

Another critical aspect to consider for application of a membrane reactor in the FPS is the rapid transient capability. The requirement is to deliver required hydrogen in less than 5 s when a step change in the electrical load is changed from 10

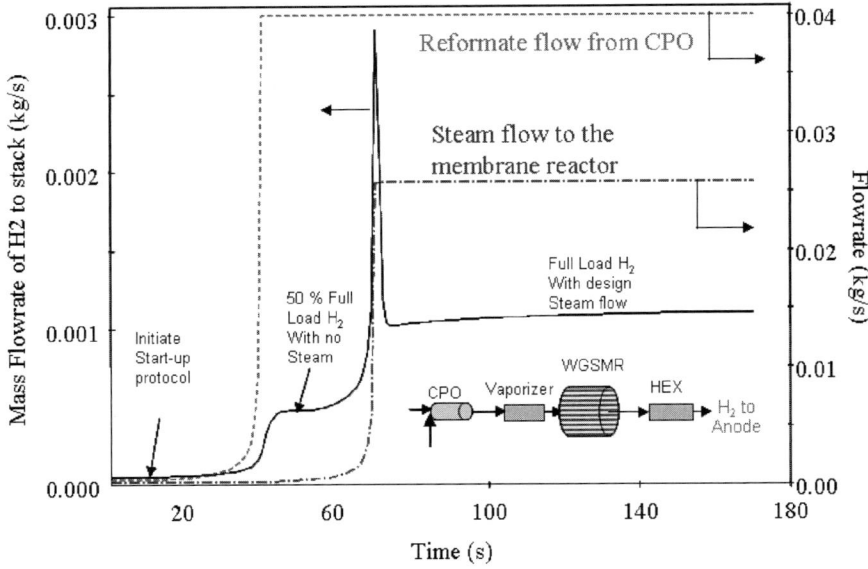

Fig. 14.8 Model based start-up simulation of the membrane reactor module

to 90%. Modeling results demonstrated less than 5 s load following capability, and that the compactness and low thermal mass of the reactor were essential to meeting this stringent requirement. The compact fuel processor established its new state relatively quickly, corresponding to the fast dynamics of reaction and transport of produced hydrogen through the support and membrane. Mass transport dynamics of the membrane were assumed to be relatively faster than through the support, hence pseudo-steady state approximation was made. Similar dynamics were evaluated for increasing the reformate flow from 10% power to full power (90% up-transient). First order response to the fuel flow change was observed with the system establishing its new state in less than 5 s. These simulations were carried out independent of the system, which might respond slower in terms of the reformate flow change.

Design Specifications Based on Systems Analysis

Once the overall system efficiency is established along with the subsequent subsystems efficiency, the performance requirements from the WGSMR are established. To achieve the 68% FPS efficiency, a membrane reactor should enable near 85% hydrogen recovery. The membrane reactor size for a varying permeance through the reactor, with steam as sweep gas in a counter flow configuration with nearly 40% steam by volume at the sweep gas exit point, is identified in Fig. 14.9. Although higher membrane permeance can be achieved with thinner membrane and higher temperatures, the selectivity to hydrogen is at stake. Based on a plot shown in Fig. 14.9, the design engineer can decide the size of the reactor for a given thickness

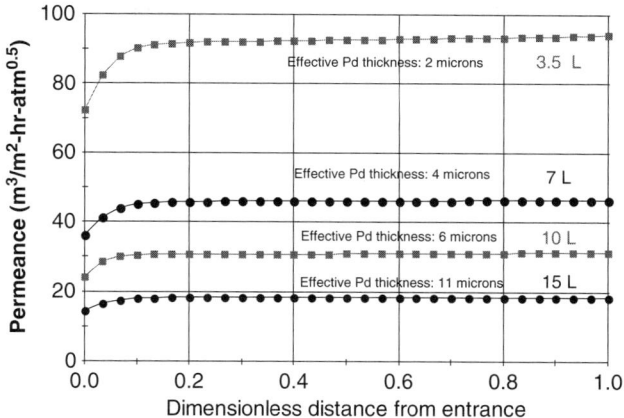

Fig. 14.9 Permeance requirements for a given volume reactor to achieve maximum system efficiency for set conditions

of the membrane or an experimentalist can target the required permeance from the membrane to fit into the system design requirements.

Practical Issues

The advanced, physics-based component and system level model development and simulations can be used as a virtual prototyping for the design engineer to perform different trade-offs and optimize the overall fuel cell power plant for performance, size, and cost. The models developed in this work, although aimed for a gasoline on-board fuel processor with a CPO for rapid startup and transients, are fuel flexible and can be applied to any type of fuel and reformer (i.e. catalytic steam reformer, autothermal, and CPO). In these cases, the overall system efficiency and membrane size will vary from the predicted values in this study. With this analysis we predicted that a Pd alloy membrane WGSR will give an $\approx 60\%$ system size reduction and an $\approx 45\%$ system cost reduction for the FPS relative to the baseline system and a 3.5% efficiency increase. What is now needed is the positive identification and life testing under realistic conditions of robust, reliable, palladium alloy membranes with sufficiently high permeance as well as a route to their cost effective, commercial-scale production.

Conclusions

Contrary to what might be assumed, the modeling work here has shown that the maximum efficiency of a WGSMR based fuel cell power system typically is not reached at maximum hydrogen recovery efficiency, provided the heating value of the retentate is captured as thermal energy elsewhere in the system. Furthermore, the FPS system efficiency has a broad maximum, the location of which is a function of the membrane reactor size and design. The WGSMR size is strongly dependent on the permeance of the palladium alloy membrane under operating conditions. Since the WGS catalyst activity is very high, most of the reactor volume is devoted to hydrogen recovery. The heat and mass transfer aspects of the reactor design, such as the membrane tube diameter and flow configuration, are other aspects that can affect hydrogen recovery and efficiency.

Acknowledgments The majority of the modeling effort was performed under US Department of Energy Contract DE-FC36-02AL67628, Mod. #8, formerly DE-FG04-2002AL67628 "On Board Vehicle, Cost Effective, Hydrogen Enhancement Technology for Transportation PEM Fuel Cells."

References

1. Vanderspurt TH, Wijzen F, Tang X, Leffler MP. US Patent Application 2003 0186805 A1, 2003.
2. Vanderspurt TH, Wijzen F, Tang X, Leffler M, Willigan RR, Newman CA, et al. US Patent Application 2003 0235526 A1, 2003.
3. Mardilovich PP, She Y, Rei MH, Ma YH. AIChE J. 1998;44(2):310.
4. Ma YH, Mardilovich PP, She Y. US Patent 6,152,987, 2001.
5. Ma YH, Engwall EE, Mardilovich IP. Composite palladium and palladium-alloy membranes for high temperature hydrogen separations. ACS Fuel Chem Div Prepr. 2003;48(1):333.
6. Engwall EE, Mardilovich IP, Ma YH. Transport resistance for oxide diffusion barriers on porous metal supports used in composite Pd and Pd-alloy membranes. Prepr Symp – Am Chem Soc Div Fuel Chem. 2003;48(1):390–1.
7. http://www.eere.energy.gov/hydrogenandfuelcells/pdfs/33098_apx.pdf.
8. Zaman J, Chakma A. J. Memb Sci. 1994;92:1–28.
9. Saracco G, Specchia V. Catal Rev Sci Eng. 1994;36(2):305–84.
10. Basile A, Criscuoli A, Santella F, Drioli E. Gas Sep Purif. 1996;10(4):243–54.
11. Criscuoli A, Basile A, Drioli E. Catal Today. 2000;56:53–64.
12. Criscuoli A, Basile A, Drioli E, Loiacono OJ. Memb Sci. 2001;181:21–7.
13. Meyer AP, Scheffler GW, Margiott PR. US Patent US5503944 A, 1996.
14. Way JD. Palladium/copper alloy composite membrane for high temperature hydrogen separation from coal derived gas streams. DOE report, 2002.
15. Rao F, Block MJ, Way JD. Desalination. 2002;147:411–6.
16. Rao F, Way JD, McCormick RL, Paglieri SN. Chem Eng J. 2003;93:11–22.
17. Subraminian PR, Laughlin DE. Phase diagrams. 2002 ASM Int'l.
18. Howard BH, Killeyer RP, Rothenberger KS, Cugini AV, Morreale BD, Enick RM, et al. J Memb Sci. 2004;241:207–18.
19. Opalka SM, Huang W, Wang D, Flanagan TB, Løvvik OM, Emerson SC, Shc Y, and Vanderspurt TH, "Hydrogen interactions with the PdCu ordered B2 alloy," J. Alloys Compd., 446–447 (2007) 583–87.
20. Völkl J, Alefeld G. Diffusion of hydrogen in metals. In: Alefeld G, Völkl J, editors. Hydrogen in metals. Vol. I. Berlin: Springer-Verlag; 1978. pp. 321–48 (vol. 28 of topics in Applied Physics).
21. Piper J. J Appl Phys. 1966;37:715.
22. Huang W, Opalka SM, Wang D, and Flanagan TB, "Thermodynamic Modeling of the Cu-Pd-H system," CALPHAD, 2007;31(3):315–29.
23. Kulprathipanja A, Alptekin GO, Falconer JL, Way JD. J Memb Sci. 2005;254:49–62.
24. Kulprathipanja A, Alptekin GO, Falconer JL, Way JD. Ind Eng Chem Res. 2004;43:4188–98.

25. Gupta N, Gummalla M, Dardas Z, Zhu TL, Sun J, Vincitore A, "Dynamic modeling and simulation of water gas shift subsystem in the fuel processing train for PEM fuel cells." 2nd Topical conference on fuel cell technology, 2003 Spring National Meeting, AIChE, March 30–April 3, 2003, pp. 87–93.
26. Gummalla M, Olsommer B, Gupta N, Dardas Z. Physics based simulations of water gas shift membrane reactor. 2nd Topical conference on fuel cell technology, 2003 Spring National Meeting, AIChE, March 30–April 3, 2003, pp. 249–254.
27. http://www.psenterprise.com/gPROMS.
28. http://www.aspentech.com/brochures/aspenplus.pdf.
29. http://yoric.mit.edu/abacuss2/abacuss2.html.
30. She Y. Ph.D. thesis, Worcester Polytechnic Institute, Worcester, MA, 2000.

Part III
Porous, Silica and Zeolite Membranes

Chapter 15
Zeolite Membranes

Weishen Yang and Yanshuo Li

In the present chapter, an outline of the preparation and characterization of zeolite membranes is given. Emphasis is then placed upon the zeolite membranes for energy and fuel applications, including clean fuels from fossil energy sources, biofuels from renewable and sustainable energy, hydrogen energy, energy-efficient and cost-efficient separations, membrane reactors and solar energy utilization.

Since the mid-1990s, owing to the potential molecular sieving action, controlled host-sorbate interactions and high thermal and chemical stability, the preparations, characterizations and applications of membranes, films and coatings of zeolite and zeolite-like materials (in the present article, in short called "zeolite membranes") have been extensively investigated. Some excellent reviews [1–8] and book chapters [9] on the preparation and characterization of zeolite membranes appeared during the last several years. Focusing on the topics of this book and limited the length, this present review attempts to emphasize the application of zeolite membranes for both the production of tomorrow's clean fuels and the establishment of energy-efficient process. For more general information on the subjects of preparations, characterizations and otherwise applications of zeolite membranes, the above cited reviews and book chapters are recommended.

Preparation of Zeolite Membranes

Although zeolite membranes have some comparabilities to zeolite powders, they indeed require new synthesis strategies and new synthesis techniques as opposed to traditional crystalline zeolite powders synthesis. The most used strategies are in situ synthesis [10], secondary (seeded) growth synthesis [11] and vapor phase transport synthesis [12]. Various techniques have been applied, such as conventional heating, microwave heating [13, 14], synthesis under centrifugal field [15], moved-synthesis [16], and so on. Diversified methods through different combinations of above

W. Yang (✉)
State Key Laboratory of Catalysis, Dalian Institute of Chemical Physics, Chinese Academy of Sciences, Dalian 116023, China
e-mail: yangws@dicp.ac.cn; URL:http://yanggroup.dicp.ac.cn

A.C. Bose (ed.), *Inorganic Membranes for Energy and Environmental Applications,*
DOI 10.1007/978-0-387-34526-0_15, © Springer Science+Business Media, LLC 2009

synthesis strategies and techniques have been reported to prepare different kinds of zeolite membranes. Often, pre-treatment of supports [17] and post-treatment of membranes [18] are carried out to improve the qualities of the as-synthesized membranes.

Characterization of Zeolite Membranes

Zeolite membranes are commonly characterized by SEM, XRD. TEM, EPMA, SEM-EDX, TEM-EDS, and Nitrogen adsorption are also used to study the morphology, microstructure and composition of zeolite membranes. Usually single gas permeation, mixed gas separation, pervaporation and vapor permeation are performed to evaluate the properties of zeolite membranes. Recently, some novel characterization techniques have been applied. Infrared reflectance measurement was used to characterize membrane thickness [19]. Fluorescence confocal optical microscopy was used to image the grain boundary structure of zeolite membranes [20]. FTIR-ATR method was used to characterize the T-O vibration of zeolite membranes [21].

Applications of Zeolite Membranes

In the middle of the last century, the original form of zeolite membranes were synthesized by dispersing the zeolite crystals in polymer membrane matrixes, which were used for gas separation and pervaporative alcohol/water separations. In the last few decades, the researches of polycrystalline zeolite membranes that supported on ceramic, glass, or metal substrates have grown into an attractive and abundant field. Their applications for gas separation, pervaporation, membrane reactors, sensors, low-k films, corrosion protection coatings, zeolite modified electrodes, fuel cells, heat pumps et al. have been wildly explored. In the following text, the applications of supported polycrystalline zeolite membranes for energy and fuels will be presented.

Tomorrow's Fuels

One of the greatest challenges facing the global society of today and tomorrow involves the production of adequate, clean, sustainable energy for all people. It is clear that new technologies are required, nevertheless adequate energy can be provided by intelligent use of solar, geothermal and fossil energy sources. These new technologies require integration of knowledge and expertise from science, engineering, and economics. New technologies are also needed for the management and use of wastes involved in energy production, including greenhouse gases. Zeolite membranes, although an end branch of material science, are expected to make contributions in the coming transition from non-sustainable energy system to a sustainable future.

Clean Fuels from Fossil Energy Sources

At the beginning of the twenty-first century, along with the coming of the peaking of world oil production, oil production from conventional reservoirs begin to decline, creating a gap between supply and demand. Alternative energy sources are needed to fill the gap. Among which, coal and natural gas, which have already played an important role in electricity generation, industrial manufacture and residential utilization, are expected to be the dominant choices because of its relative abundant reserves and mature infrastructure. In addition, with much attention paid to the protection of public health and the environment, environmentally friendly liquid fuels based on natural gas and coal are desired when gasoline and diesel fuel are substituted by them. When the times of need arose, Fischer-Tropsch (F-T) synthesis, an "old" technology that began in the 1920s, staged a comeback. Coal-to-liquids (CTL) and gas-to-liquids (GTL) process can realize the conversion of coal and natural gas to high value liquid fuels.

A common feature of all catalysis for F-T synthesis, whether they are cobalt or iron based, is that the catalytic activity is reduced due to the oxidation of active species. Under the typical reaction conditions, this oxidation may be caused by water, which is one of the primary products in the F-T process. On the other hand, at low partial pressure water can also help to increase the product quality by increasing the chain growth probability. Thus, in situ removing some of the water from the product and keeping the water pressure at an optimal value may improve the catalysis activity and promote the reaction rate. Zhu and coworkers [22] have evaluated the potential separation using NaA zeolite membrane to in situ removal of water from simulated F-T product stream. High selectivity for water removal from CO, H_2 and CH_4 were obtained. This result opened an opportunity for in situ water removal from F-T synthesis under the reaction conditions.

Espinoza et al. [23] also studied the effect of zeolite membranes in Fischer-Troscher reactors. Mordenite, ZSM-5 and silicalite membranes on stainless steel supports were used to evaluate their functions of in situ partial removal of the reaction water.

Besides producing mixed-hydrocarbons (ultra-clean diesel), F-T process can also selectively produce mixed-alcohols (oxygenated fuel). The addition of mixed-alcohols into gasoline can effectively reduce HC and CO emissions. However, before directly used as fuels or blended with conventional fuels, the water content in the as-produced F-T mixed-alcohols must be reduced below 0.5wt.%. This dehydration step is essential but difficult since most of the contained alcohols form azeotropies with water. In our group, we studied the dehydration performance of microwave synthesized NaA zeolite membrane toward F-T produced mixed-alcohols [24, 25]. The membrane also showed excellent pervaporation performance toward dehydration of simulated F-T produced mixed-alcohols. The permeate consisted of only water and little methanol (< 10%) in all the range of feed composition. This result confirmed that NaA zeolite membrane based pervaporation (or vapor permeation) process could be an effective technology for dehydration of F-T produced mixed-alcohols.

Biofuels from Renewable and Sustainable Energy

The substitution of oil by coal and natural gas to produce fuels is only a makeshift effort because as non-regenerable fossil energy sources, coal and natural gas will ultimately be depleted. The perceived risk of running out of conventional fossil fuels led to numerous programs in developing renewable sources. More recently, the risks associated with CO_2 emissions and global warming have again spurred interest in renewable energy. With the search for alternative renewable energy sources, biofuels are fast becoming a viable solution. Bio-ethanol is a readily available, clean fuel that is defined as "carbon neutral" under the Kyoto Protocol, and by replacing some of automotive gasoline with biomass ethanol, a reduction in CO_2 emissions can be expected.

Ethanol permselective membrane system has been used for the extraction of ethanol from fermentation broth. However, both membrane distillation and polymeric silicon rubber membranes showed low separation factors of ethanol and were invalid in this case. M. Nomura et al. [26] investigated the continuous extraction of ethanol from ethanol fermentation broth through a silicalite-1 membrane. From 4.73wt.% ethanol concentration of broth, the permeate ethanol concentration was 81.0wt.%. In our group, we have also investigated the potentiality of silicalite-1 membrane for alcohol extraction from aqueous solution [27].

The same as F-T produced mixed-alcohols, low purity bio-ethanol extracted from fermentation broth must be refined into high purity fuel grade ethanol. The pervaporation dehydration pilot plant based on NaA zeolite membrane was set up by Mitsui Engineering & Shipbuilding Co., Ltd. (MES) in 1999. Recently, a pilot-scale NaA zeolite membrane based vapor permeation dewatering installation has been set up in our group with a handling capacity of 250 L/D. This installation can continuously produce 225 L 99.7wt.% ethanol per day. Meanwhile the permeate is nearly pure water.

An alternative way to utilize renewable biomass is the so-called biomass-to-liquids (BTL) process. Parallel with GTL and CTL, BTL is essentially a variant of the F-T process that uses biomass as a feedstock. Hydrophilic zeolite membrane can also play an important role in the downstream dehydration process.

Hydrogen Energy

In the span of less than 150 years, the world has successfully transitioned from wood to coal, to petroleum, to increasing contributions from natural gas, hydro, nuclear energy and, most recently, renewable sources. Right now we may be standing on the brink of the next big energy transition – "Hydrogen Economy". This future "hydrogen Economy" features hydrogen as an energy carrier (like electricity) in the stationary power, transportation, industrial, residential and commercial sectors. In the near- to mid-term, hydrogen will likely be produced by steam reforming natural gas or coal gasification process and possibly other carbon reserves such as biomass. However, H_2 usually coexists with other light gases (containing CO_2, CH_4, CO, H_2O, etc.) when it is produced from industrial processes such as gasification reaction and steam reforming reaction. To use these hydrogen-rich gas streams as fuels requires gas separation. For separation of hydrogen from gaseous streams, membranes can provide an attractive alternative to PSA and cryogenic separation. In regard to separations under steam reforming reaction conditions (i.e. with the presence of water at temperatures between 500–800 K and pressures of 40–70 atm), existing technologies are inefficient or unable to operate. However this can potentially be performed using zeolite membrane processes because they can operate over a wide temperature and pressure range and in chemically challenging environments. Studies in separation of H_2 have been performed by many groups in the past decade, as summarized in Table 15.1. Some attractive works are highlighted as follows.

Hong et al. [29] exploited silylation reactions with proper silane precursors to engineer the pore size of zeolite membrane for hydrogen separation. As a result, H_2/CH_4 separation selectivity can be increased more than 50% from 35 to 59.

Lai et al. [30] prepared ZSM-5 zeolite membrane using a template (TPA^+) free synthesis gel. In this way, calcinations, which often results in cracks in zeolite film, were unnecessary. Another advantage was that expensive and toxic templates were not applied in the synthesis gel, which made the preparation process more cost effective and environmentally friendly.

Generally, zeolite membranes are synthesized on tubular supports, and used as tubular-type modules. Packing density (i.e. membrane separation area/module volume ratio) of the tubular module was low as it was compared with that of the polymeric membranes. Xu et al. [32] synthesized of NaA zeolite membrane on a ceramic hollow fiber with an outer diameter of $400\,\mu m$, a thickness of $100\,\mu m$ and an average pore radius of $0.1\,\mu m$. The quality of the as-synthesized NaA zeolite membrane held a He/N_2 separation factor of 3.66 and He permeance of $10.1 \times 10^{-8}\ mol/(m^2.s.Pa)$.

Table 15.1 H_2 separation data

Gases	Membr.	H_2 Permeance $[10^{-8}\ mol/(m^2\ sPa)]$	Selectivity	Ref.
H_2/CH_4	SAPO-34	5.0–17	4.5–13.5	[28]
H_2/CH_4	SAPO-34	—	59	[29]
H_2/CH_4	ZSM-5	12.0	231	[30]
H_2/CO_2	ZSM-5	—	235	[29]
H_2/CO_2	ZSM-5	12.0	57	[30]
H_2/N_2	ZSM-5	12.0	109	[30]
H_2/N_2	NaA	0.64	4.6	[31]
H_2/N_2	NaA	213	3.2	[13]
H_2/N_2	NaA	17.1	5.6	[14]
$H_2/n\text{-}C_4H_{10}$	SAPO-34	3.2	450	[28]
$H_2/n\text{-}C_4H_{10}$	NaA	28.6	106	[32]
$H_2/n\text{-}C_4H_{10}$	Sodalite	11.4	>1,000	[33]
H_2/CO_2, CO, CH_4	ZSM-5	67.8	16.6	[34]

Aiming at practical applications, we used microwave heating (MH) technique to prepare NaA zeolite membrane [13]. The permeance of the NaA zeolite membrane synthesized by MH was four times higher than that of the NaA zeolite membrane synthesized by conventional heating, while their permselectivities were comparable.

We also developed a novel synthesis method ("in situ aging – microwave synthesis" method) for microwave synthesis of zeolite membranes without seeding, which was adapted to the scale-up of the membrane synthesis to integrated supports (e.g. multi-channel tube modules and hollow fibers) [14].

Welk et al. from Sandia National Laboratories (SNL) (USA) [34] have intensively studied the potential applications of zeolite membranes in reforming streams separation. ZSM-5 zeolite membranes synthesized by them showed excellent performance for hydrogen separation from light gases mixture. It was reported that after a single pass through the ZSM-5 zeolite membrane, the H_2 purity was enriched from 76% (simulated reforming stream) to >98%.

The use of fossil resources (natural gas, coal, petroleum, etc.) to produce hydrogen will emit carbon dioxide (a main greenhouse gas). Therefore, to utilize fossil resources for cleaning hydrogen production facing the "Hydrogen Economy" is essentially dependent upon the associated R&D effects on CO_2 capture and sequestration. Membranes are considered to be an economical technology for capture of CO_2. Large research efforts particularly in Japan were devoted to studying zeolite membranes for CO_2 separation [35].

Energy-Efficient Process

Energy-Efficient and Cost-Efficient Separations

In chemical manufacturing processes, separations account for 43% of the energy consumed and up to 70% of the capital costs. For bulk filtration processes, the energy use is low because this physical separation is simple. The more the separation reaches the molecular level, the more sophisticated the separation technology needed and the higher the energy use.

Distillation, the separation of liquids by evaporation and condensation, is the most used technique. The thermodynamic efficiency of distillation is, however, very low and is in the order of 10%. Besides the high energy use, in many sectors of the industry about 40–70% of the investment and operational costs are used for separation technology. Membrane separation (a membrane is a partially selective barrier for transport of matter) is seen as one of the most promising and energy-efficient separation technologies. Recently, Veerle Van Hoof et al. [36] performed an economic analysis comparing different processes for the dehydration of isopropanol/water. It was concluded that the Distillation-PV hybrid process using NaA zeolite membranes is the most interesting process from the economic point of view.

Beside isopropanol/water, there are many other systems that are conventionally achieved by distillation, which is inferior to membrane separation both in operating

Table 15.2 Performances of water selective zeolite membranes

Feed Solution [a]	Membr.	WaterFlux (kg/m²h)	Selectivity	Temp. (°C)	Ref.
Water/Methanol	NaA	0.57	2, 100	50	[36]
Water/Ethanol	NaA	2.15	10, 000	75	[36]
Water/2-Propanol	NaA	1.76	10, 000	75	[37]
Water/Acetone	NaA	0.91	5, 600	50	[37]
Water/2-Propanol	NaA	0.13	8, 000	25	[38]
Water/2-Propanol	NaA	0.79	4, 000	70	[38]
Water/Ethanol	NaA	0.64	10, 000	70	[14]
Water/Ethanol	NaX	0.89	360	75	[37]
Water/Ethanol	NaY	1.59	130	75	[37]

a: All the feed solution contained 10wt.% of water.

costing and energy consumption. However, polymeric membranes are not always generically applicable. Systems based upon zeolite membranes have a large potential for energy savings in the process industry. Table 15.2 summarizes pervaporation results of water/alcohol mixture by hydrophilic zeolite membranes.

Although pervaporation has gained widespread acceptance in the chemical industry as an effective process for separation of azeotropic mixtures, the application to separate organic liquid mixtures such as aliphatic/aromatic hydrocarbon or alcohol/ether is still very limited because of low selectivity due to swelling of polymer membranes. Zeolite membranes have the advantage of being chemically and mechanically robust; they can withstand severe operating conditions and not undergo physical and chemical changes as a result of exposure to organic liquids and elevated temperatures. Thus zeolite membranes are expected to be inherently superior for the separation of organic mixtures. Results of pervaporation separation of organic mixtures by zeolite membranes are summarized in Table 15.3. As shown, high perm-selectivity for methanol over other solvents with lower polarity were obtained by NaA, NaX and NaY zeolite membranes. Furthermore, high benzene selectivity was observed for benzene/p-xylene, benzene/cyclohexane and benzene/n-hexane separation through FER and NaY zeolite membranes, respectively.

Table 15.3 Organic mixtures separation results by zeolite membranes

Feed solution	Memb.	Flux (kg/m²h)	Selectivity	Temp. (°C)	Ref.
Methanol/ETBE	NaA	0.21	1, 200	50	[37]
Methanol/Benzene	NaX	1.25	24	50	[37]
Methanol/Benzene	NaY	0.62	1, 400	50	[37]
Methanol/MTBE	NaX	0.26	320	50	[37, 39]
Methanol/MTBE	NaY	1.70	5, 300	60	[37, 39]
Benzene/n-hexane	NaY	0.2	98	120	[39]
Benzene/cyclohexane	NaY	0.3	190	150	[39]
Benzene/p-xylene	FER	0.2	600	30	[40]

Table 15.4 Comparison of xylene isomers separation by zeolite membrane and polymeric membranes

Feed mixtures	Membr.	p-Xylene Flux	Sep. Factor	Temp. ($^\circ$C)	Ref.
p-/o-xylene: 0.45 / 0.35 kPa	Silicalite-1	34.3×10^{-3} kg/m^{2}h	500	200	[41]
p-/o-xylene: 50 / 50 mol%	Silicalite-1	137×10^{-3} kg/m^{2}h	40	50	[42]
p-/o-xylene: 0.31 / 0.26 kPa	Silicalite-1	0.97×10^{-8} mol/m^{2}sPa	25	200	[43]
p-/o-xylene: 0.30 / 0.27 kPa	Silicalite-1	8.3×10^{-8} mol/m^{2}sPa	225	200	[44]
p-/o-xylene: 0.27 / 0.59 kPa	Silicalite-1	30×10^{-8} mol/m^{2}sPa	16	390	[45]
m-/p-xylene: 48.9 / 51.1vol%	Dinitro-benzene	2.33 kg.μm/m^{2}h	2.39	30	[46]
p-/o-xylene: 50 / 50wt%	Copoly-imides	4.2 kg.μm/m^{2}h	1.33	65	[47]

All the above mentioned high perm-selectivity of zeolite membranes can be attributed to the selective sorption into the membranes. Satisfactory performance can be obtained by defective zeolite membranes. Xylene isomers separation by zeolite membranes compared with polymeric membranes are summarized in Table 15.4. As shown, zeolite membranes showed much higher isomer separation performances than that of polymeric membranes. Specially, Lai et al. [41] prepared b-oriented silicalite-1 zeolite membrane by a secondary growth method with a b-oriented seed layer and use of trimer-TPA as a template in the secondary growth step. The membrane offers p-xylene permeance of 34.3×10^{-3}kg/m^2.h with p- to o-xylene separation factor of up to 500. Recently, Yuan et al. [42] prepared silicalite-1 zeolite membrane by a template-free secondary growth method. The synthesized membrane showed excellent performance for pervaporation separation of xylene isomers at low temperature (50°C).

Hydrogen as a potentially clean fuel can be produced from diverse feed stocks using a variety of process technologies. At present, most of the hydrogen is made from fossil fuels and largely used in the oil refinery plant. Refineries already have a shortage of affordable hydrogen even though they have large volumes of off-gas streams that contain considerable amount of hydrogen. So recovery of hydrogen from off-gas refinery streams is resources and cost saving. Lin et al. have done a lot of work on the multicomponent hydrogen/hydrocarbon separation. Two types of zeolite membrane were investigated (i.e. secondary-grown MFI zeolite membrane prepared without template and in situ synthesized MFI zeolite membrane prepared with template). In the case of the latter, the zeolite membrane showed excellent separation properties for rejection of hydrogen from the hydrogen/hydrocarbon mixture at 25°C. Hydrogen permeation rate is almost zero, while the hydrocarbon permeation rate is $2 \sim 4 \times 10^{-4}$ mol/m^2.s. The zeolite membrane outperforms the microporous carbon membrane in terms of both selectivity and permeance for hydrocarbons over hydrogen.

Membrane Reactor

Membrane reactors (MRs) as a concept, dates back to 1960s and a large number of papers have been published on this multidisciplinary vibrant subject at the frontier between catalysis, membrane science and chemical engineering. In such an integrated process, the membrane is used as an active participant in a chemical transformation for increasing the reaction rate, selectivity and yield, thus saving energy. Zeolite MRs have attracted intense research activities during the last decade. Under the development of zeolite membrane science, zeolite membranes and layers with sufficient quality and reliability to be used as MRs have been successfully prepared in many groups. Since the current subject is chiefly focused on the applications of zeolite membranes in energy field, hereinafter, the applications of zeolite membrane reactors in hydrogen fuel production and storage will be given into particulars.

During the past two decades, the CO_2 reforming of methane has attracted much attention from both industrial and environmental sectors. In the environmental aspect, both CO_2 and methane are undesirable greenhouse gases, and the reforming reaction provides a way for CO_2 and CH_4 utilization. From the industrial viewpoint, the conversion of CH_4 and CO_2 into value-added products is desirable since they are cheap and abundant. Synthesis gas (i.e. $CO+H_2$) can be used in chemical energy transmission systems or utilized in the F–T reaction. Liu et al. [48] applied NaA zeolite membrane in a membrane reactor (NiO/La_2O_3-γ-Al_2O_3 as catalyst) for CH_4/CO_2 reforming. The CH_4 and CO_2 conversions were investigated over a fixed-bed and a zeolite membrane reactor under similar conditions as a function of reaction temperature. Above 400°C, especially at 700°C, the conversions of CH_4 and CO_2 over the membrane reactor were ca. 73.65 and 82.44mol% respectively; whereas those over the fixed-bed reactor were 45 and 52mol%. This indicates that the removal of H_2 and CO via membrane separation promoted CH_4 and CO_2 conversions. Meanwhile, the problem of catalyst deactivation due to coking can be partly resolved by adopting a membrane reactor. However, this process is highly endothermic and requires a lot of energy. Therefore, thermal arrangements (e.g. coupling with exothermic reactions) are of great significance when this process is expected to be actually applied.

A major goal for establishing a hydrogen energy economy is to gain consumer acceptance for fuel cell powered vehicles in the transportation sector. Hydrogen storage is a "critical path" technology that will facilitate the commercialization of fuel cell powered vehicles. The use of chemical cycles (such as borazine-cyclotroborazane and benzene-cyclohexane) to store hydrogen has attracted remarkable interest. These systems have been looked into previously and were found to be lacking in terms of catalyst needs as well as the pressure and temperature excursions required for charging and discharging. New approaches having new strategies to overcome these limitations would be attractive, which includes membrane reactor. Jeong et al. [49] applied FAU zeolite membrane for the catalytic dehydrogenation of cyclohexane in a membrane reactor packed with a Pt/Al2O3 catalyst. This zeolite membrane reactor simultaneously removed hydrogen and benzene from cyclohexane, which thus increased the conversion.

Other Applications

The sun is the primary source for most forms of energy found on Earth. Solar energy is clean, abundant, widespread, and renewable. Various technologies have been developed to utilize solar energy. Zeolite membranes or films also take part in this capture.

Some frontier research works on the artificial photosynthetic assemblies with zeolite membranes have been done by Kim and Dutta et al. from the Ohio State University [50].

Although adsorptive processes have been extensively studied for gas separation, catalysis, it is only recently that they have been proposed for heat management. M. Tather et al. [51] developed a mathematical model for a novel arrangement proposed in order to cope with the drawbacks originating from the inefficient heat and mass transfer in adsorption heating pumps with Zeolite 4A used as the adsorbent. L. Bonaccorsi et al. [52] have successfully synthesized zeolite coatings on metal supports with thickness ranging from few to several tens of microns, which had important technical applications in adsorption heat pumps.

Conclusions

Within the last decade, remarkable progresses have been made in the preparation of supported zeolite membranes in laboratory scale. In 1999, Mitsui Engineering & Shipbuilding Co. Ltd. put the first commercial unit of NaA zeolite membrane on the market, which is a milestone in the course of commercial production and application of zeolite membranes. In Europe, bio-ethanol dehydration plant has been set up in UK by the alliance of Smart and inocermic GmbH. In our group, organics dehydration installation based on NaA zeolite membrane with membrane area of $0.8\,m^2$ has also been set up. The pilot-scale test is undergoing. Fast and reproducible mass production of tubular NaA zeolite membranes is realized through microwave synthesis, and multi-channel supports are tested facing at a high area-to-volume ratio. Nevertheless, the only commercialized type is LTA (NaA type) zeolite membrane, and the only realized application is for organics dehydration. The Al-rich (Si/Al ratio = 1) property of LTA zeolite membrane makes it unstable from stream and traces of acid. Therefore increase of acid stability of zeolite membranes by variation of the Si/Al is expected. Despite that, LTA zeolite has a narrow pore system (Φ 0.43 nm in the Na-form); the separations of light gas mixtures of Mitsui LTA membrane and most of the literature reported LTA membrane are near to the Knudsen factors. Therefore novel methods to prepare more compact LTA zeolite membrane with real molecular sieving effects are still highly demanded. MFI zeolite membrane, due to its shape selective properties, wide Si/Al ratio, and high thermal stability (up to $500°C$), is promising in gas mixture separation and isomer separation. Nowadays, ZSM-5 membrane can meet DOE's 2010 targets for H_2 separation easily, except flux. The development of high area-to-volume modules can compensate the shortage of flux and has been done in Germany. In the foreseeable future, urged by the demand of a safe, efficient, renewable and sustainable energy system and promoted by the continuous fruitful efforts from academic and industrial circles, the fascinating potential advantageous properties of zeolite membranes are convinced to be brought into actual effects for energy and fuel applications.

Acknowledgments This work was supported by National Science Fund for Distinguished Young Scholars of China (20725313), and the Ministry of Science and Technology of China (Grant No. 2005CB221404).

References

1. Caro J, Noack M, kolsch P, Schafer R. Micropor Mesopor Mater. 2000;38:3–24.
2. Tavolaro A, Drioli E. Adv Mater. 1999;11:975–96.
3. Chiang AST, Chao KJ. Phy Chem Solids. 2001;62:1899–910.
4. Jansen J, Koegler JH, Bekkum H, Calis HPA, Bleek CM, Kapteijn F, et al. Micropor Mesopor Mater. 1998;21:213–26.
5. Matsukata M, Kikuchi E. Bull Chem Soc Jpn. 1997;70:2341–56.
6. Bein T. Chem Mater. 1998;8:1636–53.
7. Noack M, Kolsch P, Schafer R, Toussaint P, Caro J. Chem Eng Technol. 2002;25:221–30.
8. Lin YS, Kumakiri I, Nair BN, Alsyouri H. Separ Purif Method. 2002;31:229–379.
9. Nair S, Tsapatsis M. In Auerbach SM, Carrado KA, Dutta PK, editors. Handbook of zeolite science & technology. New York: Marcel Dekker; 2003. pp. 867–919.
10. Noble RD, Falconer JL. Catal Today. 1995;25:209–12.
11. Xomeritakis G, Nair S, Tsapatsis M. Micropor Mesopor Mater. 2000;38:61–73.
12. Kikuchi E, Yamashita K, Hiromoto S, Ueyama K, Matsukata M. Micropor Mater. 1997;11:107–16.
13. Xu XC, Yang WS, Liu J, Lin LW. Adv Mater. 2000;12:195–8.
14. Li YS, Chen HL, Liu J, Yang WS. J Memb Sci. 2006;277:230–9.
15. Tiscareño-Lechuga F, Tellez C, Menendez M, Santamaria J. J Memb Sci. 2003;212:135–46.
16. Richter H, Voigt I, Fischer G, Puhlfürß P. Sep Purif Technol. 2003;123:95.
17. Berg AWC, Gora L, Jasen JC, Makkee M, Maschmeyer Th. J Memb Sci. 2003;224:29–37.
18. Yan YS, Davis ME, Gavalas GR. J Memb Sci. 1997;123:95–103.
19. Nair S, Tsapatsis M. Micropor Mesopor Mater. 2003;58:81–9.
20. Bonilla G, Tsapatsis M, Vlachos DG, Xomeritakis G. J Memb Sci. 2001;182:103–9.
21. Li YS, Liu J, Yang WS. J Memb Sci. 2006;281:646–57.
22. Zhu W, Gora L, Berg AWC, Kapteijn F, Jansen JC, Moulijn JA. J Memb Sci. 2005;253:57–66.
23. Espinoza RL, du Toit E, Santamaria J, Menendez M, Coronas J, Irusta S. Stud Surf Sci Catal. 2000;130:389.
24. Li YS, Chen HL, Huang AS, Liu J, Lin YS, Yang WS. Inorganicmembranes. Proceedings of Eighth International Conference on Inorganic Membranes, 2004, pp. 613–7.
25. Li YS, Chen HL, Liu J, Li HB, Yang WS. Sep Purif Technol. 2007;57:140–6.
26. Nomura M, Bin T, Nakao S. Sep Purif Technol. 2002;27:59–66.
27. Chen HL, Li YS, Yang WS. J Memb Sci. 2007;296:122–130.
28. Poshusta JC, Tuan VA, Falconer JL, Noble RD. Ind Eng Chem Res. 1998;37:3924–9.
29. Hong M, Falconer JL, Nobel RD. Ind Eng Chem Res. 2005;44:4035–41.
30. Lai R, Gavalas GR. Micropor Mesopor Mater. 2000;38:239–45.
31. Aoki K, Kusakabe K, Morooka S. J Memb Sci. 1998;141:197–205.
32. Xu XC, Yang WS, Liu J, Lin LW, Stroh N, Brunner H. Chem Commun. 2000;7:603–4.

33. Xu XC, Bao Y, Song CS, Yang WS, Liu J, Lin LW. Micropor Mesopor Mater. 2004;75: 173–81.
34. Welk ME, Nenoff TM, Bonhomme F. Stud Surf Sci Catal. 2004;154:690–4.
35. Kusakabe K, Hasegawa Y. J Memb Sci. 2002;208:415–8.
36. Van Hoof V, Van den Abeele L, Buekenhoudt A, Dotremont C, Leysen R. Sep Purif Technol. 2004;37:33–49.
37. Kita H, Asanuma H, Tanaka K, Okamoto K, Kondo M. Abstr Pap – Am Chem Soc 1997;214:269-PMSE, part 2.
38. Jafar LJ, Budd PM. Micropor Mater. 1997;12:305–11.
39. Kita H, Fuchida K, Horita T, Asamura H, Okamoto K. Sep Purif Technol. 2001;25:261–8.
40. Nishiyama J, Ueyama K, Matsukata M. Micropor Mater. 1997;12:293–303.
41. Lai Z, Bonilla G, Diaz I, Nery, JG, Sujaoti K, Amat MA, et al. Science. 2003;300:456–60.
42. Yuan WH, Lin YS, Yang WS. J Am Chem Soc. 2004;126:4776–7.
43. Keizer K, Burggraaf AJ, Vroon ZAEP, Verweij H. J Memb Sci. 1998;147:159–72.
44. Sakai H, Tomita T, Takahashi T. Sep Purif Tech. 2001;25:297–306.
45. Hedlund J, Sterte J, Anthonis M, Bons A, Carstenson B, Corcoran N, et al. J Micropor Mesopor Mater. 2002;52:179–89.
46. Ishihara K, Matsui K, Fujii H. Chem Lett. 1985;32:1663–6.
47. Schleiffelder M, Staudt-Bickel C. React Funct Polym. 2001;49:205–13.
48. Liu BS, Gao LZ, Au CT. Appl Catal A Gen. 2003;235:193–206.
49. Jeong BH, Sotowa KI, Kusakabe K. J Memb Sci. 2003;224:151–8.
50. Kim Y, Das A, Zhang H, Dutta PK. J Phys Chem B. 2005;109:6629–932.
51. Tather M, Tantekin-Ersolmaz B, Erdem-Senatalar A. Micropor Mesopor Mater. 1999;27: 1–10.
52. Bonaccorsi L, Proverbio E. Micropor Mesopor Mater. 2004;74:221–9.

Chapter 16
Advanced Materials and Membranes for Applications in Hydrogen and Energy Production

Structural and Stability Aspects of Amorphous Materials

Balagopal N. Nair, Y. Ando, and H. Taguchi

General criteria for selection of materials for the processing of hydrogen separation membranes are discussed. Performance and stability standards required for applications in high temperature membrane reactors have been focused. The correlations between pore structure and stability issues of membranes made of amorphous materials, specifically silica membranes are discussed in detail.

The production of hydrogen using steam-reforming technology is relevant now more than ever, as newer applications based on the use of hydrogen gas emerge. Steam reforming of methane, run at rather high temperatures to get maximum efficiency out of the equilibrium-limited reaction, is generally used to produce hydrogen. It is well known that the application of a membrane-based reformer would enable the reaction to run efficiently at temperatures as low as 773 K [1–4]. In order to achieve high efficiency, the membrane must be capable of separating the product H_2 gas with high separation power (permeation and separation). Hence the key points in the selection of the membrane material are separation, permeation and stability at high temperature.

Chemical stability of the membrane material is another key point. In order to run the reforming reaction at a low temperature without significant coke formation, the amount of steam in the reforming mixture must be reasonably high. Hence the membrane material must be stable against the water vapor corrosion. However, water vapor is not the only corrosive medium inside the reformer. Gases such as CO, CO_2 as well as H_2 can also react with membrane materials and destroy their microstructure. Reactions between catalyst/catalyst-additives and membrane materials are possible as well. All these possibilities demand membrane materials with

B.N. Nair (✉)
R&D Center, Noritake Company LTD., Miyoshi, Aichi 470-0293, Japan
e-mail: bnair@n.noritake.co.jp

A.C. Bose (ed.), *Inorganic Membranes for Energy and Environmental Applications*,
DOI 10.1007/978-0-387-34526-0_16, © Springer Science+Business Media, LLC 2009

extremely good chemical and morphological stability. Another deciding factor is the cost of the membrane material. The key points should be summarized as follows:

1. Good H_2 separation efficiency (permeation, separation);
2. Thermal and hydrothermal stability;
3. Stability against reaction with gases, impurities, catalysts/chemicals;
4. Cost.

Different materials have distinct advantages and disadvantages with regard to each of these points. In order to facilitate an easy comparison we have organized candidate materials into five groups. The approximate temperature ranges of application of these materials are schematically shown in Fig. 16.1. Though there are reports detailing membrane properties outside this range and about membrane materials outside these categories, the classifications in Fig. 16.1 are representative of hydrogen selective membrane materials. Membranes based on liquids, polymers and organic/inorganic hybrids are not included in this list as the temperature of stability of these groups of materials are normally much lower than 400°C and therefore unsuitable for use in methane reforming reactors – the most relevant application in the context of this manuscript.

Several publications on the processing of membranes based on these materials could be found in the literature [5–28]. The selection of membrane material for a given application could be divided in to two parts: Screening of materials based on bulk properties and screening based on thin film properties. In the former case, intrinsic material properties such as stability and conductivity will decide the outcome of the research work. In the latter case, the defect free formability of thin film will be the deciding part. The method of film formation as well as the quality of the support substrates could become important in this respect. In supported membranes, material stability and membrane performance are very much related. The most important issue – the application of membranes in high temperature environments – is therefore the study of the structure of the membrane/material and its correlation with the stability/durability.

In the latter part of this chapter we have addressed some of the above-mentioned issues in more detail, giving specific attention to amorphous membrane materials. Yet it must be emphasized that this manuscript is not a review of publications

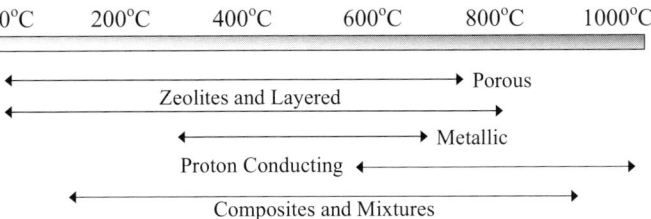

Fig. 16.1 Temperature range of application of various hydrogen selective membranes/materials. Schematic prepared based on reported membrane performance or material stability data

concerning the development of hydrogen separation membranes from amorphous materials. A number reviews on this subject, including ours, could be found in the literature [5, 11]. In this manuscript we will address some of the fundamental issues that have not been fully addressed in the literature so far.

Membranes Based on Amorphous Materials

Majority of the microporous membranes (except zeolites and layered materials) so far reported in the literature are made of amorphous materials [5–12]. Some amorphous materials tend to densify or transform to their stable crystalline form. For example, microporous membranes made of materials like alumina [13] and titania [14] are sometimes reported in the literature but normally are unstable at high temperatures. Crystallization will destroy the microporous structure of these ceramics and will therefore decrease their separation performance. Materials like silica tend to undergo densification at high temperatures. Hence the key requirement in the application of amorphous membrane materials, in methane reforming reactors, is the delay in the onset of crystallization or densification to temperatures >600°C. Because of the difficulties in achieving this condition, the applications of such membranes are presently limited to mild operating conditions. New understanding related to pore-structure stability, as detailed in the latter part of this article, and developments related to composite materials should assist in the processing of membranes with better performance and stability.

Structure and Stability

The crystallization or densification of microstructure is one of the main problems of amorphous microporous materials. Figure 16.2 shows a rough representation of this situation. Typical microporous silica membranes made by sol-gel dip coating process usually have poor thermal stability and at high temperature densification of matrix is common. Materials like alumina and zirconia usually crystallize at high temperature. Carbides and nitrides are better candidates in this regard [29, 30]. Figure 16.3 shows the microporosity values of SiC and Si_3N_4 materials. It is shown that microporosity could be retained in these materials at temperatures well above 800°C. It can be concluded that, as far as thermal stability is concerned, non-oxide materials offer opportunities to make stable microporous membranes.

The hydrothermal stability of the materials is more important as far as the use of membranes in methane reforming reactor is concerned. In general, improving the hydrothermal stability of membranes is difficult owing to the metastable nature of porous, particularly microporous structures and their tendency to change in the way of surface area reduction. Yet, recent reports [33, 34] show that improvements have been made in the hydrothermal stability of membranes based on silica, a material

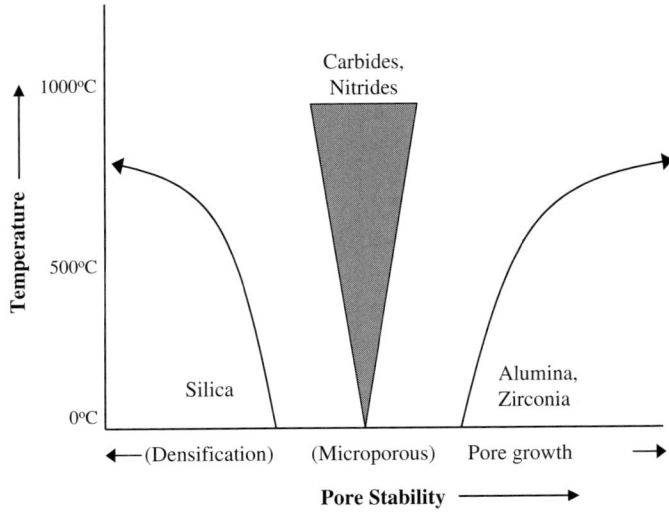

Fig. 16.2 Schematic showing stability problems associated with 'Group A' porous materials

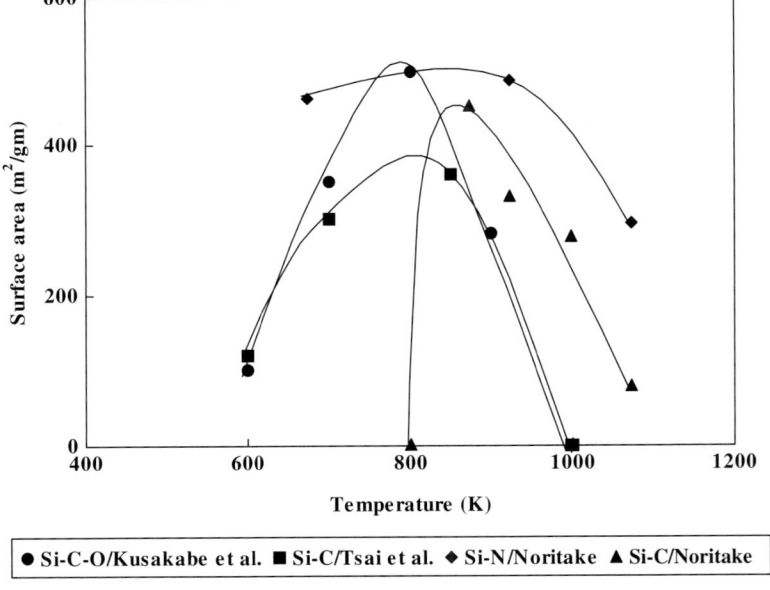

● Si-C-O/Kusakabe et al.   ■ Si-C/Tsai et al.   ◆ Si-N/Noritake   ▲ Si-C/Noritake

Fig. 16.3 Surface area values of microporous non-oxide ceramics vs. heat-treatment temperature. Calculated pore size values were roughly 0.5 nm for the "Noritake" samples. Si-C-O/Kusakabe et al. – Data redrawn from [31]. Si-C/Tsai et al. – Data redrawn from [32]. Lines fitted to the points are meant for guidance only

well known for its reactivity with water. In order to rationalize this, understanding the pore structure of silica membranes could be important.

Pore Structure of Silica Membranes

A representation of pore structure of silica membranes is shown in Fig. 16.4. As shown, the pores of silica membranes are probably formed, as in β-cristobalite, by the 5, 6, 7, or 8 membered rings of Si-O [35]. The presence of solubility sites, shown in Fig. 16.4, with an opening of 0.3 nm is the most likely reason for the large ideal selectivity obtained for He (0.26 nm) and H_2 (0.289 nm) molecules against molecules such as N_2 (0.364 nm). Based on their studies, Oyama et al. state that molecules with sizes smaller than 0.3 nm permeate easily through silica membranes and larger molecules apparently permeate with difficulty.

Based on the gas permeation measurement results on a variety of silica membranes, we have also reported similar findings [7]. We reported that the pore size distribution of sol-gel silica membranes appear to be bimodal with majority pores with sizes around 0.3 nm and some pores with much larger sizes formed due to the opaqueness of silica polymer clusters. Figure 16.5 shows the permeation results of silica membranes. The four membranes used in the study had been tailored with different porosity values. In spite of the differences in porosity values between the membranes, all the membranes showed ability to separate He from other molecules but the ability to separate between molecules of sizes greater than 0.3 nm was poor. Figure 16.6 shows a schematic representation of the separation behavior exhibited by silica membranes in comparison to a zeolite ZSM-5 membrane [7, 16, 36].

Pervaporation and adsorption studies performed in our group [36, 37] with methanol and methyl tertiary butyl ether (MTBE) molecules, showed that the silica membrane pores allowed the MTBE molecule (~4.4 Å) to permeate, though

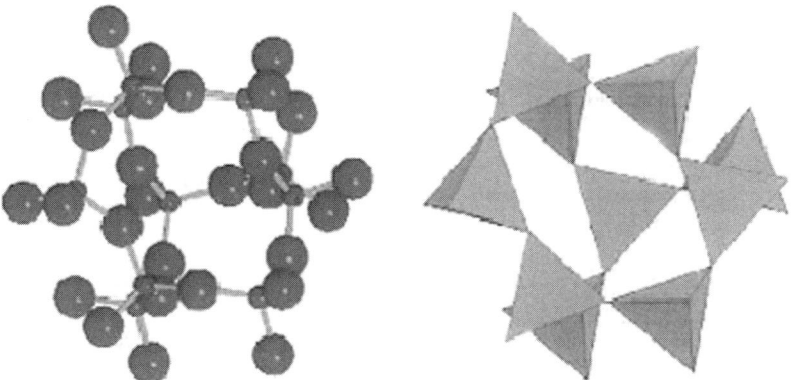

Fig. 16.4 Pore structure of silica membranes could be represented using this schematic of solubility site in β-cristobalite. Reproduced with permission from [35]. Copyright 2004 Elsevier

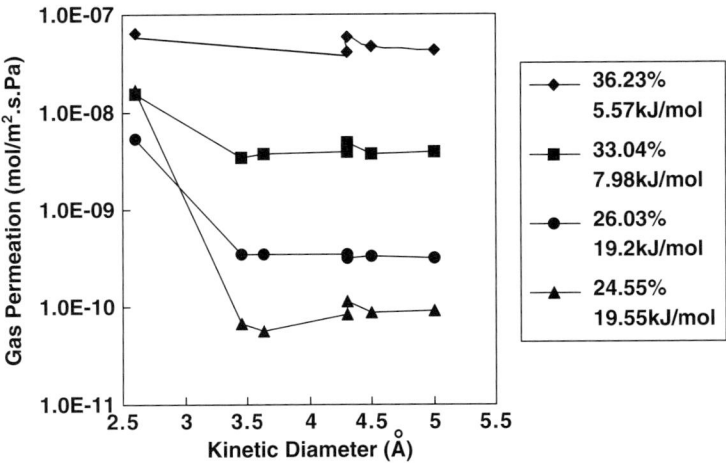

Fig. 16.5 Permeation of gas molecules with different sizes through silica membranes with different values of porosity (% value in legend) and apparent activation energy for helium permeation (kJ/mol value in legend). Redrawn with permission from [7]. Copyright 1998 Wiley-VCH

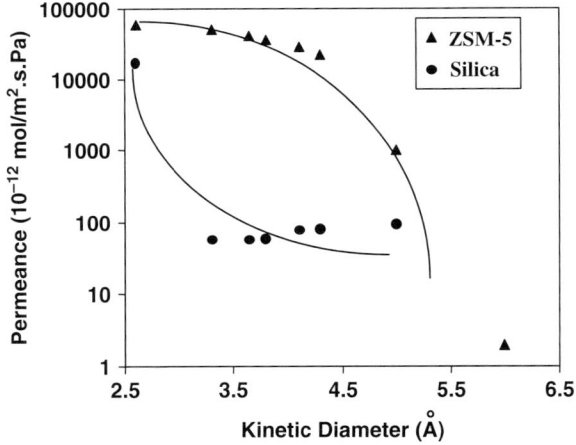

Fig. 16.6 Comparison of pore sizes of silica membrane with zeolite ZSM-5 membrane. Lines fitted to the points are meant for guidance only. Data are from [7] and [16]

with extreme difficulty [37]. In short, our studies showed that the pores of silica membranes allow molecules as large as MTBE molecules to fit into them but the permeation rate is so low that for all practical separation purposes the pore size could be considered as ~0.3 nm. The control of the pore size of silica membrane was found difficult in our studies. Incorporation of template molecules [38, 39], however, could control the pore size of silica membranes. Raman et al. reported that the size of the largest of the alcohol molecule that could fit into the pores of the silica membranes increased from 0.38 nm to 0.41 nm and 0.45 nm with an increase in the template molecular size incorporated with the tetra ethyl ortho silicate (TEOS) precursor [38]. These membranes could be specifically advantageous for the separation/production of hydrogen from higher hydrocarbons. It could be expected that the incorporation of the templates helped to increase the number of higher membered rings in the silica matrix and thereby the active pore size of the membrane.

Control of Structure by Chemical Modifications

Apart from the engineering of pore size, the mixed precursor route also allows the control of chemical structure of silica. Avnir et al. [40] have reported on the improvements in the properties of silica by incorporating suitable molecules/ions inside the pore structure of silica gels. The processing of such catalytically active membranes could be useful for a variety of fuel cell related separations and purification applications. Schwertfeger et al. [41] and Komarneni et al. [42] have reported the application of such mixed precursor routes for the production of hydrophobic silica porous structures. Figure 16.7 shows the change in water adsorption of silica gels with change in mixing ratio of the precursors $Si(OME)_4$ and $MeSi(OMe)_3$.

The mixed precursor route seems to allow a certain degree of control in the processing of silica structures with controlled hydrophobicity. The relation between hydrophobicity and H_2O permeation of microporous membranes is not clear at this moment. But recent results reported in literature should give a good starting point for further understanding of this subject. Tsuru et al. [43] showed that the H_2/H_2O selectivity values of silica membranes have good correlation with the He/H_2 selec-

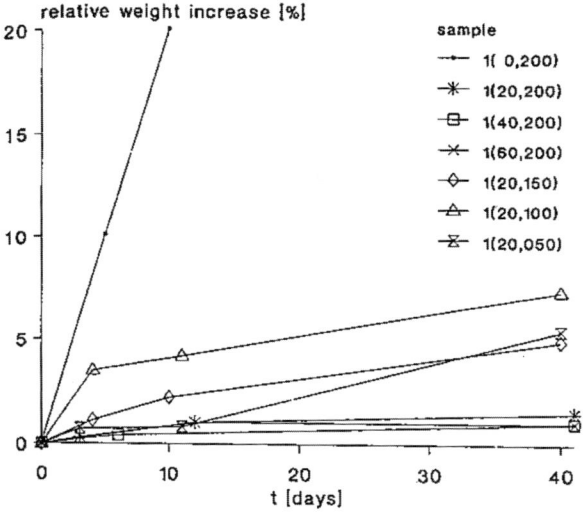

Fig. 16.7 The change in water adsorption of silica gels with changes in precursor chemistry. The ratio of precursors $Si(OME)_4$ and $MeSi(OMe)_3$ were controlled to make the samples. Reproduced with permission from [41]. For details of samples see the original article. Copyright 1992 Elsevier

tivity values. However, no correlation to H_2/N_2 selectivity values were apparent in their studies. The authors suggested that H_2O permeates through the active pores of the membranes through which apparently He and H_2 molecules also permeated. Seemingly N_2 molecules failed to permeate through these active pores of the membrane. Details of the hydrothermal stability of the membranes were not discussed in their paper. Nakao et al. reported [34] that their CVD silica membranes showed extremely high H_2/H_2O separation factors of about 100. These silica membranes reportedly showed reasonably good hydrothermal stability as well.

Pore Structure and Stability

The cause and effect relation between hydrothermal stability and water permeation is not apparent. This is a point where more attention is necessary in the future studies. It is possible that the larger H_2/H_2O separation factor or in other words the low permeation of H_2O through the membrane is the reason for the better hydrothermal stability of the membranes. Kinetic limitations have indeed shown to enhance the stability of protective coatings and coated materials. On the other hand, it should also be reasoned that the inherent hydrophobic structure of the material is the source of the improvement in hydrothermal stability and for the reduction in H_2O permeation by limiting the number of sites where H_2O could be adsorbed (as in Fig. 16.7). The results discussed so far clearly indicate that the improvement in hydrothermal stability of the membranes is linked to the improvement in porous structure (structure and quantity of surface exposed to H_2O) of the membranes itself.

Irrespective of its implications on stability, the results reported by Tsuru et al. clearly indicate that He/H_2 separation factor should be a guiding tool for the understanding of the structure of the membranes. We have previously reported the differences in He/H_2 separation factors between sol-gel and CVD made silica membranes [36]. We proposed that the study of the difference in permeation behavior should give an understanding of the structural differences between apparently porous (sol-gel) and dense (CVD) membranes. Recent results have confirmed the importance of this understanding. It should be possible to extend the theoretical model of permeation reported by Oyama et al. [35], incorporating molecules like H_2O with a lesser degree of vibrational freedom.

The study of the porous structure as discussed in the previous paragraphs would allow researchers to tailor more durable membranes for H_2 separation. Making changes in nanostructure of the silica based porous material could be important in this respect. The partial substitution of "Si" or "O" in the pore structure is one possible direction (see Fig. 16.8). For example, it is reported that the partial substitution of "O" with "N" could increase the stiffness of the network [44]. This stiffness could change the activation enthalpy required by the gas molecules to diffuse through the network, but will increase the resistance of the network against any degradation. The

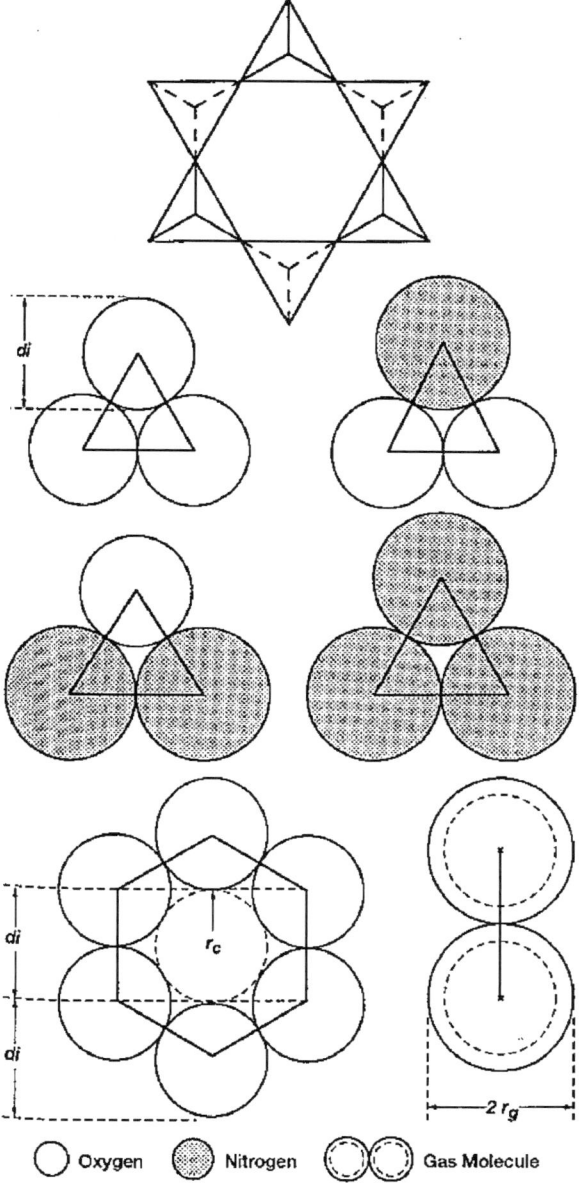

Fig. 16.8 Idealized diagram of the limiting orifice through which molecules permeate in SiO_xN_y. Reproduced with permission from [44]. Copyright 1995 Am Ceram Soc

balancing of these two factors could be difficult but should be a way to tailor high performance membranes with stability and durability. Further improvements in the performance of the membranes should be possible by incorporating novel ideas of architecture [45] or mechanism of permeation [46] of the membranes.

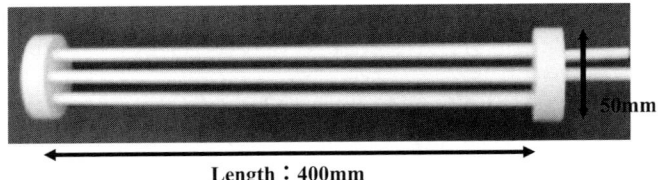

Length : 400mm

Fig. 16.9 Photograph of a high temperature stable silica membrane module. The membrane module is fully made of ceramic materials (Courtesy: Noritake Company LTD., Aichi, Japan)

Fabrication of Large Area Membranes and Modules

The processing of gas separation membranes with large membrane area (>100 cm^2) is seldom reported in the membrane literature. The processing of high quality membranes with top layer thickness values <1 μm from amorphous membrane materials in large areas is obviously a difficult task. Excellent control on the membrane making process as well as on the substrate quality is required to achieve good reproducibility. In addition to this, the fabrication of membrane modules requires selection of other ceramic components and sealing materials with matching thermal expansion behavior. Recently, we fabricated amorphous silica-based membrane modules with total membrane area of around 0.05 m^2 on alumina substrates (Fig. 16.9). The characteristics of some of the components of the module and the separation performance are described elsewhere [47, 48]. The exercise of design and fabrication of the module has taught us that the selection of materials for the module fabrication is as difficult and important as the selection of membrane materials detailed in this chapter.

Acknowledgments The research work on silica membranes was carried out as a part of the R&D Project for High Efficiency Hydrogen Production/Separation System using Ceramic Membranes funded by NEDO, Japan. The research work on Si-N based membrane was partly funded by Chubu Power Company, Aichi, Japan. We would like to thank our colleagues S. Suzuki, T. Suzuki, M. Yokoyama, K. Miyajima and Y. Yoshino for developing the of Si-N and Si-O membrane technology partly described in this paper. We also thank Dr. S. Gopalakrishnan, Dr. M. Nomura, Dr. Y. Iwamoto and Prof. S. I. Nakao for the fruitful discussions.

References

1. Itoh N. A membrane reactor using palladium. AIChE J. 1987;33:1576.
2. Zaman J, Chakma A. Inorganic membrane reactors. J Memb Sci. 1994;92:1–28.
3. Armor JN. Membrane catalysis: where is it now, what needs to be done. Catal Today. 1995;25:199–207.
4. Tsuru T, Wada S, Izumi S, Asaeda M. Catalytic membrane reactors for methane reforming using hydrogen perm-selective silica membranes. J Memb Sci. 1998;149:127.
5. Lin J, Kumakiri I, Nair BN, Alsyouri H. Microporous ceramic membranes: review. Sep Purif Methods. 2002;31(2):229.

6. Uhlhorn RJR, Keizer K, Burggraaf AJ. Gas transport and separation with ceramic membranes. Part II. Synthesis and separation properties of microporous membranes. J. Memb Sci. 1992;66:271–87.

7. Nair BN, Keizer K, Okubo T, Nakao SI. Evolution of pore structure in microporous silica membranes: sol-gel procedures and strategies. Adv Mater. 1998;10(3):249–52.

8. Larbot A, Julbe A, Guizard C, Cot L. Silica membranes by the sol-gel process. J Memb Sci. 1989;44:289.

9. Brinker CJ, Sehgal R, Hietala SL, Deshpande R, Smith DM, Loy D, et al. Sol-gel strategies for controlled porosity inorganic materials. J Memb Sci. 1994;94:85.

10. Suda H, Haraya K, Hydrogen separation with carbon membranes. Membrane. 1995;30(1): 7–12.

11. Nair BN, Okubo T, Nakao S. Structure and separation properties of silica membranes: review. Membrane. 2000;25(2):73–85.

12. Burggraaf AJ, Cot L, editors. Fundamentals of inorganic membrane science and technology. The Netherlands: Elsevier Science BV; 1996; pp. 331–50.

13. Chang CH, Gopalan R, Lin YS. Comparative study on thermal and hydrothermal stability of alumina, titania and zirconia membranes. J Memb Sci. 1994;91:27–45.

14. Sekuli J, ten Elshof JE, Blank DHA. A microporous titania membrane for nanofiltration and pervaporation. Adv Mater. 2004;16(17):1546–50.

15. Vercauteren S, Keizer K, Vansant EF, Luyten J, Leysen R. Porous ceramic membranes: preparation, transport properties and applications. J Porous Mater. 1998;5:241–58.

16. Vroon ZAEP. Ph.D. thesis, University of Twente, Enschede, The Netherlands, 1995.

17. Yamanaka S, Hattori M. Inorganic phosphate materials. Tokyo: Kodansha/Elsevier; 1989; pp. 131–55.

18. Armor JN, Farris TS. The unusual hydrothermal stability of Co-ZSM-5. Appl Catal B. 1994;4, L11–17.

19. Uemiya S. Metal membranes for hydrogen separation. Membrane. 1995;30(1):13–9.

20. Iwahara H. Technological challenges in the application of proton conducting ceramics. Solid State Ionics. 1995;77:289–98.

21. Norby T, Larring Y. Mixed hydrogen ion-electronic conductors for hydrogen permeable membranes. Solid State Ionics. 2000;136–137:139–48.

22. Nowick AS, Du Y. High temperature protonic conductors with perovskite-related structures. Solid State Ionics. 1995;77:137.

23. Balachandran U, Ma B, Lee T, Chen L, Song SJ, Dorris SE. Development of dense ceramic membranes for hydrogen production and separation. Proceedings of the ICIM-8, Cincinnati, OH, 2004, pp. 163–6.

24. de Lange RSA. Ph.D. thesis, Uni. Twente, Enschede, The Netherlands, 1993.

25. Yoshida K, Hirano Y, Fujii H, Tsuru T, Asaeda M. Hydrothermal stability and performance of silica-zirconia membranes for hydrogen separation in hydrothermal conditions. J Chem Eng Jpn. 2001;34(4):523–30.

26. Julbe A, Guizard C, Larbot A, Cot L, Giroir-Fendler A. The sol-gel approach to prepare candidate microporous inorganic membranes for membrane reactors. J Memb Sci. 1993;77: 137–53.

27. Kusakabe K., Membrane separation for hydrogen utilization. Membrane. 2005;30(1):2–6.

28. Hasegawa Y, Kusakabe K, Morooka S. Selective oxidation of CO in hydrogen rich mixtures by permeation through a Pt-loaded Y-type zeolite membrane. J Memb Sci. 2001;19:1–8.

29. Yokoyama M, Miyajima K, Nair BN, Taguchi H, Ando Y, Nagaya S, et al. Development of SiN based ceramic membrane. Proceedings of the ICIM-8, Cincinnati, OH, 2004, pp. 561–5.

30. Miyajima K, Ando Y, Yokoyama M, Nair BN, Taguchi H, Nagaya S, et al. Development of Si-N based ceramic membranes. Memb News. 2004;66(12):17–21.

31. Kusakabe K, Li ZY, Maeda H, Morooka S. Preparation of thermostable amorphous Si-C-O membrane and its application to gas separation at elevated temperature. J Memb Sci. 1995;103:175–80.

32. Lee LL, Tsai DS. Synthesis and permeation properties of silicon-carbon based inorganic membranes for gas separation. Ind Eng Chem Res. 2001;40:612–6.

33. Prabhu AK, Oyama ST. Highly hydrogen selective ceramic membranes; application to the transformation of greenhouse gases. J Memb Sci. 2000;176(2):233.

34. Nomura M, Ono K, Gopalakrishnan S, Sugawara T, Nakao SI. Preparation of a stable silica membrane by a counter diffusion chemical vapor deposition method. J Memb Sci. 2005;251(1–2):151–58.

35. Oyama ST, Lee D, Hacarlioglu P, Saraf RF. Theory of hydrogen permeability in nonporous silica membranes. J Memb Sci. 2004;244(1–2):45–53.

36. Nair BN. Ph.D. thesis, Uni. Tokyo, Tokyo, Japan, 1998.

37. Nair BN, Keizer K, Suematsu H, Suma Y, Ono S, Okubo T, et al. Synthesis of gas and vapor molecular sieving silica membranes and analysis of pore size and connectivity. Langmuir. 2000;16(10):4558–662.

38. Raman NK, Brinker CJ. Organic template approach to molecular sieving silica membranes. J Memb Sci. 1995;105:273–9.

39. Raman NK, Anderson MT, Brinker CJ. Template based approaches to the preparation of amorphous nanoporous silicas. Chem Mater. 1996;8:1682–701.

40. Avnir D. Organic chemistry within ceramic matrices: doped sol-gel materials. Acc Chem Res. 1995;28:328–34.

41. Schwertfeger F, Glaubitt W, Schubert U. Hydrophobic aerogels from $Si(OMe)_4/MeSi(OMe)_3$ mixtures. J Non-Cryst Solids. 1992;145:85–9.

42. Liu C, Komarneni S. Nitrogen and water sorption properties of ethyl substituted silica aerogels and xerogels. Mat Res Soc Symp Proc. 1995;371:217–22.

43. Tsuru T, Yamaguchi K, Yoshioka T, Asaeda M. Methane steam reforming by microporous catalytic membrane reactors. AIChE J. 2004;50(11):2794–805.

44. Ogbuji LUJT. The $SiO_2 - Si_3N_4$ interface. J Am Ceram Soc. 1995;78(5):1272–84.

45. Nair BN, Suzuki T, Yoshino Y, Gopalakrishnan S, Sugawara T, Nakao SI, et al. An oriented nanoporous membrane. Adv Mat. 2005;17(9):1136–40.

46. Kawamura H, Yamaguchi T, Nair BN, Nakagawa K, Nakao SI. Dual ion conducting membrane for high temperature CO_2 separation. J Chem Eng Jpn. 2005;38(5):322–8.

47. Yoshino Y, Suzuki T, Yamada D, Nair BN, Taguchi H, Itoh N. Development of tubular substrates, silica based membranes and membrane modules for hydrogen separation at high temperature. J Memb Sci. 2005;267(1–2):8–17.

48. Yoshino Y, Suzuki T, Nair BN, Taguchi H, Nomura M, Nakao SI, et al. Silica embrane modules for hydrogen separation at high temperature. Proceedings of the ICIM-9, Lillehammer, Norway, June 2006, pp. 143–8.

Chapter 17
Catalytic Dehydrogenation of Ethane in Hydrogen Membrane Reactor

Jan Galuszka, Terry Giddings, and Ian Clelland

Abstract The effect of a hydrogen permselective membrane (H-membrane) reactor on catalytic dehydrogenation of ethane was assessed using a fixed bed conventional reactor and a double tubular H-membrane reactor. A 5.0wt.% $Cr_2O_3/\gamma - Al_2O_3$ catalyst prepared by incipient wetness impregnation of a $\gamma - Al_2O_3$ (BET surface area = 50 m^2/g) support was used at 555°C and 600°C. Although about 40% of H_2 produced during dehydrogenation of ethane in the membrane reactor passed through the membrane, only moderate enhancement in ethane conversion was observed. The slow processes on the catalyst surface are thought to counterbalance the positive effect of membrane assisted hydrogen removal. Also, decreased selectivity to ethylene due to enhanced carbon formation in the membrane reactor led to faster deactivation of the catalyst. A strategy for commercialization of catalytic dehydrogenation of ethane through the development of a better hydrogen membrane might require a reevaluation.

Introduction

Olefins are important building blocks for a variety of petrochemicals and fuels. A technology for inexpensive dehydrogenation of light paraffins to olefins would permit flexibility and innovation in the design of new process schemes [1]. Dehydrogenation is strongly endothermic and since it increases the number of moles, high pressure has an adverse effect on the process [2]. An effective catalyst is necessary to achieve high per-pass conversion and selectivity. Despite considerable work, only chromia-alumina and platinum-alumina systems gained commercial viability [1–4].

Dehydrogenation of paraffins extracts a mole of hydrogen from each molecule and introduces an olefinic double bond in the conversion to olefins. Since this reaction is reversible, conversion is limited by thermodynamics [2]. Carrying out the process in a membrane reactor having hydrogen-permselective walls would allow

J. Galuszka (✉)
Natural Resources Canada, CANMET Energy Technology Centre-Ottawa, 1 Haanel Drive,
Ottawa, Ontario, Canada K1A 1M1
galuszka@NRCan.gc.ca

A.C. Bose (ed.), *Inorganic Membranes for Energy and Environmental Applications*,
DOI 10.1007/978-0-387-34526-0_17, © Springer Science+Business Media, LLC 2009

continuous removal of hydrogen. This should substantially increase overall yield and selectivity by equilibrium displacement combining reaction and separation in one step. This idea has gained significant interest in the research community and since the early 1990s a significant number of relevant reports have been published. Membrane-assisted catalytic dehydrogenation of ethane and propane gained the most attention as these processes were considered to be good model reactions to demonstrate the combined power of catalysis and membrane reactors.

Early contributions [5, 6] claimed substantial influence of a membrane reactor on the kinetics of catalytic dehydrogenation of ethane as increases between 6 and 8 times above the thermodynamically allowed ethane conversion were reported for the membrane assisted runs. It was observed, however, that a long contact time was necessary to get these results.

Weyten et al. [7, 8] studied dehydrogenation of propane in a ceramic and a palladium membrane reactor. They concluded that Le Chatelier's principle governs the operation of a dehydrogenation membrane reactor, and therefore the principle of operation is nothing more than using the equilibrium constant. Also it was stated that good membrane performance and long contact time were necessary to see the benefits of a membrane reactor that at its best gave propane conversion about two times above the equilibrium. Schafer et al. [3] compared performance of chromia-alumina and platinum-alumina catalysts for dehydrogenation of propane in a ceramic membrane reactor. Although their ceramic membrane had a relatively good hydrogen permeance and separation, the increase in propane yield in a membrane reactor was only about 5% for both catalysts. Since significantly faster catalyst coking was observed in a membrane-supported dehydrogenation, it was concluded that special catalysts would need to be developed for a membrane assisted catalytic dehydrogenation of paraffins.

Dittmeyer et al. [9] reviewed the applications of a palladium membrane reactor to catalytic dehydrogenation of paraffins. Based on simplified simulation, it was concluded that a beneficial effect of hydrogen removal through the membrane could be offset by strong adsorption of the product olefin on the catalyst surface that would eliminate the active surface sites and effectively slow down the forward reaction rates. Also, it was observed that the long time stability and reactivity of the palladium membranes would require more research.

We have made significant advances in hydrogen separating ceramic membrane (H-membrane) technology [10]. The purpose of this work is to demonstrate its application to catalytic dehydrogenation of ethane and assess the feasibility of enhancement of olefin production in a membrane reactor.

Experimental

Catalyst

A 5wt.% Cr_2O_3/Al_2O_3 catalyst was prepared by incipient wetness impregnation method. Alfa Aesar γ-alumina powder, 99.997% purity, with a nominal particle size of 110 μm and a BET surface area of 50 m^2/g was dried overnight at 140°C then cooled in a desiccator. A 10% chromium (III) nitrate monohydrate (from Fisher Scientific Company) solution was prepared in deionized water. Typically about 140 mL of this solution was added to about 50 g of γ-alumina under constant stirring. The mixture was heated to 70°C and evaporated over 24 h then calcined at 700°C for 24 h. The catalyst was pelletized and sized to 0.60–0.85 mm particles for reactor testing.

Membrane Preparation

H-membranes were prepared on a 60 cm long α-alumina porous tube with 10 mm OD. The macropores of the initial tube were reduced to about 50 nm by coating the tube with α-alumina followed by γ-alumina that formed a top layer of about 1–2:m thick. The silica layer was deposited on the γ-alumina layer by chemical vapour deposition (CVD) using tetraethyl-orthosilicate (TEOS). The highest He permeance (He was used instead of H$_2$ for safety reasons) of the silica membrane at 500°C was 32.0 cm^3(STP)/cm^2 · min ·atm. He/N$_2$ selectivity was 17.6, greatly exceeding the Knudsen separation of 2.6 [10]. The reported H$_2$/He permeance ratios (separation) were $\geq$ 1 [11], therefore H$_2$ permeance of this membrane is either equal to or higher than the He permeance. The performance of H-membranes used in this study was more moderate. He permeance was between 3.6 and 4.2 cm^3 (STP)/cm^2 · min ·atm and He/N$_2$ separation was between 6 and 10. The permeance and separation of the H-membrane were stable for the duration of the dehydrogenation runs.

Reactor Testing

The catalyst performance was assessed in a fully automated reactor system operated either with a conventional fixed-bed or a membrane reactor module. The schematic of the system is shown in Fig. 17.1. An on-line Hewlett-Packard (HP) gas chromatograph (GC) equipped with TCD/FID detectors was used to analyze streams containing reagents and products. All streams were fed through mass flow controllers and reagent stream contained ethane diluted in 80% N$_2$ fed at weight hourly space velocity (WHSV) between 1 and 1.4. Nitrogen contained in the feed streams was used as internal standard for calculating conversion, reaction product yields and selectivities.

The membrane reactor was a double tubular module containing a quartz shell and H-membrane tube, which was packed with catalyst as shown in Fig. 17.2. The enameled endings of the membrane were sealed with high temperature O-rings. Argon was used as sweep gas and both shell and feed sides were controlled independently by back-pressure regulators and analyzed separately by an online GC. Both these lines were equipped with separate automatic flow measuring devices that are essential for proper evaluation of reaction conversion, yield and selectivity in a membrane reactor. The sample loop of the GC analyzer was also controlled by a back-pressure regulator and the flow was measured electronically.

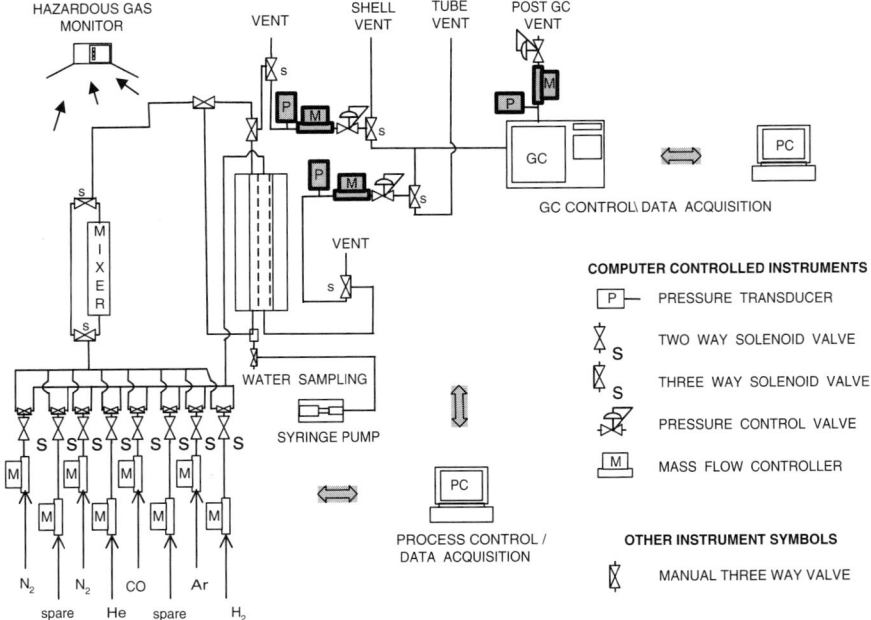

Fig. 17.1 Schematic of membrane reactor system

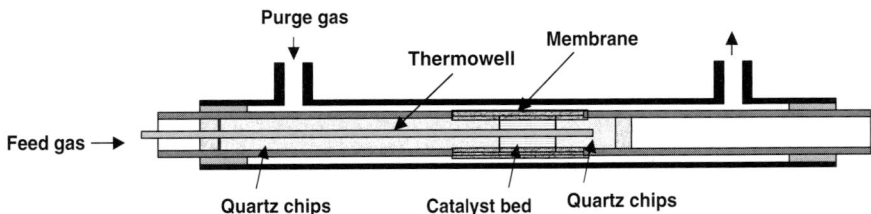

Fig. 17.2 Schematic of membrane reactor showing the placement of catalyst

For comparison, similar experiments were conducted using a conventional fixed-bed reactor module. The membrane reactor module was replaced with a quartz tube and the reactor effluent was analyzed for products and reactants. In some cases, a conventional reactor was simulated by conducting experiments in the membrane reactor without sweeping by blocking off the shell side of the reactor.

The reactor modules were charged with about 1.0 g of catalyst. The catalyst bed was maintained between 555°C and 600°C. After each dehydrogenation experiment, the temperature was lowered to 450°C and the catalyst was regenerated in situ for 2 h in a stream of air flown at 100 mL/min followed by a reduction in a diluted stream of hydrogen.

Results and Discussion

Conventional Reactor

The performance of $Cr_2O_3/\gamma - Al_2O_3$ catalyst for the dehydrogenation of ethane was first tested in a conventional fixed bed reactor at 555°C and 600°C. Figure 17.3 and Table 17.1 show the results for both temperatures. The catalyst gave a relatively steady performance at 555°C but conversion decreased more than 5% during about 2.5 h on stream at 600°C due to catalyst deactivation. These experiments were run in duplicate. Figure 17.3 shows a good reproducibility for both temperatures indicating the effectiveness of the catalyst regeneration.

The experimental parameters were adjusted to give ethane conversions below those dictated by the thermodynamic equilibrium. For the equilibrated systems it is difficult to know the amount of catalyst effectively participating in the dehydrogenation reaction. This may pose a problem if a conventional reactor run is used as a reference for assessing the effect of a membrane only.

Membrane Assisted Dehydrogenation

The results of the membrane assisted dehydrogenation of ethane on $Cr_2O_3/\gamma - Al_2O_3$ at 555°C and 600°C are shown in Figs. 17.4 and 17.5 and Table 17.1. The membrane reactor was operated with co- and counter-current sweep modes. For comparison, these figures also show the runs done in a conventional reactor. Since the feed and the shell streams were analyzed separately, the calculated ethane conversion and ethylene selectivity and yield were adjusted for the amount of ethane and ethylene slippage through the membrane. On average, about 5% of ethylene and

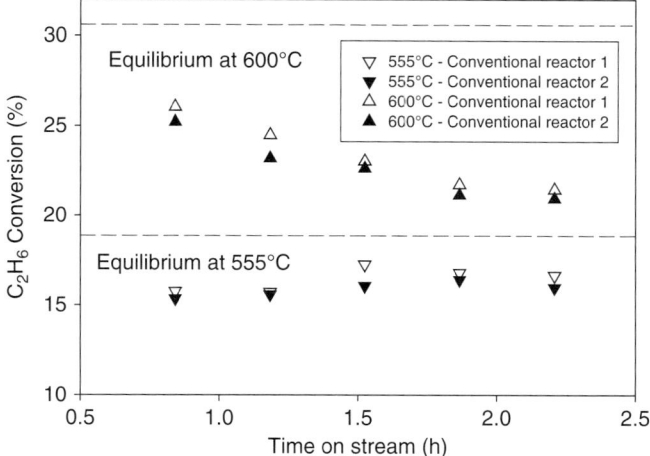

Fig. 17.3 Ethane conversion in a conventional fixed bed reactor

Table 17.1 Summary of ethane dehydrogenation in conventional and membrane reactors

Temperature (°C)	Reactor type[1]	Average conversion (%)	Average selectivity (mole %)				Average yield (mole %)				Average hydrogen ratio (%)	
			C_2H_4	CH_4	$C_3/C_3^=$	C	C_2H_4	CH_4	$C_3/C_3^=$	C	Shell	Tube
555	C	16.2	90.9	0.21	0.6	8.3	14.7	0.03	<0.1	1.4	–	–
	M/Co	16.6	97.2	0.40	<0.1	2.3	16.1	0.07	<0.1	0.8	40	60
	M/Ct	18.5	85.5	0.35	<0.1	14.1	15.8	0.06	<0.1	2.6	11	89
600	C	23.0	84.4	0.70	<0.1	14.8	19.4	0.16	<0.1	3.4	–	–
	M/Co	24.8	84.9	0.78	<0.1	14.3	21.1	0.19	<0.1	3.5	42	58
	M/Ct	25.2	81.5	0.76	<0.1	17.7	20.6	0.19	<0.1	4.4	15	85

NOTE:[1] C = Conventional; M = Membrane; Co = Co – current; Ct = Counter – current

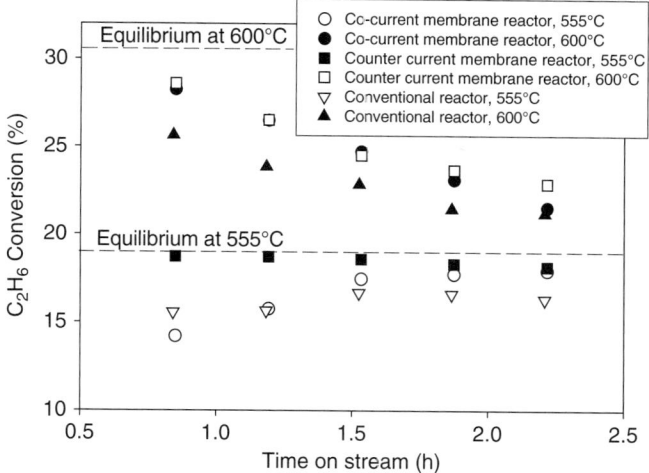

Fig. 17.4 Comparison of ethane conversion in conventional and membrane reactor operated with co- and counter-current sweep

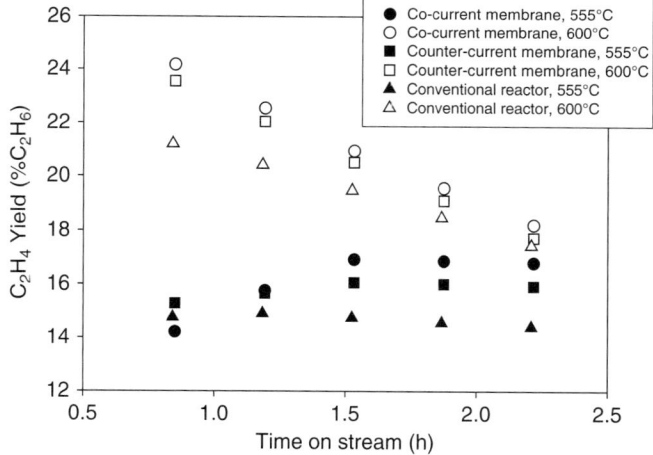

Fig. 17.5 Ethylene yield in different modes of operation

about 14% of ethane passed through the membrane to the shell side of the membrane reactor.

Clearly a membrane reactor increased ethane conversion and ethylene yield. However, it is also evident that the increase is rather moderate $<5\%$ and depends on the method of sweeping. The counter-current sweep gave initially higher conversion at 555°C, but at longer time on stream this difference disappeared. At 600°C both modes of sweeping produced no significant differences. However, the co-current sweep gave consistently higher C_2H_4 yield for both temperatures although the increase was rather moderate, not exceeding 4%. The catalyst deactivated faster in a membrane reactor at 600°C. As a result, all 3 modes gave very similar ethylene yields after about 2.5 h on stream.

Table 17.1 summarizes and compares the results of the ethane dehydrogenation runs performed in a conventional and a membrane reactor. The ethylene selectivity was negatively affected by a counter current sweep in the membrane assisted runs for both temperatures. Selectivity to carbon increased significantly, especially at 555°C. Since selectivity to ethylene is below 100%, the ethylene yield is a more suitable parameter for assessing the influence of a membrane reactor on the extent of ethane dehydrogenation. Clearly a membrane reactor with a co-current sweep gives the highest, although very modest, < 2% ethylene yield increase similarly to an earlier report on dehydrogenation of propane [3].

Co-Current and Counter-Current Configurations

Table 17.1 gives the distribution of total hydrogen that is produced during catalytic dehydrogenation of ethane between the feed and the shell of a membrane reactor. On average, about 40% of hydrogen was removed through a membrane in the co-current sweep at 555°C and 600°C. The counter-current sweep was much less effective in hydrogen removal as only about 11% passed through the membrane at 555°C and only about 15% at 600°C.

Figure 17.6 shows a probable explanation of the differences in the effectiveness of a membrane for removing hydrogen from the reaction environment imposed by the different sweep modes. A counter-current configuration has two segments of a high hydrogen pressure gradient ΔP located at the inlet and at the outlet of the membrane reactor. However, these ΔP pressure gradients are in opposite directions. At the outlet of the membrane reactor the hydrogen concentration on the feed side must be the highest as the extent of the dehydrogenation reaction is also the highest. Since the sweep stream is fresh, the hydrogen ΔP is high and the removal of hydrogen through the membrane is very effective. Consequently, the concentration of hydrogen in the sweep stream increases to its maximum. This hydrogen is carried by the sweep stream to the inlet of the membrane reactor entering another segment of a high hydrogen ΔP. However, this time the concentration of hydrogen is high on the sweep side and minimal on the feed side as the extent of the dehydrogenation reaction at the inlet of the reactor is low. Accordingly, the hydrogen contained in the sweep stream reenters the feed side of the membrane reactor that suppresses the dehydrogenation of ethane and effectively lowers the effectiveness of the membrane.

A co-current sweep is clearly more agreeable with the natural extent of ethane dehydrogenation. A fresh sweep stream enters at the inlet of a membrane reactor. As the concentration of hydrogen increases along the catalyst bed, hydrogen is removed through a membrane. At the outlet of the membrane reactor, the concentration of hydrogen is high on the sweep side and the feed side minimizing hydrogen back diffusion into the reaction zone. Thus the co-current sweep promotes the effectiveness of a membrane in removing hydrogen and provides a better kinetic

COUNTER-CURRENT SWEEP

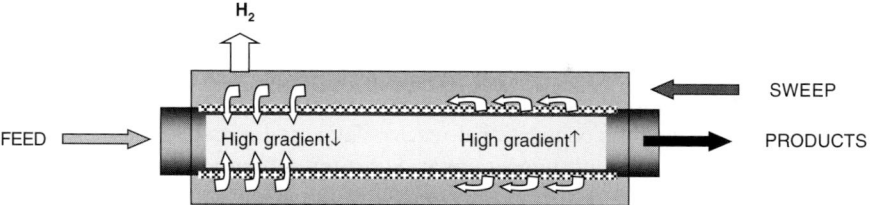

CO-CURRENT SWEEP

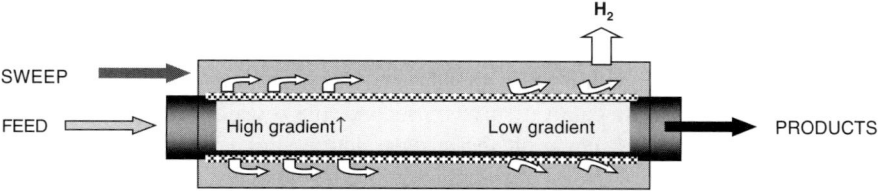

Fig. 17.6 Hydrogen concentration gradients in membrane reactor operated in co- and counter-current sweep modes

compatibility between the catalyst and the membrane as we confirmed experimentally.

Compatibility of Hydrogen Removal and Ethane Conversion

About 40% of the total hydrogen that was produced during dehydrogenation of ethane in a membrane reactor was removed through a membrane in the co-current mode of sweeping at both temperatures. Therefore, it would be interesting to correlate that amount of hydrogen with the experimentally observed gains in the conversion of ethane in the membrane assisted runs.

If the side reactions are ignored, the overall dehydrogenation rate of ethane can be described by Eq. 17.1. This equation is derived from a simple formula for the equilibrium constant given in Eq. 17.2 that predicts a beneficial effect of hydrogen removal [12] for dehydrogenation processes.

$$r = \overrightarrow{k} \left[p_{C_2H_6} - \left(\frac{p_{C_2H_4}\, p_{H_2}}{K_p} \right) \right] \qquad (17.1)$$

$$K_p = \left(\frac{p_{C_2H_4}\, p_{H_2}}{p_{C_2H_6}} \right) \qquad (17.2)$$

Eq. 17.2 theoretically stipulates that the more hydrogen is removed the higher conversion of ethane could be achieved. Assuming that the partial pressure of H_2 was somehow reduced by 40% in a conventional reactor and using the thermodynamic equilibrium constants for $555°C$ and $600°C$ given in Table 17.2, the calculated equilibrium ethane conversions were 23.9% and 37.8% giving respectively about 5% and 7% conversion gains.

Using Eq. 17.1 and the conventional reactor data, the forward reaction rate constant was estimated for both temperatures. Substituting this rate constant and the membrane reactor data to Eq. 17.1 the ethane dehydrogenation reaction rates were calculated. Then the average ethane conversions in the membrane assisted runs were estimated. As can be seen from Table 17.2, these calculated expected values of ethane conversions are significantly higher than the conversions observed experimentally stipulating that the amount of hydrogen removed through the membrane should warrant higher conversions if the kinetics were governed by a power rate law.

Catalytic dehydrogenation of alkanes involves at least three steps. First, alkane must be adsorbed on the catalyst surface. Second, the process of alkane dehydrogenation to alkene takes place on the surface active sites [2, 12, 13], and third, hydrogen and alkene are released from the surface through desorption [14, 15]. The catalyst involvement has somehow been ignored in a good portion of the reports on catalytic dehydrogenation of alkanes in the membrane reactors (e.g. [5, 8, 16, 17]). Recently, this possibility was considered and partial blocking of the catalyst surface by a strong adsorption of reactants (r), products (p) or coke (c) precursors was contemplated. Different forms of a general rate Eq. 17.3 having a denominator D accounting for these inhibitions were modeled [9, 14,18–21].

$$r = \frac{\overrightarrow{k}\left[p_{C_2H_6} - \left(\frac{p_{C_2H_4}\,p_{H_2}}{K_p}\right)\right]}{D(r, p, c)} \tag{17.3}$$

However the interpretation of a physical meaning of that derived surface related numerous parameters is rather discretionary since any discussion on the kinetics of a process without actually identifying its rate determining step is questionable.

Using a rate equation that allowed two rate-determining steps adsorption and surface reaction derived by Masson et al. [21] and cited by Zwahlen et al. [18], we obtained a very good fit to the experimental data for a conventional reactor at $555°C$ where deactivation of Cr_2O_3/Al_2O_3 catalyst was not significant. The estimated standard deviation was $<5\%$ and there was no dependence on the partial pressure of hydrogen. Nevertheless, we are just satisfied to conclude that a potential beneficial effect of hydrogen removal through the membrane during catalytic ethane dehydrogenation seems to be greatly diminished by a kinetic limitation from the slow reactions occurring on the catalyst surface.

Contrary to earlier stipulations [7, 8], the operation of a membrane reactor for catalytic ethane dehydrogenation seems to be more complex than Le Chatelier's principle and the matching membrane performance with the production rate of

Table 17.2 Experimental and calculated ethane conversions

| Temperature (°C) | Reactor[1] type | Average conversion (%) | | Equilibrium | | $\vec{k}$ (mol · g^{-1} · hr^{-1} · atm^{-1}) | Average H$_2$ ratio (%) | |
		Actual	Expected (calculated)	40% H$_2$ removed	()		Shell	Tube
555	C	16		23.9 (18.9)		0.0597	–	–
Kp = 0.0104 atm	M	17	22.9				40	60
600	C	23		37.8 (30.7)		0.0702	–	–
Kp = 0.0317 atm	M	25	30.1				42	58

NOTE: [1]C = Conventional, M = Membrane

hydrogen could be of secondary importance. Actually, removal of ethylene could be more beneficial than removal of hydrogen. Thus the development of a high-flux-super-tight hydrogen membrane for a commercial realization of catalytic ethane dehydrogenation might not be justifiable.

Conclusions

Catalytic dehydrogenation of ethane on $Cr_2O_3/\gamma-Al_2O_3$ catalyst seems to be kinetically controlled by reactions on the catalyst surface. These reactions substantially reduce the influence of hydrogen produced during dehydrogenation on the overall reaction rate. Consequently, the removal of hydrogen through a hydrogen permselective membrane only modestly improves ethane conversion and ethylene yield. The substantial increases of the conversions in the membrane assisted dehydrogenation of paraffins, claimed in earlier reports, are most likely related to the quality of the membranes and the experimental specifics including partial loss of feed through a membrane, lack of material balance and not accounting for coke formation on the catalysts and the palladium membranes [22].

Commercialization of catalytic dehydrogenation of ethane in a membrane reactor could possibly occur either by developing a catalyst that makes the rate determining step sensitive to the partial pressure of hydrogen or developing an ethylene permselective membrane.

Acknowledgments Financial support provided by the Consortium on the Conversion of Natural Gas (CCNG) and Federal Program on Energy Research and Development (PERD) is gratefully acknowledged.

References

1. Sanfilippo D, Cattech. 2000;4(1):56.
2. Weckhuysen BM, Schoonheydt RA, Catal Today. 1999;51:223.
3. Schafer R, Noack M, Kölsch P, Stör M, Caro J. Catal Today. 2003;82:15.
4. Bhasin MM, McCain JH, Vora BV, Imai T, Pujabo PR. Appl Catal. 2001;221:397.
5. Champangnie AM, Tsotsis TT, Minet RG, Wagner EJ. Catalysis. 1992;134:713.
6. Gobina E, Hughes R. J Memb Sci. 1994;90:11.
7. Weyten H, Keizer K, Kinoo A, Luyten J, Leysen R. AIChEJ. 1997;43(7):1819.
8. Weyten H, Luyten J, Keizer K, Willems L, Leysen R. Catal Today. 2000;56:3.
9. Dittmeyer R, Hollein V, Daub KJ. Mol Catal. 2001;173:135.
10. Galuszka J, Liu D. Stud Surf Sci Catal. 2001;136:363.
11. deVos RM. Thesis Enschede, 1998, ISBN 90 365 11410.
12. Lugo HJ, Lunsford JHJ. Catalysis. 1985;91:155.
13. Gorriz OF, Cadus LE. Appl Catal. 1999;180:247.
14. Raich BA, Foley HC. Appl Catal. 1995;129:167.
15. Gascon J, Tellez C, Herguido J, Menendez M. Appl Catal. 2003;248:105.
16. Gobina E, Hou K, Hughes R. Catal Today. 1995;25:365.
17. Gobina E, Hou K, Hughes R. Chem Eng Sci. 1995;50:2311.

18. Zwahlen AG, Agnew JB. Ind Eng Chem Res. 1992;31:2088.
19. Sheintuch M, Dessau RM. Chem Eng Sci. 1995;51(4):535.
20. Sznejer G, Sheintuch M. Chem Eng Sci. 2004;59:2013.
21. Masson J, Bonnier JM, Delmon BJ. Chim Phys Chim Biol. 1979;76:458.
22. Galuszka J, Pandey R, Ahmed S. Catal Today. 1998;46:82.

Index

A

Acid gas removal (AGR) system, 174–175
Acidic phosphates, 68
Air separation process, using ITM, 29
Air Separation Unit (ASU), 159
Alkanes, catalytic dehydrogenation of, 306
All-ceramic composite membrane, 71
Amorphous materials membranes, 287–293
    chemical modification to control structure,
        291–292
    fabrication, 294
    modules and area, 294
    pore structure, 289–293
    stability, 287–289, 292–293
    structure, 287–293
Annealing, in Pd-Cu alloy membranes, 212
Artificial photosynthetic assemblies, 282

B

BCN membranes, 114–116, 120
Bergius Process, 159
Bi-metal multilayer (BMML) deposition,
    239, 244
BIMEVOX, 55
Bio-ethanol, 276
Biofuels, from renewable and sustainable
    energy, 276
Biomass-to-liquids (BTL) process., 276
Brownmillerites, 6, 39–40, 55–56
BSCF membrane reactors, 61–62

C

Carbon dioxide (CO_2), routes for mitigation, 27
Catalytic membrane reactors (CMR), 95
Catalytic poisoning, and dense metal
        membrane degradation, 185–187
Cell stack subsystem (CSS), 255–256,
    264–265

Ceramatec membranes, *see* Ceramic
    membranes
Ceramic Autothermal Recovery (CAR)
    process, 28
Ceramic composite membranes, mixed
    conduction in, 71, 74
Ceramic membranes, 201–202
    composite membrane, 71
    concept, 70–71
    conductivity measurement, 73
    devices for UHP production, 67–79
    evaluation, 73–76
    materials selection, 71–72
    mixed conduction in ceramic composite, 74
    options for H_2 separation, 69–70
    scale up, 76–77
    stability, 74–76
    thin membrane fabrication, 76
Ceramic–metal joining and sealing technology,
    9
Chemicals, GOX use in production of, 32
Clean fuels, from fossil energy sources, 275
Coal
    biomass gasification process, 107
    gasifiers, 159–160
    hydrogen from, 157–158
Coal-to-liquids (CTL) process, 275–276
Cold gas cleaning technologies, 162–163
Composite membrane
    evaluation of mixed conduction in, 74–75
    thermogravimetry of, 74–75
Composite Pd/Pd alloy membranes
    characterization, 244–248
    deposition on porous substrates, 241–244
        electroless deposition, 242–243
        porous substrates, 243–244
    flux and long-term stability, 248–250
    hydrogen permeation, 240–241
    mass transfer resistances, 250

Composite Pd/Pd alloy membranes (*cont.*)
    microscopic surface and microstructure
        analysis, 245–248
    Pd-hydrogen system, 240–241, 251
    permeation flux measurements, 244–245
    support type, 250
    thermodynamics, 240–241
    thickness, 248–250
    See also Palladium alloy membranes
Corrosive decay, and dense metal membrane
    degradation, 187–194
Cryogenic technology
    energy use for power plant, 34
    for oxygen production, 4–5, 23–24
CVD silica membranes, 292

D

Dense composite membranes
    containing metallic elements, 127–135
    definition, 125
    hydrogen separation using, 125–150
    materials for fabrication, 127–135
    membrane flux measurements, 146–149
    metal alloys use in, 131–132
    metal lattice parameters and, 130–132, 134
    permeability and diffusivity of hydrogen,
        127–131
    process integration and scale-up, 155–170
    solubility of hydrogen and, 129–130,
        132–134
    surface and interface effects, 135–142
    surface catalysis for, 142–146
    variation in, 125–127
Dense metal membrane
    configuration, 183–184
    degradation mechanism, 185–194
        catalytic poisoning, 185–187
        corrosive decay, 187–194
        WGSMR operation impact, 190–194
    fabrication method, 183–185, 194
    materials, 178–180
    palladium, 178–179
    palladium alloys, 179–180
    permeability, 180–183
    super permeable metals, 180
    transport mechanism, 180–183
Differential thermal analysis (DTA), 83–84,
    86, 87, 89–91
Distillation-PV hybrid process, 278
Double exchange mechanism, 56

E

Electroless palladium-plating bath, 242
Energy cycles, ITM oxygen integration with,
    12–13
Ethane
    catalytic dehydrogenation
        with co-current sweep, 303–305
        compatibility and ethane conversion,
            305–308
        in conventional and membrane reactor,
            301–303
        with counter-current sweep, 303–305
        in hydrogen membrane reactor,
            297–308
    oxidative dehydrogenation, 62
Ethanol permselective membrane system, 276

F

FAU zeolite membrane, 282
Fick's Law, 146, 181
Fischer-Tropsch (F-T) synthesis, 157, 159,
    275–276, 281
Fluorite solid electrolyte, electronic
    conductivity, 57
Fossil energy sources, clean fuels from, 275
Fuel cell power plants
    description, 255–256
    modeling and analysis, 258–267
    palladium alloys for long life and
        performance, 256–258
    sub-systems in, 255–256
    system level models, 259
    WGSMR based, 259–267
        design analysis, 261–263
        design specifications, 266–267
        fuel processing subsystem, 263–264
        start-up time and transient capability,
            265–266
        system, 264–265
Fuel processing subsystem (FPS), 254, 256,
    263–268

G

Gaseous oxygen (GOX), in production of
    chemicals and steel, 32
Gasification, 107
    of carbonaceous solids, 174
    and integrated membrane technologies,
        173–178
Gas separation membranes, integration into
    IGCC plant, 175–177, 194
Gas-to-liquids (GTL) process, 275–276

H

Heat recovery steam generator (HRSG), 174
Henry's law, 241
High temperature membrane technology,
     see Mixed-conducting perovskite
     reactor
Hydrogen
    advanced membrane reactor for, 108–109
    from biomass, 108–109
    from coal, 108–109, 157–158
    economy, 276, 278
    energy, 276–278
    industrial sources of, 157
    membrane technology, 68–70, 194–195
    MPEC membrane for production of,
        107–121
    permeability, 178–179, 220–222
        measurements, 77–79
        in Pd-Cu alloy membranes, 207, 211,
            221–222
    production
        palladium alloy membranes for,
            201–216
        for power, 158
        from solid fuels, 107–121
        for synthetics fuels, 158–159
    separation
        experiments, 77–79
        technologies, 155, 277
    transport mechanism through dense metal
        membranes, 180–183
Hydrogen membrane reactor, 163–164
    catalyst, 297–300
    ethane catalytic dehydrogenation in,
        297–308
    membrane preparation, 299
    system, 300
    testing, 299
Hydrogen separation membranes
    applications, 156, 158–160, 201, 220
        coal gasifiers, 159–160, 220
        hydrogen production for synthetic fuels,
            158–159
        IGCC plants, 156, 159
        process description for hydrogen from
            coal, 157–158
        WGS reactors, 155–157, 160
    for hydrogen production and CO_2
        sequestration, 156, 160
        impact of impurities and gas cleaning
            methods, 161–162
    integration into IGCC plant, 155–156, 159,
        175–177, 194

materials for, 285–287
    performance and testing, 163–166, 222,
        235–236
    plant design issues, 169
    reactor, 163–164
    scale-up, 164, 166–169
    temperature range for application of, 286
Hydrogen transport membranes (HTM), 143,
    161, 163, 169–170, 185

I

Integrated Gasification Combined Cycle
    (IGCC) system, 67–68, 79, 155–156
    gas separation membranes integration in,
        175–177, 194
    hydrogen separation membranes integration
        into, 156, 158, 194
    process with CO_2 capture, 175
    process without CO_2 capture, 174–175
Integrated membrane technologies, and
    gasification, 173–178
Ion transport membranes (ITM)
    acceptability, 48
    application, 23
    architecture, 17, 41–42
    evolution of, 5
    flux of oxygen through, 12, 16, 23
    gas separation applications, 3–24
    materials, 5–7, 23, 38–41
    membrane stability and mechanical issues,
        7–9
    modules
        architecture, 17, 23
        concepts, 44–47
        sealing and joining techniques, 9–10
    operating principles, 10–11
    performance enhancements, 10–11
    sealing, 9–10, 38, 42–44
    technology
        applications, 31–32
        costs, 33, 35–36
        general aspects of, 29–31
        oxidation reactions, 36–38
        for oxygen production, 5–6
        pure oxygen production, 32–33
        viability of, 27–49
    thin film membrane devices, 16–17
    viability
        economic, 47–48
        technological, 48
Ion transport membranes (ITM) oxygen
    integration with energy cycles, 12–13, 23

Ion transport membranes (ITM) oxygen (*cont.*)
   module
      architecture, 17, 23
      commercial production, 18–20
      concepts, 44–47
      design, 16–18
      sealing and joining technique, 9–10
   technology
      commercialization timeline, 22
      cost and performance target, 14–15
      design, construction, commissioning
         and testing aspect, 15–16
      RD&D, 14–16
      validation and benefits, 19, 21
Isocompositional heating and cooling
   technique, 9
ITC technology, 34, 37
ITM Oxygen Cooperative Research and
   Development Agreement, 5

K
Knudsen diffusion, 216

L
Langmuir Adsorption Isotherm, 146–147
Lanthanum calcium ferrite (LCF) perovskite
   system, 9
Lattice matched system, 136, 139
Light alkanes
   OITM reactors for selective oxidation of,
      58–63
   oxidative coupling of methane, 58–60
   partial oxidation of methane, 60–62
   selective oxidation of, 58–63
Liquefied oxygen (LOX), 32
LSGF-BSCF membrane, 57
LTA zeolite membrane, 282

M
Membrane reactors (MRs) systems, 281
   modeling and analysis, 258–267
      design specifications, 266–267
      fuel processing subsystem, 263–264
      reactor design analysis, 261–263
      start-up time and transient capability,
         265–266
      system, 264–265
      system level models, 259
      WGSMR modeling, 259–261
Membranes
   based on amorphous materials, 287–293
   concepts and reactor configurations, 53–55
   hydrothermal stability of, 287, 292
   materials, 55–58

material selection criteria for, 285–287
   See also specific membranes
Mesoporous oxides, 68
Mesoporous phosphates, 68
Metal membrane fabrication method, 183–185,
   194
   layered membrane, 184
   thin films on porous substract, 183–184
   unsupported thin foils, 183
Metal-organic chemical vapor deposition
   (MOCVD), 241
Methane
   oxidative coupling of, 58–60
   partial oxidation of, 60–62
   reforming reactors, 287
Methyldiethylamine (MDEA), for gas
   cleaning, 162, 169
MFI zeolite membrane, 280, 282
Microporous membranes, 287
Microstructure control, with magnesia, 97–99
Mixed-conducting perovskite membranes, 95
Mixed-conducting perovskite reactor
   architecture, 101–104
      multilayer membrane, 104
      porous layer elaboration, 101–104
   for high-temperature applications, 95–105
   microstructure, 97–102
      control with magnesia, 97–99
      effects on oxygen permeation rates,
         99–102
   theoretical models, 96–97
Mixed ionic and electronic conducting
   membrane (MIECM), 29
   material, 53–57
   mechanism of OCM in, 58–59
   oxygen permeation fluxes of, 56
   reactor, 61–62
Mixed protonic-electronic conducting (MPEC)
   membrane
   BCN membranes, 114–116, 120
   fabrication, 114–115
   hydrogen flux measurement in high
      pressure permeation unit, 113–114
   hydrogen permeation data, 115–116
   for hydrogen production from solid fuels,
      109–121
   materials, 110
   perovskite membrane, 115–116
   SCY membrane, 117–120
   simulation results for hydrogen transport
      in, 117–120
   transport model for, 110–113

Multi-channel tubular monoliths, hexagonal packing of, 45
Multi Electrode Assembly (MEA) approach., 96

N

NaA zeolite membrane, 275–279, 281–282
Natural gas combined cycle (NGCC) plant, 37
Nernst–Einstein formula, 12
Niobium, 125–126
Non cryogenic air separation technology, for oxygen production, 4
Nonfaradaic electrochemical modification of catalytic activity (NEMCA), 54
Nonwetting, phenomena of, 135

O

On-board fuel processing system, performance requirements, 254
Organic mixtures, separation by zeolite membrane, 279
Oxidative coupling of methane (OCM), 58–60
Oxidative dehydrogenation of ethane (ODE), 62
Oxidative dehydrogenation of propane (ODP), 62–63
Oxide ceramic materials, perovskite crystal structure of, 136–137
Oxygen
    in bulk production of chemicals and steel, 32
    conducting membranes, 42
    cryogenic distillation for production, 4
    flux expression for separation, 12
    large-scale production, 33
    permeable membrane, 57
    use in industry, 31–32
Oxygen-ion transport membranes (OITM)
    concepts and reactor configurations, 53–55
    development of, 53
    materials, 55–58
    OCM and, 58–60
    ODP and, 62–63
    POM and, 60–62
Oxygen-ion transport membranes reactors (OITMRs)
    for selective oxidation of light alkanes, 54, 58–63
Oxygen-permeable perovskite-type ceramic membranes, 11
Oxygen permeation rates, microstructure effects on, 99–102
Oxygen Transport Membranes (OTM), 29

P

Palladium
    in dense metal membrane, 178–179
    deposition on porous substrates, 241–244
        electroless deposition, 242–243
        porous substrates, 243–244
    hydrogen system, 240–241
    membranes, 178–179, 203
        characterization, 244–248
        flux and long-term stability, 248–250
        mass transfer resistances, 250
        microscopic surface and microstructure analysis, 245–248
        permeation flux measurements, 244–245
        support type, 250
        thickness, 248–250
        See also Palladium alloy membranes
    thermodynamics and hydrogen permeation, 240–241
Palladium alloy membranes
    characterization, 224, 257
    in dense metal membrane, 179–180
    economic constraints, 205–206, 212–214
    fabrication and testing, 224
    gas mixture effects, 226–235
        CO/CO$_2$ effects, 226–228, 235
        sweepgas effects, 233–236
        WGS mixture effects, 228–231, 235
        WGS mixture effects including H$_2$S, 230–234
    high temperature permeation test, 224–225
    hydrogen
        permeation, 225–226, 257–258
        production, 201–218, 219–237
    materials for, 223
    module, 224–225
    Pd-Cu alloy membrane, 205–209, 211–217, 221–236
    performance, 235–236
    preparation, 223–225, 235
    self-supporting, 205–206
    vaccum deposit palladium membrane, 206–207, 212–214
    See also Composite Pd/Pd alloy membranes
Palladium cermets, 125, 136–139
Palladium-copper (Pd-Cu) alloy membranes, 206–209, 211–216, 257–258
    annealing, 212
    characterization, 224
    deposition on
        plastic, 210
        silicon, 211

Palladium-copper (*cont.*)
  fabrication and testing, 224
  gas mixture effects, 226–235
    CO/CO_2 effects, 226–228, 235
    sweepgas effects, 233–236
    WGS mixture effects, 228–231, 235
    WGS mixture effects including H_2S,
      230–234
  high temperature permeation test, 224–225
  hydrogen
    permeation testing, 209, 214–217,
      225–226
    separations, 221–238
  materials for, 223
  module, 224–225
  performance, 235–236
  plastic backing removal in, 209
  See also Composite Pd/Pd alloy membranes
Palladium-gold (Pd-Au) alloy membranes
  characterization, 224
  fabrication and testing, 224
  gas mixture effects, 226–235
    CO/CO_2 effects, 226–228, 235
    sweepgas effects, 233–236
    WGS mixture effects, 228–231, 235
    WGS mixture effects including H_2S,
      230–234
  high temperature permeation test, 224–225
  hydrogen
    permeation, 225–226
    separations, 221–239
  materials for, 223
  module, 224–225
  performance, 235–236
  See also Composite Pd/Pd alloy membranes
Palladium-silver (Pd-Ag) alloy membrane
  system, 180, 257
Paraffin, dehydrogenation of, 297–298
Parker gas separation, 28
Partial oxidation of methane (POM), 60–62
Perovskites, 6, 9, 28, 30, 39–40, 49, 55–57, 68,
  71–74
Polymer Electrolyte Membrane (PEM) fuel
  cell power system, 253
  See also Fuel cell power plants
Polymeric membranes, xylene isomers
  separation by, 280
Polymer membranes, 201–202
Porous exclusion membranes technology, 155
Power conditioning subsystem (PCS), 255
Power plant
  cryogenic and ITC technology based, 34
  ITM based membrane process viability, 37
  STAG based, 34
  Zero Emission Power Plants (ZEPP), 37
Pressure-driven hydrogen separation
  membranes, 69, 71, 79
Pressure Swing Adsorption (PSA) technology,
  4, 28, 35, 107, 109, 155, 169, 277
Pure oxygen production, 32–33
Pyrochlores, 68, 70

R

Reactor configurations, and membrane
  concepts, 53–55
Rectisol system., for gas cleaning, 156, 162,
  169
Renewable and sustainable energy, biofuels
  from, 276

S

SCY membrane, 117–120
Selexol system., for gas cleaning, 156, 161,
  162, 169, 174–175
Self-supporting palladium alloy membranes,
  207, 218
Sieverts' law, 148–149, 202, 241, 245, 260
Sieverts thermodynamic relation, 181
Silicalite-1 zeolite membrane, 276, 280
Silica membranes
  chemical modifications to control structure,
    291–292
  gas permeation through, 289–290
  module, 294
  pore structure, 289–291
Sintered composite membranes, 77
Size exclusion membranes technology, 155
Solid electrolyte potentiometry (SEP), 54
Solid electrolyte (SE) membrane
  materials, 53–56
  reactors, 62
Solid fuels, hydrogen production from,
  107–121
Solution-diffusion theory, 127
Solvent-based recovery methods, 156
STAG based power plant, 34
Steel, GOX use in production, 31–32
Super permeable metals, 180, 184
Surface oxygen exchange kinetics, 11
Synthetic fuels, hydrogen for production of,
  158–159

T

Technology Development Unit (TDU), 14–15
Temperature-programmed desorption (TPD)
  technique, 60

Thermal management subsystems (TMS), 255–256
Thermogravimetric analysis (TGA), 84, 86, 91
Thermogravimetry, of composite membrane, 74–75
Thin membrane fabrication, 76, 78
5 TPD sub-scale engineering prototype (SEP) facility, 14–16
Transient liquid phase (TLP) method, 9
Tubular configuration, principle of compression sealing of, 43

U
Ultra-high purity hydrogen (UHP) production, ceramic membrane devices for, 67–79
Un-supported palladium alloy membranes, *see* Palladium alloy membranes

V
Vaccum deposit palladium membrane, 206–207, 212–214
Vacuum Swing Adsorption (VSA) technology, 28, 159

W
Wagner theory, 96
Warm-gas clean-up systems, 175–176
Water-gas shift (WGS)
    catalysis, 156, 160–161, 253–254, 259–260, 268
    feed streams, 163–166
    membrane reactors, *see* WGS membrane reactor (WGSMR)
    mixtures, 155, 161, 225, 228–235
    reaction, 157, 170, 176, 178–179, 234, 253, 260–262
    reactors, 158, 160–161, 169, 173, 253, 263
Water management subsystems (WMS), 255–256
Water selective zeolite membranes, performances of, 279
Wetting, phenomena of, 135
WGS membrane reactor (WGSMR), 173, 176–177, 253–254
    based fuel cell power plants, 259–267
    design analysis, 261–263

design specifications, 266–267
fuel processing subsystem, 263–264
palladium alloys for long life and performance, 256–258
start-up time and transient capability, 265–266
system, 264–265
operation impact on corrosion, 190–194
performance requirements, 254

X
Xylene isomers, separation by membranes, 280

Y
Yttria stabilized zirconia (YSZ), 55, 60, 62–63
Yttrium-doped barium cerate
    barium stoichiometry, 87
    decomposition in carbon dioxide, 83–94
    HTXRD analysis, 84–89, 91
    modeling, 86–87, 89–91
    sample preparation, 84
    thermal analysis, 86–90
    zirconium substitution, 91–94
        kinetics, 92–93
        undoped, 91
        yttrium-doped, 91–92

Z
Zeolite membrane
    application
        biofuels from renewable and sustainable energy, 276
        clean fuels from fossil energy sources, 275
        hydrogen energy, 276–278
    artificial photosynthetic assemblies, 282
    characterization, 274
    energy-efficient process, 278–281
        and cost-efficient separations, 278–280
        membrane reactor, 281
    organic mixtures separation by, 279
    peparation, 273–274
    xylene isomers separation by, 280
Zeolite MRs, 281
Zero Emission Power Plants (ZEPP) system, 37, 38
ZSM-5 membrane, 275, 277–278, 282, 290

Printed in the United States of America